HAWAI'I'S

SEA CREATURES

A GUIDE TO HAWAI'I'S
MARINE INVERTEBRATES

TEXT AND PHOTOGRAPHS
BY JOHN P. HOOVER

ADDITIONAL PHOTOGRAPHS
BY
SCOTT JOHNSON
MIKE SEVERNS
AND MEMBERS OF THE UNDERWATER
PHOTOGRAPHIC SOCIETY OF HAWAI'I

MUTUAL PUBLISHING

CONTENTS

Copyright ©1998
by John P. Hoover

No part of this book may be reproduced
in any form or by any electronic or mechanical
means including information storage and retrieval
devices or systems without prior written permission
from the publisher except that brief passages may
be quoted for reviews.

All rights reserved

Library of Congress Catalog Card Number
98-68053

ISBN 1-56647-220-2 softcover
ISBN 1-56647-235-0 casebound

First Printing, March 1998
1 2 3 4 5 6 7 8 9

Design by Angela Wu-Ki

Mutual Publishing
1215 Center Street, Suite 210
Honolulu, Hawaii 96816
Ph (808)732-1709
Fax (808)734-4094
Email: mutual@lava.net
URL: http://www.pete.com/mutual

Printed in Korea

INTRODUCTION

As increasing numbers of snorkelers and divers take to Hawai`i's waters, the need has arisen for an illustrated guide to Hawaiian marine invertebrates. This book attempts to fill that gap. It describes over 500 species of sponges, corals, anemones, worms, molluscs, crustaceans, echinoderms, and other animals occurring in the Islands from the shoreline to scuba diving depths. A one-volume work such as this cannot possibly be complete (Hawaiian molluscs alone would fill a book four times this size); it can, however, cover the larger and more conspicuous animals and introduce the fascinating world of marine invertebrates to the reading public.

Almost all the photographs in this book are of living animals in their natural Hawaiian habitats. Occasionally an animal was moved to make it easier to photograph, but seldom far from where it was found. Location and depth information is given for all pictures and every effort has been made to supplement the pictures with accurate and readable text, including notes for aquarists wherever practical. Animals are arranged in broad evolutionary groups (phyla) beginning with the simplest and ending with the most complex. An exception is made for the wormlike invertebrates, which are combined in a single section. Within phyla, animals are arranged by classes, orders, and genera.

Common names, Hawaiian names (when available), and full scientific names are given for each species. In cases where common names did not previously exist, they have been invented for the convenience of readers, usually based on the scientific name or on some obvious characteristic of the animal.

Little more needs to be said about the arrangement and conventions of this book. In the main it follows the classification scheme in Ruppert and Barnes' *Invertebrate Zoology, 6th ed.*, 1994. The geographic range "Indo-Pacific" denotes tropical waters from East Africa to Hawai`i. Although some information presented here is based on observations made by the author and his associates, most comes from published sources. The bibliography lists all publications consulted during the preparation of this book except encyclopedias, dictionaries, and other general reference books.

To ensure accuracy, many of the animals illustrated were photographed, collected, and submitted to specialists for identification. In several instances, species new to science or previously unknown in Hawai`i were discovered this way. The scientists who helped make this book possible are listed in the Acknowledgments along with their specialties.

ACKNOWLEDGMENTS

As someone once said, the secret of success is to surround yourself with people who know more than you do. This book could not have been put together without the help and support of many persons who gave freely of their time, expertise and resources. Among those listed below, special thanks go to Cory Pittman for critically reviewing the entire manuscript (his field and laboratory knowledge of Hawaiian marine invertebrates is unsurpassed), and to Ron Holcom and Darrell Takaoka, who almost on demand located rare and unusual animals I was unable to find myself.

Linda & Ken Bail, Bubbles Below Scuba Charters, Kaua`i (diving); Patrice Belcher, B.P. Bishop Museum Library (reference); Julie Brock, University of Hawai`i (bryozoans, annelid worms); Alexander Bruce, Queensland Museum (shrimps); Steve Buck, Kāneohe Marine Corps Air Station (diving, photos); Dale R. Calder, Royal Ontario Museum (hydroids); Roy L. Caldwell, University of California at Berkeley (mantis shrimps); Bruce Carlson, Waikīkī Aquarium (photos); Peter Castro, California State Polytechnic University, Pomona (trapeziid crabs); Norton T.M. Chan, Waikīkī Aquarium (specimen collecting); Lisa Choquette, Dive Makai Scuba Charters, Kailua-Kona (diving); Steve Coles, B.P. Bishop Museum (corals, diving); Alain Crosnier, Muséum national d'histoire naturelle (penaeid shrimps); Jerry Crow, Waikīkī Aquarium (scyphozoans, cubozoans); Peter J.F. Davie, Queensland Museum (crabs); Ralph DeFelice, B.P. Bishop Museum (sponges, diving); Charles Delbeek, Waikīkī Aquarium (soft corals, specimen collecting); Dan Dickey, Austin, Texas (diving, specimen collecting, observations); Marilyn Dunlap, University of Hawai`i (sea hares); John L. Earle (molluscs, specimen collecting, diving); Lucius G. Eldredge, B.P. Bishop Museum (crustaceans); Daphne G. Fautin, University of Kansas (anemones); Darryl L. Felder, University of Southwestern Louisiana (ghost shrimps); G. Curt Fiedler, University of Hawai`i (crustaceans); David B. Fleetham, Kīhei, Maui (photos); Terrence M. Gosliner, California Academy of Sciences (opisthobranchs); Dave Gulko, University of Hawai`i (cnidarians); Michael G. Hadfield, University of Hawai`i (sea hares); Eileen Herring, University of Hawai`i Hamilton Library (reference); Eric Hochberg, Santa Barbara Museum of Natural History (octopuses); Ron Holcom, Honolulu, Hawai`i (specimen collecting, observations, photos); John N.A. Hooper, Queensland Museum (sponges); Carol N. Hopper, Waikīkī Aquarium (worm snails); Hal Ing, SnapShot Photo, Honolulu, Hawai`i (photography, diving); Scott Johnson, Kwajalein, Marshall Islands (molluscs, photos); Jerry & Lori Kane, Honolulu, Hawai`i (photos); E. Alison Kay, University of Hawai`i (molluscs); Derek W. Keats, University of the Western Cape (coralline algae); Brian Kensley, Smithsonian Institution (lobsters); Gretchen Lambert, California State University, Fullerton (tunicates); Keith Leber, Honolulu, Hawai`i (editing); Christopher L. Mah, California Academy of Sciences (sea stars, brittle stars); James E. Maragos, East-West Center Program on Environment (corals); Jennifer & Lynn Mather, University of Lethbridge (cephalopods); Margaret McFall-Ngai, Kewalo Marine Laboratory, Honolulu *(Euprymna)*; Karen Medeiros, Hawaii Medical Library; Patsy A. McLaughlin, Western Washington University (hermit crabs); Colin L. McLay, Canterbury University (sponge crabs); Mark & Cary Moehlman, University of Hawai`i (diving); Robert B. Moffitt, National Marine Fisheries Service, Honolulu (lobsters); Anthony Montgomery, Waikīkī Aquarium (black corals); Richard Mooi, California Academy of Sciences (sea urchins); Michael D. Miller, The Slug Site slugsite.tierranet.com (photos); Bruce C. Mundy, National Marine Fisheries Service, Honolulu (photos, observations); Norm and Linda Nelson, Camp Pecusah, Maui; A. Todd Newberry, University of California, Santa Cruz (tunicates); Leslie J. Newman, Smithsonian Institution (polyclad flatworms); William A. Newman, Scripps Institution of Oceanography (barnacles); Junji Okuno, Natural History Museum and Institute, Chiba (shrimps); Bob and Tina Owens, Kailua-Kona, Hawai`i (diving, photos); Gustav Paulay, University of Guam (echinoderms); David L. Pawson, Smithsonian Institution (sea cucumbers); Cory C. Pittman, Fairfield, Washington (specimen collecting, diving, observations, review of entire ms.); Joseph Poupin, Laboratoire d`Océanographie, Ecole Navale du Poulmic, Brest (hermit crabs); John E. Randall, B.P. Bishop Museum (photos); Stewart B. Reid (octopuses); Ed Robinson, Ed Robinson's Diving Adventures, Kīhei, Maui (photos); David R. Schrichte, Honolulu, Hawai`i (photos); Pauline Fiene Severns, Kīhei, Maui (opisthobranchs, specimen collecting, diving); Mike Severns, Mike Severns Diving, Kīhei, Maui (photos, diving); Holger Schramm, Waikīkī Aquarium (photos); Susan Scott, Honolulu, Hawai`i (medical, photos); Tom Shockley, Dive Makai Scuba Charters, Kailua-Kona, Hawai`i (diving); Jan Short, B.P. Bishop Museum (reference); Keoki Stender, Honolulu, Hawai`i (corals, photos, observations); Marcia A. Stone, Honolulu, Hawai`i (proofreading and TLC); Darrell Takaoka, Honolulu, Hawai`i (specimen collecting, observations); Chris Takahashi, Honolulu, Hawai`i (specimen collecting); Vici Tate, Kīhei, Maui (specimen collecting); Craig Thomas, Wahiawa General Hospital (medical); Mary K. Wicksten, Texas A & M University (shrimps).

WHAT IS AN INVERTEBRATE?

Any animal without a backbone is an **invertebrate**. Over 90 percent of animals fall in this category, including the insects, crabs, shrimps, snails, slugs, clams, octopuses, worms, sea stars, sea urchins, sponges and a host of lesser-known creatures. Invertebrate animals exist in enormous variety, especially in the sea.

Animals with backbones are called **vertebrates.** Although more familiar to us, vertebrates form only a small minority of Earth's living creatures. They include mammals, birds, fishes, reptiles and amphibians. Being vertebrates ourselves, we humans tend to lump the invertebrate animals together in a group. In truth, however, they form no natural group; the various types of invertebrates are as different from each other as we are from them. The word "invertebrate," therefore, is mostly a term of convenience.

WHAT IS A PHYLUM?

Just as vertebrate animals are organized around a backbone, or vertebral column, other groups of animals have their own distinct body types. Insects, for example, have jointed legs and a hard exterior skeleton. So do crabs, lobsters, and shrimps. These and other creatures with jointed

Top: Most people recognize fishes as animals; fewer realize that the subtly-hued coral gardens through which they swim are also animals—colonies of small limestone-secreting invertebrates. (Ornate Butterflyfish, *Chaetodon ornatissimus*, Cauliflower Coral, *Pocillopora meandrina*, and Lobe Coral, *Porites lobata*. Molokini Islet, Maui. 20 ft.

Right: Marine invertebrates often have unfamiliar forms. The spiral feathery structures are the twin tentacle crowns of a worm that lies buried in coral. Each tiny pit in the coral contains a tiny flower-like animal, or coral polyp. (Christmas Tree Worm, *Spirobranchus giganteus*, in Lobe Coral, *Porites lobata*). Mākua, O`ahu. 90 ft.

Right: The phylum Arthropoda consists of animals with hard shells and jointed appendages, such as this Banded Spiny Lobster *(Panulirus marginatus)* photographed at the Lāna`i Lookout, O`ahu.

Bottom: Animals of the phylum Mollusca have soft, legless bodies with or without a shell. This large red Spanish Dancer sea slug *(Hexabranchus sanguineus)* was photographed at Pūpūkea off O`ahu's north shore.

appendages and external skeletons (there are many) form a grouping of animals called a **phylum**—in this particular case, the vast phylum Arthropoda (meaning "jointed legs").

Another enormous grouping, the phylum Mollusca, consists of soft-bodied animals that are often protected by a hard outer shell of calcium carbonate. Snails, clams, oysters, slugs, octopuses, and squids are all molluscs.

Vertebrates—birds, fishes, amphibians, reptiles, and mammals—belong to the phylum Chordata. All chordates have a backbone-like structure. A few, however, possess this structure for only part of their life cycle and are therefore considered invertebrates. Tunicates, or sea squirts, which often resemble sponges to the casual observer, are a good example. A tunicate's "backbone" exists only during the larval stage and is visible only through a microscope.

As seen above, an animal's external appearance can be misleading. No one would ordinarily suppose sea squirts to belong to the same phylum as human beings. Careful dissection of their larvae, however, shows this to be the case. Barnacles are another example. Superficially they resemble molluscs; when their larvae are examined they are seen to be arthropods.

Right: Invertebrate animals of three phyla are visible in this picture. White-Margin Nudibranchs (phylum Mollusca) feed on a sponge (phylum Porifera), which is carried as camouflage by a Sleepy Sponge Crab (phylum Arthropoda). Pūpūkea, Oʻahu. 20 ft.

Bottom: Marine invertebrates can be difficult to identify by external appearance. This tan colonial tunicate (phylum Chordata) resembles an encrusting sponge (phylum Porifera) but is actually an evolutionarily advanced animal belonging to the same phylum as human beings. Lānaʻi Lookout, Oʻahu. 15 ft.

There are 30-33 named phyla (depending on whom you consult). Except for the phylum Chordata, all consist entirely of invertebrate animals. Although members of some phyla (principally the Chordata, Arthropoda, and Mollusca) have adapted completely to life on dry land, almost all the rest are aquatic, populating oceans, rivers and streams, or the moist inner bodies of larger animals.

 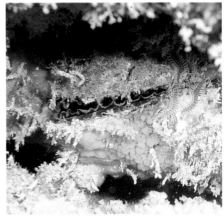

Day Octopus *(Octopus cyanea)*   Cliff Oyster *(Spondylus violacescens)*

Classes of animals within a phylum can vary enormously. Compare the intelligent, mobile octopus with its eight prehensile arms (phylum Mollusca, class Cephalopoda) with the headless, brainless Cliff Oyster (phylum Mollusca, class Bivalvia) that lives fixed in place under a ledge, its shell covered with sponges and worms.

WHAT IS A CLASS?

The phyla of the world are divided into **classes**. Mammals, birds, reptiles, amphibians, bony fishes, and cartilaginous fishes all form separate classes within the phylum Chordata. Classes within the phylum Arthropoda include insects (class Insecta); spiders (class Arachnida); and lobsters, crabs and shrimps (class Crustacea). Molluscs are divided principally into snails and slugs (class Gastropoda); clams, oysters and other bivalves (class Bivalvia); and octopuses and squids (class Cephalopoda).

Classes in turn are divided into **orders**, orders into **families**, and families into **genera** (singular: **genus**). Each descending level represents a group of more closely related animals. Last is the **species**, the basic unit of biological classification. The **subspecies** is sometimes used, especially in older works, to denote regional or other variations within a species.

WHAT IS A SPECIES?

In general, a species can be defined as an interbreeding population of animals or plants. Species of larger animals are usually recognizable by external characteristics. For example, there are about 20 species of common shallow-water hermit crabs in Hawai`i. Most have distinct color patterns. With a little effort anyone can learn to tell one of these hermit crab species from another. Many of the smaller species of invertebrate animals, however, require a microscope to identify with confidence; others, such as tunicates, are classified on the basis of internal anatomy and usually require dissection.

WHAT'S IN A NAME?

Sea cucumbers *(Holothuria cinerascens)* line the walls of a sponge-filled crevice at the Lāna`i Lookout, O`ahu. The English name for these animals comes from their shape. The scientific name combines *holothurion*, the Greek word for "sea cucumber," and *cinerascens*, Latin for "ash-colored." This book calls the species "Ashy Sea Cucumber." In Hawaiian it is traditionally called **loli pua**. **Loli** means "sea cucumber" and **pua** means "flower," in reference to the radiating tentacles, visible at left.

WHAT'S IN A NAME?

The formal, scientific names of animals consist of two parts: the genus and the species. This "binomial nomenclature," invented by the great Swedish naturalist Linnaeus (1707-1778), is used in the naming of all animals and plants. The genus is always capitalized, while the species is entirely in lower case, as in *Octopus ornatus*. Both names are italicized.

Binomial names are often, but not always, composed of descriptive Latin or Greek words. *Octopus* is a genus of the family Octopodidae. Both names mean "eight arms." The species name *ornatus* indicates precisely which octopus we mean, in this case the Ornate Octopus. Following the two-part name is the name of the biologist(s) who first formally described the species and the year in which the description was published. Thus we have *Octopus ornatus* Gould, 1852. The authorship is considered part of the full scientific name and functions as a kind of "hyperlink," enabling researchers to more easily find information about the animal. Parentheses around the author's name indicate that the originally assigned genus has been changed.

It has often happened (especially in the last century) that two or more scientists, working in different parts of the world, independently "discovered" and named the same animal. Rarely did they give it the same name. Confusion resulted, with some species of animals receiving a dozen or more published scientific names. The rules of scientific nomenclature state that, within certain limits, only the first published name for a species is valid. Later (invalid) names are known has "synonyms." Great progress has been made in recent years toward sorting out synonyms and valid names. That work is still going on, and some of the scientific names in this book will undoubtedly be revised in the future. Similarly, older books may use a scientific name different from the one given here. Classification of animals (and plants) is always subject to change.

Scientific names can be intimidating to nonscientists. They look difficult and hard to pronounce, and they frequently do not connect in any obvious way with the actual organism. Nevertheless, if one wants to communicate precisely about plants and animals, scientific names have no substitute. This book tries to make the scientific names more meaningful by providing translations or explanations whenever practical.

Common or popular names present a dilemma: most invertebrates have none. As a convenience to readers unfamiliar with Latin or Greek, English-language names have been "invented" in this book when necessary. Readers with a more than passing interest in the animals, however, are urged to learn their scientific names.

In the Hawaiian language, animal names are often in two parts: a general name coupled with a specific descriptor (similar to the genus and species of a scientific name). Thus **loli** (sea cucumber) and **pua** (flower) join to form **loli pua**, a sea cucumber with radiating, flower-like tentacles. Many species or groups had more than one Hawaiian name, which often differed from island to island. Unfortunately, by the time anyone thought to record Hawaiian names and fishing lore, much of the old culture and knowledge had been lost. In many cases only the general name has survived. Many of these general Hawaiian names remain in common use, especially for edible animals. Translations have been provided wherever practical.

ISOLATION AND ENDEMISM

Scientists believe that most tropical marine life—even that of the tropical Atlantic and Caribbean—originated in the ancient warm seas around Indonesia and the Philippines. More species of marine animals are found in these waters than anywhere else, and their number decreases markedly as one moves away. Shallow-water animals and plants have spread slowly from this "center of dispersal," moving from island to island or along the shores of continents.

Pacific island groups tend to lie close together, forming a series of stepping stones along which animals can disperse. The Hawaiian Islands, however, are an exception. Discounting tiny Johnston Atoll lying to the south, they are separated from all other Pacific islands by distances of more than 1,000 miles. This gap—greater than the distance between any other Pacific islands and their neighbors—is probably the most important factor influencing the composition of Hawai`i's marine fauna.

Almost all marine animals begin life as minute drifting larvae. Carried varying distances by ocean currents, they eventually settle to the ocean floor or other suitable habitat to mature. The ancestors of most Hawaiian reef and shore animals arrived here the same way as drifting larvae. But so isolated were these island waters that only species with long-lasting larval stages survived the journey. Those with brief larval stages died in the open sea. Distance acted as a filter.

Crossing the gap was only the first challenge. Having arrived in Hawaiian waters, a species still had to find favorable habitat and suitable food. Lacking these it would quickly perish. To reproduce, it had to arrive in numbers sufficient for males and females to mature at the same time and find each other. Because of these winnowing effects, far fewer marine species occur in

The Hawaiian Pom-Pom Crab *(Lybia edmondsoni)* occurs only in the Islands. A similar anemone-bearing species from which the Hawaiian species probably descended occurs elsewhere in the tropical Indo-Pacific. Lāna`i Lookout, O`ahu. 30 ft.

Hawai`i than in Indo-Pacific locations such as French Polynesia, Micronesia, Australia, Thailand or even far-flung East Africa. Indeed, whole groups of animals common in these areas are absent from the Hawaiian shallow-water fauna. Crinoids and basket stars, for example, are not found in the main Hawaiian Islands at depths attainable by scuba divers, and soft corals are so poorly represented that they may seem entirely missing to the casual observer.

Isolation, however, works two ways. Although impoverishing Hawai`i's fauna on one hand, it has enriched it on the other. The great distance between Hawai`i and other islands made possible the emergence of many new species. About 20 percent of Hawaiian marine invertebrates and 24 percent of Hawaiian fishes are unique to the Islands. Most probably evolved from Indo-Pacific ancestors that arrived by chance in the distant past. A few may be "relics," species once widespread that died out everywhere else. The occurrence of unique species in a limited geographical area is called **endemism**.

Isolation encourages endemism because small, confined species populations are easily affected by genetic changes. Favorable mutations establish themselves quickly in the gene pool, organisms become progressively better adapted to their environment and, given enough time, often evolve characteristics distinct from those of their ancestors. The little anemone-bearing Pom-Pom Crab *(Lybia edmondsoni)* is a good example. Although similar to its Indo-Pacific cousin *(L. tessellata)*, it differs sufficiently in color pattern and other details to be considered a separate Hawaiian species. Only a few places in the world have a percentage of endemic animals comparable to Hawai`i's. Easter Island, also greatly isolated, is one; the Red Sea region, isolated not by distance but by geology, is another.

Gaskoin's Cowry *(Cypraea gaskoini)*, one of Hawai`i's beautiful endemic cowries. Kahe Point, O`ahu. 20 ft.

In Hawai`i the Elegant Hermit Crab *(Calcinus elegans)* has orange bands on its dark brown legs. Elsewhere in the Indo-Pacific its bands are bright turquoise blue. Hekili Point, Maui. 3 ft.

Among the best known of Hawai`i's endemic sea animals are nine species of cowries, whose shells are prized by collectors worldwide. Over the years, however, the list of endemic cowries has changed considerably. Some "endemics" have been dropped as their wider geographic range became known; others, newly discovered, have been added. Any list of Hawaiian endemics is likely to change in the same way.

In addition to the endemics, there exist a number of Hawaiian sea animals that differ distinctly in size, color, or other details from members of the same species elsewhere. Hawaiian Tiger Cowries, for example, grow up to six inches long, dwarfing those from any other part of the Indo-Pacific. Large or small, however, these animals are in other respects much the same and all are assigned to the species *Cypraea tigris*. Similarly, Elegant Hermit Crabs *(Calcinus elegans)* from Hawai`i sport bright orange bands on their dark brown legs; everywhere else the bands are bright blue. Unable to find other significant differences, zoologists stop short of declaring these Hawaiian hermits a separate species.

The Red Pencil Urchin *(Heterocentrotus mammillatus)* occurs throughout the tropical Indo-Pacific but is abundant only in Hawai`i. Here it is seen between colonies of Blue Octocoral *(Anthelia edmondsoni)*, an endemic soft coral. Honokōhau, Hawai`i. 3 ft.

Another characteristic of Hawaiian reefs is the abundance of certain animals that are uncommon everywhere else. Red Pencil Urchins *(Heterocentrotus mammillatus)* occur sparsely throughout the Indo-Pacific but in Hawai`i can be astonishingly plentiful, adding bright red splashes of color to our reefs. Perhaps their predators and competitors had brief larval stages and were left behind.

The careful observer is sure to find many other differences between Hawaiian marine animals and their counterparts elsewhere in the Indo-Pacific. In addition, the relatively small number of species in the Islands and the somewhat simplified ecosystems make Hawai`i ideal in many respects for the study of other aspects of marine biology and ecology.

WHERE TO LOOK FOR MARINE INVERTERBRATES

Coral grows abundantly at O`ahu's prime snorkeling spot, Hanauma Bay Nature Park, although most is in deeper water accessible only to divers.

Hawai`i's undersea environment differs somewhat from the tropical Indo-Pacific norm. The main islands are geologically young, surrounded by slightly cooler waters than most other tropical island groups, and subject on the northwestern sides to enormous seasonal surf. As a consequence, Hawai`i's underwater landscapes often appear barren. Rocky bottom with scattered low corals is typical. Although fish are usually abundant, invertebrate animals may seem scarce. You need to know where to look.

Along sheltered leeward coasts and in a few protected bays where conditions are favorable for coral growth, living coral cover approaches 100 percent. This coral reef environment is excellent habitat for invertebrate life. On O`ahu, coral reefs occur at Hanauma Bay, Kāne`ohe Bay, and Kahe Point; on Maui at Honolua Bay, Āhihi-Kīna`u, Olowalu, and the protected inner crater of Molokini Islet; on the Big Island at Hilo Bay, Puakō, Hōnaunau, Kealakekua Bay and many other spots along the Kona coast. Most of these sites are accessible from shore and offer good diving and snorkeling.

Above: Rocky coastlines such as this near the Lāna`i Lookout on O`ahu's southeastern shore are full of submarine walls, caves and crevices—rich hunting grounds for marine invertebrates. This is scuba diving territory; the seas here are generally too rough and deep for snorkeling, but during calm weather many animals can be found in splash pools at low tide.

Left: Excellent snorkeling in the rocky shore habitat can be had along O`ahu's north shore during the summer. Here snorkelers explore the west side of Waimea Bay.

Equally good but quite different habitat occurs along Hawai`i's exposed rocky shores and cliffs. Although coral cover in this environment is generally sparse, walls, crevices, undercuts, caves, lava tubes and canyons harbor invertebrates not often found on coral reefs. The sheer back walls of Molokini Islet, Maui, and Lehua Rock, Ni`ihau, offer dramatic diving of this type; numerous locations off Kaua`i, Lāna`i, and the Big Island are also memorable. Many of these sites require a boat for access. Excellent shore diving (and some snorkeling) in rocky habitat are to be had at Pūpūkea, Mākua, the Hālona Blowhole, and the Lāna`i Lookout, all on O`ahu, and at "Black Rock" at Kā`anapali, Maui. A combination of rocky shore and coral reef habitats occur in many places along the Kona coast of the Big Island.

Shallow reef flats such as this one near Black Point, O`ahu, are home to many invertebrate animals. They are best explored at low tide when turbulence is at a minimum.

Other Hawaiian habitats good for "critter hunting" are sand and rubble bottoms, shallow reef flats (such as at Kewalo and Black Point, Oahu), tide pools, brackish (anchialine) pools, mangroves, and even harbors. (The boat channel at Magic Island, O`ahu, is a harbor-like environment where diving is allowed.) Each of these areas has its special animal and plant inhabitants; all are worth exploring.

Although emphasis in this book is on the larger invertebrates living on rocky coasts and coral reefs in clear water, animals from other habitats are not omitted. The reader—whether diver, snorkeler, surfer, or beachgoer—stands an excellent chance of finding almost any animal he or she encounters, or something very much like it, in these pages.

Marine invertebrates generally are most active at night. Night diving and snorkeling, however, are not for everyone. Fortunately, many animals can be seen during the day by peering into crevices and caves with the aid of a good underwater light. A good naturalist/dive guide can be invaluable in knowing where to look. Years of experience have given these professionals a wealth of local knowledge and a keen eye for spotting well-camouflaged animals. If you have trouble finding some of the creatures in this book, avail yourself of their services. Dive charter companies that have been especially helpful to the author are listed in the Acknowledgments.

Hawai`i desperately needs more Marine Life Conservation Zones such as O`ahu's Hanauma Bay, where all marine life is protected.

For most of human history the sea and its creatures have been considered inexhaustible. People have taken what they needed and more, and there has always been enough. Today, however, our population is too large and our technology too efficient for this to continue. Fish stocks are plummeting. On the reef, the big fish are mostly gone and the small ones are in danger of following. Marine invertebrates, too, are at risk. Almost all black coral at sport diving depths has been taken for the jewelry industry, and many sports divers are stripping our shores of lobsters and cowries. Even the lowly limpets are vanishing.

The time has come to restrain ourselves. Most of us truly do not need the living animals that we take from the sea. Removing them impoverishes the environment and lessens the diving or snorkeling experience of others. **Don't remove anything from the reef that you do not truly need.** Let the creatures be. If you keep an aquarium, take only a few animals that you know you can feed and care for. Return them where you got them when done. If you move a stone to see what is underneath, replace it carefully. If you must collect live shells, take only one—not every one that you see. If you fish, practice catch and release. Support the creation of marine parks so that animals will have places to mature and reproduce undisturbed. Scientists estimate that a miminum of 10 percent of Hawaii's coastline should be protected, and we are nowhere close to that.

Finally, if you need a challenge and a goal to make diving fun, try finding, identifying and perhaps photographing, all the animals in this book. Learn how they live, what they eat, what their predators are, how they protect themselves. In the process, you are almost certain to find something new—a behavior, an association, or even a new species. There is still an enormous amount to learn about life under the sea, and discovery is the greatest thrill of all.

SPONGES
PHYLUM PORIFERA

Sponges of different colors compete for space on the wall of a cave. At least 8 of the species described in this chapter are visible. Lāna'i Lookout, O'ahu. 20 ft.

Sponges are often called the simplest multicelled animals; in truth, they are surprisingly complex. Living filters with no mouths, stomachs, gills, hearts or other organ systems, they draw water in through a multitude of small pores, strain out plankton and organic particles, and return filtered water to the sea. Because they appear not to move, early naturalists considered sponges to be plants. Only after they were found to pump water were sponges recognized as animals.

The basic life functions of sponges are carried out by individual cells rather than by tissues or organs. In this respect sponges differ from all other multicelled animals. There are many types of sponge cells. Some secrete spicules and protein fibers that form a skeleton, others beat the tiny hairlike flagella that move water through the system, and yet others strain out food particles or have reproductive functions. Some cells roam freely within the sponge and certain amoeba-like ones can transform into any other kind of cell as needed. Additional cell types exist whose functions are unknown. Although apparently simple, the inner workings of sponges are sophisticated, elegant and not yet completely understood. Zoologists sometimes debate whether these diverse assemblages are colonies of individual animals or individuals in their own right (most today lean toward the latter view), and none can really explain how they organize themselves. In a famous experiment replicated many times, a soft sponge is pressed through fine silk to completely separate the cells. Incredibly, the individual cells eventually recombine into a new sponge. Recent research reveals that sponges are not as fixed in structure and sedentary as once thought. Some species can slowly "crawl" amoeba-fashion by absorbing skeletal elements on one side and depositing them on the other.

Most sponges are marine and live in relatively shallow water on hard substrates; some are adapted to soft bottoms. Free-standing species often form tubes, vases, barrels or amorphous masses. Encrusting sponges coat hard surfaces. Boring sponges excavate chemically into coral limestone or mollusc shells. Sponge surfaces may be hard, soft, spiny, rough or slimy. These animals occur in almost every conceivable color. Vivid reds, oranges, purples, yellows and blues are common, as well as dull browns and grays. Since many sponges are toxic, the brighter colors may serve to warn grazing fish away.

Sponge "bodies" are usually composed of a nonliving skeleton, an amorphous gelatinous substance (the mesohyl) containing numerous living cells, and a skinlike covering. Together these form a complex system of inner chambers and canals open to the surface. Water enters through numerous small openings (incurrent pores, or ostia) and exits through a lesser number of large ones (excurrent pores, or oscules). The latter are often raised above the surface of the sponge. Special chambers lined with tiny whiplike flagella function as pumping stations to force the water through the increasingly small spaces where microscopic organic matter, much of it extremely fine, is filtered out by specialized feeding cells called collar cells (choanocytes).

Most sponge skeletons are composed of a fibrous protein, spongin, stiffened by spicules of silica. Some, however, lack spicules, the familiar "bath sponge" for example.) Other sponge skeletons are purely mineral, formed from fused spicules of calcite or silica. Sponges may be so variable in shape and color, even within a species, that microscopic examination of the spicules and other structures is necessary to identify them with certainty.

Like many marine invertebrates, sponges are usually hermaphroditic, combining both sexes in a single organism. Because their gametes mature at different times, individual sponges cannot fertilize themselves. Typically they release sperm into the surrounding water to be captured by other sponges and conveyed to eggs held within. Fertile eggs hatch into free-swimming larvae that settle in a suitable spot and become new sponges. Sexual reproduction, however, is generally a less common means of propagation and dispersal than either simple fragmentation (in which a bit of the animal breaks off and grows into a new sponge), or budding (where tiny new individuals grow from a mature one).

Sponges form close associations with other organisms. Algae living in or among the cells of some species produce excess sugars that help feed their host. The algae benefit too from this arrangement, gaining a safe place to live and utilizing sponge waste products as "fertilizer." Shrimps and brittle stars, often in considerable numbers, may inhabit the inner spaces of sponges; zoanthids and stinging scyphozoan polyps may colonize the outer surface. Some crabs carry sponges on their carapaces as camouflage or possibly as chemical protection from predators.

Because they manufacture compounds not found in other animals or plants, sponges are of great interest as a source of pharmaceuticals. Some of their products have antibiotic or anticancer properties. Others irritate human skin. For this reason, and also because of their often sharp spicules, it is best not to handle these animals.

Sponges separate easily into three evolutionary groups. Those of the enormous class Demospongiae have skeletons either of spongin, silica spicules, or both. Sponges of the relatively small class Calcarea have skeletons composed of calcium carbonate spicules. The deep sea "glass sponges" (class Hexactinellida) have pure silica skeletons. (The "stony sponges," which have skeletons formed of spongin and silica spicules resting on a hard base of calcium carbonate, are sometimes placed in a fourth class, the Sclerospongiae. Today, however, most specialists place them in the Demospongiae.) There are probably about 15,000 species of sponges worldwide, of which only 30 percent have been formally described. An ancient and successful group, sponges constitute an evolutionary branch of their own that has given rise to no other forms of life.

In Hawai`i most sponges form irregular masses or thin encrustations. The large free-standing vase, tube and barrel formations conspicuous on Caribbean and Indo-West Pacific coral reefs are almost completely absent. Colorful sponges, however, often blanket the walls of caves and crevices along Hawai`i's volcanic shores, especially in areas of constant surge or current. Sponges are also common under stones or rubble and on docks, pilings or mangrove roots in protected locations such as Pearl Harbor, Honolulu Harbor and the quiet backwaters of Kāne`ohe Bay, O`ahu. Existing literature records 84 marine sponge species in Hawai`i, mostly from shallow-water habitats. Many more remain to be discovered. Sponge identification can be extraordinarily difficult and some of the names presented here are tentative.

In ancient Hawai`i sponges were known as **hu`akai** ("foam of the sea") or **`ūpī** (from the word meaning "to squeeze"). With the exception of a hard cave sponge (noted below), they were not generally used as tools, food or medicine.

CLASS CALCAREA

Sponges of the class Calcarea have skeletons composed of calcareous spicules. Because there is no spongin in the skeleton, calcareous sponges are somewhat brittle. They are usually small.

PINK LEUCETTA

Leucetta sp.
Order Clathrinida. Family Leucettidae
• This small, pink calcareous sponge is common on the ceilings of overhangs and caves exposed to moderate surge. It forms irregular rounded lobes 2 in. or less across. Photo: Palea Point, O`ahu. 30 ft.

WHITE LEUCETTA

Leucetta solida de Laubenfels,
1950
Order Clathrinida
Family Leucettidae
• White calcareous sponges of this somewhat doubtful species form small irregular lobes with prominent excurrent pores. Common under ledges and along shaded vertical walls in areas with good water movement, they are especially numerous on O`ahu's north and south shores. Tiny shrimps, *Discias exul*, sometimes inhabit cavities in the sponge; the large Spanish Dancer nudibranch, *Hexabranchus sanguineus* (p. 172), reportedly feeds on it. Usually no more than 2 in. long. Worldwide in warm waters. Photo: Pūpūkea, O`ahu. 30 ft.

CLASS DEMOSPONGIAE

This large class consists of sponges whose skeletons are usually made of the fibrous protein spongin strengthened by spicules of silica (a substance similar to glass or quartz). Spicules are absent in some species. About 95 percent of all known sponges are demosponges.

ACQUISITIVE SPONGE

Dysidea cf. *avara*
(Schmidt, 1862)
Order Dendroceratida
Family Dysideidae
• Sponges of this family are able to construct a skeleton of foreign matter that partially or completely replaces their own skeleton. This one uses sand grains, visible in a cobweb-like pattern on its surface. It is pinkish or dull lavender. The surface is coarse and the texture very spongy. The abbreviation "cf." indicates that this Pacific species is similar but probably not identical to *D. avara*, a Mediterranean sponge. The drug AZT used in AIDS therapy was developed from a chemical (avarol) contained in the Mediterranean species. The name means "eagerly desirous" or "greedy." Photo: Moku o Lo`e (Coconut Island), Kāne`ohe Bay, O`ahu. 3 ft. (on mangrove root)

YELLOW DACTYLOSPONGIA

Dactylospongia sp.
Order Dictyoceratida
Family Spongiidae
• This common Hawaiian sponge, apparently undescribed, forms massive encrustations up to several inches thick on the walls and ceilings of overhangs and caves exposed to current or surge. Prominent excurrent pores may form a line along a central ridge. Color is dull yellow to greenish yellow. A curious chemical reaction takes place when this sponge is preserved: placed in alcohol, it turns dark within a few minutes; the alcohol turns pink and then dark red. Some specialists place this sponge in the genus *Luffariella* (family Thorectidae). Sponges may be 12 in. across. Photo: Hālona Blowhole, O`ahu. 15 ft.

YELLOW LUFFARIELLA

Luffariella metachromia (de Laubenfels, 1954)
Order Dictyoceratida. Family Thorectidae
• The surface of this common, light yellow sponge is covered by many small points interconnected by ridges. When brightly illuminated it glistens. It occurs under overhangs and in caves in areas of strong to moderate current or surge, often with the *Dactylospongia* species above. Tiny endemic nudibranchs, *Hypselodoris andersoni* (p. 165), sometimes feed on it. When cut (not recommended) the sponge exudes copious mucus. Individual sponges are usually less than 6 in. across. Photo: Kea`au Beach Park, O`ahu. 50 ft.

Hippospongia cf. *ammata* de Laubenfels, 1954
Order Dictyoceratida
Family Spongiidae
• This sponge grows in the open on mixed sand and rock at depths of 20 ft. or more along exposed shores. The surface is usually covered with sand and detritus; the texture is supple and resilient. This and other species of the order Dictyoceratida lack hard spicules. Some make good bath sponges and are harvested commercially. The name derives from the Greek *amathos*, meaning "sand." Photo: Lāna`i Lookout, O`ahu. 20 ft.

BLACK REEF SPONGE
Spongia oceania de Laubenfels, 1950
Order Dictyoceratida. Family Spongiidae
• This sponge grows in the open on hard substrate in areas of strong surge. The surface is dull black and relatively smooth with many small excurrent pores. Although the skeleton lacks spicules, its texture is too firm for commercial use. In the late 1940s this was apparently the most common large Hawaiian sponge. The shallow reef top just north of the entrance to Hanauma Bay, O`ahu, was said to be thick with them, many less than a yard apart and larger than a human head. Few occur there now. It was also reported as common along the Kona coast of the Big Island. Known only from Hawai`i. To almost 16 in. across. Photo: Hālona Blowhole, O`ahu. 15 ft.

GRAY CACOSPONGIA

Cacospongia sp.
Order Dictyoceratida. Family Thorectidae
• Always covered with fibers and detritus, this gray-black sponge blends with its surroundings and is scarcely recognizable. Its presence is often betrayed by the White-Margin Nudibranch, *Glossodoris rufomarginata* (p. 164), which feeds on the sponge and lays coiled ribbons of eggs on its surface. The sponge is common under overhangs and in other shaded areas. Photo: Pūpūkea, Oʻahu. 30 ft.

MEANDERING SPONGE

Chondrosia chucalla de Laubenfels, 1936
Order Hadromerida
Family Chondrillidae
• This sponge is abundant on some exposed rocky shores at depths of 3-20 ft., forming low, irregular, sometimes branching or meandering encrusting lobes about 1/2 in. thick, about 1 in. wide, and several inches long. When growing in the open it is grayish black. In crevices and caves it may be gray, sometimes forming an almost continuous encrusting mat. The surface is smooth, and the flesh inside is light in color. Perhaps because it lacks either a mineral or a spongin skeleton, this sponge is often eaten by the Moorish Idol fish *(Zanclus cornutus)*. It has an Indo-Pacific distribution. Photo: Hālona Blowhole, Oʻahu. 15 ft. [The paper-thin, encrusting orange sponge is *Prosuberites oleteira* de Laubenfels, 1957 (order Hadromerida, family Clionidae).]

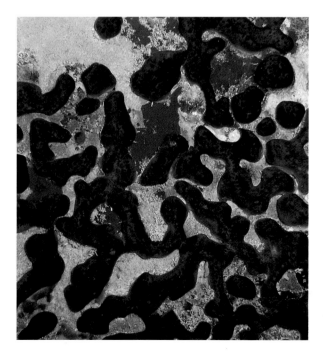

VAGABOND BORING SPONGE

Spirastrella (Spheciospongia) vagabunda Ridley, 1884
Order Hadromerida. Family Spirastrellidae
• This sponge bores into limestone rock, probably by means of acid secretions. Excurrent tubes (fistules) protruding above the substrate are the only visible parts of the organism. About 20 sponge genera in a number of different orders share a similar growth form and habitat. Sponges such as these contribute greatly to the breakdown of coral rock. Photo: Kahe Point, O`ahu. 30 ft.

BLUE SUBERITES

Suberites sp.
Order Hadromerida. Family Suberitidae
• This dark blue sponge forms hard, paper-thin encrustations on the walls of crevices and caves exposed to constant surge. Patches are rarely more than a few inches across. The species is undescribed. Photo: Lāna`i Lookout, O`ahu. 30 ft.

VARIABLE TERPIOS

Terpios zeteki (de Laubenfels, 1936)
Order Hadromerida
Family Suberitidae
• Remarkably variable in color, this sponge is usually turquoise or red but can also be green or violet. Different color individuals may grow side by side. The internal color is always yellow. The sponge grows in harbors, often forming long, irregular lobelike projections. It is believed to be an introduced species. Photo: Pearl Harbor, O`ahu. 5 ft.

YELLOW AXINYSSA

Axinyssa aculeata Wilson, 1925
Order Hadromerida
Family Suberitidae
• Smooth rounded swellings of an intense orange-yellow characterize this sponge; dark gray specimens also occur. Excurrent pores are usually at the apex of the colony. It grows in caves or wedged deep in crevices on rocky reefs in areas of good water movement. The species name means "sharp pointed," in reference to the smooth, sharply pointed spicules that make up the skeleton. Specimens may be the size of a hand but are usually smaller. Indo-Pacific. Photo: Lāna`i Lookout, O`ahu. 20 ft.

POLYP-BEARING SPONGE

Timea sp.
Order Hadromerida
Family Timeidae
• Tiny white polyps, possibly of a deep-sea scyphozoan jellyfish, live embedded in the surface of this orange encrusting sponge. It occurs on cave walls and ceilings and in crevices. The polyps' sting may protect the sponge. Similar polyps occur commensally with many different sponges in tropical seas around the world. The polyps are about 1/25 in. across and the thin encrusting sponge may cover a square foot or more. The species is undescribed. Photo: Lāna`i Lookout, O`ahu. 30 ft.

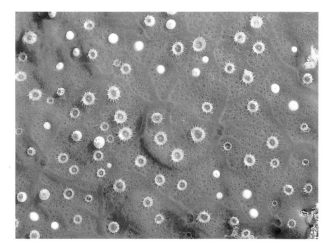

BLUE CALLYSPONGIA

Callyspongia(?) sp.
Order Haplosclerida(?)
Family Callyspongiidae(?)
• One of the few Hawaiian sponges that grow openly on the reef, this small light blue species also occurs in crevices. It is not common. Sponges of the genus *Callyspongia* have a cobweb-like surface composed of spicules. Photo: Hālona Blowhole, O`ahu. 20 ft.

PARADOXICAL SPONGE

Liosina paradoxa Thiele, 1899
Order Halichondrida. Family Dictyonellidae
• One of the few approximately tube-shape sponges in Hawai`i, this species grows openly on the reef at depths of about 25 ft. or more. It is large for a Hawaiian sponge but its tan color renders it inconspicuous. The consistency is soft and the tubes are sometimes closed at the top. Tubes are 3-4 in. high and 1-2 in. across. Indo-Pacific. Photo: Kewalo, O`ahu. 30 ft.

VIOLET CALLYSPONGIA

Callyspongia diffusa (Ridley, 1884)
Order Haplosclerida
Family Callyspongiidae
• This lavender-to-brown branching sponge is common in Kāne'ohe Bay, O'ahu, growing on mangrove roots and even among branches of coral. It is often covered with silt. Branches grow up to 5 in. long and there are typically 3-5 branches per sponge. Indo-Pacific. Photo: Moku o Lo'e (Coconut Island), Kāne'ohe Bay, O'ahu. 5 ft. (in *Montipora capitata*)

BLUE SIGMADOCIA

Sigmadocia sp.
Order Haplosclerida
Family Chalinidae
• These small, light blue sponges are common in harbors and quiet lagoons, often attached to mangrove roots. They may be the same as *S. caerulea* from the Caribbean; if so, they were probably introduced unintentionally. Individuals are usually no more than 1-2 in. across. Photo: Pearl Harbor, O'ahu. 5 ft.

AMORPHOUS CAVE SPONGE

Strongylophora sp.
Order Haplosclerida
Family Petrosiidae
• This hard stiff sponge grows deep in caves away from light, forming thick, irregular encrustations that may cover large portions of the ceiling and sides. Sometimes it forms lobes or projects into fingers. Its color is cream. It has not yet been formally described. Photo: Pūpūkea, O'ahu. 30 ft.

WAVY CAVE SPONGE • `ana

Leiodermatium sp.
Order Lithistida. Family Azoricidae
• This distinctive hard, stony sponge grows deep in caves and crevices, away from light and strong water movement. It forms erect wavy folds and compact cauliflower-size heads, usually cream or light brown in color. Known as `**ana** in old Hawai`i, it was used medicinally. A rinse or gargle made of crushed `**ana** and water was said to be effective against thrush, a yeast infection of children. Although common in Hawai`i, this sponge has probably not yet been formally described. Photo: Puakō, Hawai`i. 30 ft.

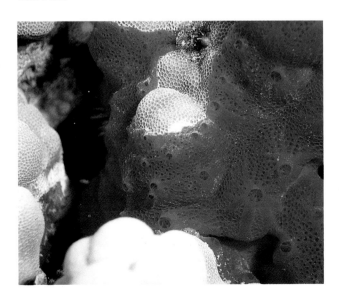

RED PHORBAS

Phorbas sp.
Order Poecilosclerida
Family Anchinoidae
• This red encrusting sponge is densely covered with pores. It is soft and easily torn. Smooth ridges sometimes radiate from the large excurrent openings. The sponge grows openly on the reef, coating the sides and undersides of dead coral and sometimes overgrowing living coral (here *Porites lobata*). It is abundant in some areas along the leeward shores of O`ahu at depths of 20-40 ft. Photo: Palea Point, O`ahu. 40 ft.

RED BORING SPONGE

Hamigera sp.
Order Poecilosclerida
Family Anchinoidae

• Like many other sponges, this species bores into coral limestone, probably dissolving it with acid secretions. The only parts of the sponge visible are soft red excurrent tubes (fistules) projecting from the surface of the rock. Sponges such as these contribute greatly to the breakdown of coral rock. Although abundant at many locations along the Wai`anae coast of O`ahu, this sponge is new to science. Photo: Kahe Point, Oahu. 20 ft.

HAWAIIAN LISSODENDORYX

Lissodendoryx hawaiiana (de Laubenfels, 1950)
Order Poecilosclerida
Family Coelosphaeridae

• This sponge is dull orange to brilliant vermilion with prominent excurrent pores that close slowly if the sponge is disturbed (taking about 5 minutes). It is soft and easily torn. This species might be a synonym of *Lissodendoryx aspera* (Bowerbank, 1875). Photo: Moku o Lo`e (Coconut Island), Kāne`ohe Bay, O`ahu.

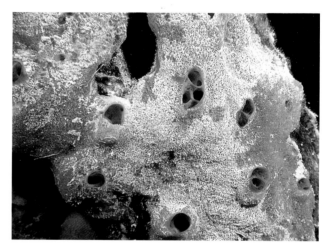

TUBULAR BIEMNA

Biemna fistulosa (Topsent, 1897)
Order Poecilosclerida
Family Desmacellidae

• The word fistula means "pipe" in Latin. This sponge encrusts vertical surfaces in quiet silty waters, often forming many soft, easily torn tubes up to several inches long. These project through the algae and other growths that typically cover the sponge. It varies from yellow to whitish. This is a toxic species that can irritate human skin. Avoid contact. Indo-Pacific. Photo: Magic Island, O`ahu. 10 ft.

13

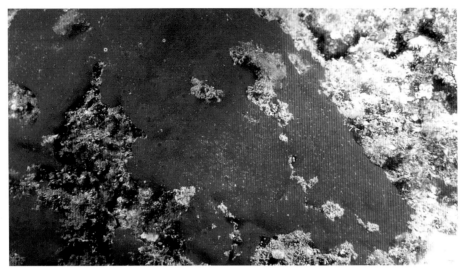

VERMILION CLATHRIA

Clathria (Microciona) sp.
Order Poecilosclerida. Family Microcionidae
• This is Hawai`i's most common encrusting sponge, ubiquitous on exposed shores and reefs to depths of at least 200 ft. It forms irregular bright red patches a few inches across, usually on vertical or near vertical surfaces, on the ceilings of overhangs and within caves. It also overgrows dead black coral skeletons. The surface is soft and smooth with slightly elevated excurrent pores. Photo: Palea Point, O`ahu. 30 ft.

RED MYCALE

Mycale (Mycale) sp.
Order Poecilosclerida. Family Mycalidae
• Much of the surface of this thick, intense red encrusting sponge is covered by algae and detritus. It occurs in quiet harbors and is common at Magic Island, O`ahu. Tiny brittle stars (*Ophiactis* sp.) live in the inner spaces, extending one or two arms from the incurrent pores to catch some of the larger particles being drawn in. This is probably helpful to the sponge (large particles might clog its pores). In return, the sponge provides a safe dwelling place for the brittle stars. Sponges of this family produce a pigmented mucus that stains or irritates human skin. Avoid contact. Other species of *Mycale*, similar in appearance, may occur in the same habitat. Photo: Magic Island boat channel, O`ahu. 15 ft.

BLACK CAVE SPONGE

Tetrapocillon sp.
Order Poecilosclerida. Family Guitarridae
• This black encrusting sponge is abundant on the sides of caves and crevices along the north shore of O`ahu. It is extremely soft and easily torn. Photo: Pūpūkea, O`ahu. 20 ft.

ORANGE STYLINOS

Stylinos sp.
Order Poecilosclerida
Family Mycalidae
• Encrusting patches of this distinctive orange sponge are common on many reefs, growing in the open. The rough-looking surface is soft and easily torn. Sponges of this family produce a pigmented mucus that stains or irritates human skin. Avoid contact. Patches are usually 2-3 in. across. Photo: Pūpūkea, O`ahu. 50 ft.

STAINING SPONGE

Iotrochota protea (de Laubenfels, 1950)
Order Poecilosclerida
Family Myxillidae
• This velvety black sponge grows on sand or rubble bottoms in sheltered bays. It stains the skin dark purple when touched. It is typically fist-size. The species name comes from the Greek word protos meaning "first." (When originally described it was thought to be the first of its genus.) Photo: Kāne`ohe Bay, O`ahu. 5 ft.

GREEN BATZELLA

Batzella sp.
Order Poecilosclerida
Family Phoriospongidae
• This vivid green encrusting sponge grows in small thin patches on the walls of caves where there is good water movement. Patches rarely exceed 1 in. diameter. The order Poecilosclerida, to which it belongs, is the largest order of living sponges. Photo: Lāna`i Lookout, O`ahu. 20 ft.

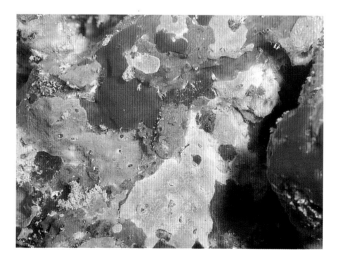

TAN PSAMMOCLEMMA

Psammoclemma sp.
Order Poecilosclerida. Family Phoriospongidae
• Like a number of other sponges, this species is able to incorporate foreign particles into its skeleton. The cobweb-like surface pattern is formed by sand grains. This sponge grows on the reef in the open. Photo: Hot water outfall, Kahe Point, O`ahu. 20 ft.

FIRE SPONGE

Tedania sp.
Order Poecilosclerida
Family Tedaniidae
• Touching this sponge can cause reddening of the skin, swelling and extreme tenderness that lasts 3-7 days. Soaking the affected area in vinegar or rubbing alcohol may relieve the pain. Hydrocortisone ointment or Benadryl might also help. The sponge is basically encrusting with elevated, chimney-like excurrent pores (oscules). If growing in the open it is bright orange-red, if in the shade it is orange or yellow. Its consistency is soft, and the flesh is easily torn. Specimens may attain the size of a hand but are usually smaller. This may be *T. ignis* from the tropical Atlantic and Caribbean. If so, it was probably introduced to Hawai`i sometime before 1950. Photo: Moku o Lo`e (Coconut Island), Kāne`ohe Bay, O`ahu. 2 ft.

CNIDARIANS
PHYLUM CNIDARIA (or COELENTERATA)

Some of the most beautiful marine animals are cnidarians. Here orange polyps of the stony coral *Tubastraea coccinea* nestle among branches of the soft coral *Carijoa riisei*. Pūpūkea, Oʻahu. 30 ft.

Cnidarians (pronounced without the "c") are simple, multicelled animals consisting basically of a stomach and a mouth. Sea anemones, jellyfishes, hydroids and corals belong to this group. All have radially symmetric, saclike or bell-shape bodies open at one end. Stinging tentacles usually surround the opening to collect food and provide protection. In many cnidarians an external skeleton cradles the soft body and positions the animal advantageously with respect to currents or light. Cnidarians may be solitary or colonial.

The basic cnidarian life cycle alternates between two adult body forms, the polyp and the medusa. Polyps are fixed in place (sessile) while medusae are free-swimming. A polyp's mouth opens upward (or away from the attached end), its tentacles resembling a flower. A medusa's mouth usually faces down, with tentacles hanging from the margin. (Medusa was a mythical Greek monster-woman who had venomous snakes hanging from her head instead of hair.) The existence of two body phases is reproductively advantageous, as will be seen below, but many cnidarians have lost the medusa stage and exist only as polyps.

The simple body cavity of cnidarians (sometimes called the coelenteron) was for many years considered their defining characteristic; for that reason they were called "coelenterates" (phylum Coelenterata). Although the older name lingers in some books, most zoologists today use the name cnidarians (from the Greek *knide*, meaning "nettle") because of the unique stinging capsules, or nematocysts, manufactured in the tentacles and other parts of these animals.

17

The cnidarians' ability to create nematocysts distinguishes them from the comb jellies of the small phylum Ctenophora (silent "c" again), which share a similar saclike radial body enclosing a central cavity.

Most cnidarians are carnivores, stunning or killing small animals that contact their deadly tentacles. The stinging mechanism consists of a tiny coiled thread tipped with a spine and coated with venom. It is stored "inside out" within the nematocyst capsule. Under the proper physical or chemical stimulus the capsule fires, everting its tiny harpoon into prey or predator. The cumulative effect of hundreds or thousands of microscopic nematocysts can be considerable. Although most cnidarians are harmless to humans, some jellyfishes, hydroids and anemones pack a powerful punch, and the infamous Sea Wasp *(Chironex fleckeri)* of northern Australia can kill a grown man in minutes. There are many types of nematocysts, however, and not all sting. Some entangle prey with long threads while others are adhesive, used by the animal to anchor itself to the substrate or to build a protective tube.

As a supplement to their prey, some cnidarians derive nourishment from sugars "leaked" by symbiotic single-celled algae (zooxanthellae) living in their tissues. In return, the zooxanthellae gain a secure, well-lit home. Through photosynthesis zooxanthellae produce about nine times more carbohydrate than they actually need, the surplus going to the cnidarian.

Cnidarians (and their close relatives the ctenophores, or comb jellies) are unique in having their cells arranged in only two layers, inner and outer, or endoderm and ectoderm. (By comparison, other multicelled animals, apart from sponges, start life with three basic layers of cells.) Filling the space between the inner and outer layers is a gelatinous substance called the mesoglea containing few or no cells; in some cnidarians, such as jellyfishes, this "jelly" layer gives shape and form to the animal.

Although lacking true organs, cnidarians possess groups of specialized cells, or tissues, for carrying out the simple life functions of digestion, reproduction, neural transmission, muscular contraction and the like. Scientists believe cnidarians were the earliest animals to evolve such tissues.

Many cnidarians secrete external skeletons, which may be flexible (most hydroids and soft corals) or rigid (stony corals). Australia's Great Barrier Reef, often called the world's largest structure, is composed primarily of the accumulated skeletons of untold billions of cnidarians cemented together by limestone- secreting algae. Many tropical islands, even entire nation states, owe their existence to these animals.

The reproductive strategies of cnidarians are interesting. Species that alternate between the polyp and medusa phases are asexual in their polyp phase. Such polyps usually reproduce by splitting in half (fission), by shedding a small piece of tentacle or foot that regenerates into a complete animal (laceration), or by budding. In the last case, a small medusa can be formed instead of a polyp. Such offspring are clones, genetically identical to their parent. Asexual reproduction is efficient and can quickly produce large colonies, but because no DNA is exchanged it provides no mechanism for evolutionary change.

Sexual reproduction in these cnidarians occurs in the medusa stage. Unlike polyps, individual medusae are always male or female, producing sperm or eggs. Fertilized eggs hatch into planktonic forms called planulae that drift in the sea for a time before settling on a suitable surface to become polyps. Although somewhat inefficient (most of the planulae perish or are eaten), sexual reproduction facilitates dispersal and produces genetic variation in the offspring, making evolution of the species possible.

Cnidarians maintain fascinating associations with other animals, often to the benefit of both. Twenty-eight species of anemonefishes live in and around ten species of large Indo-Pacific sea anemones, defending them from anemone-eating fishes. The anemones, in turn, provide safe havens for the fishes, which are resistant to the sting. Various species of small shrimps and crabs also live in anemones, and some jellyfishes and hydrozoans host small fishes in their tentacles. Some crabs carry anemones in their claws for self-defense, and several species of hermit crabs cultivate anemones on their shells, carefully removing and replacing them when they change homes. Other hermit crabs place zoanthids on their shells. The zoanthids eventually dissolve the shell and form a living covering for the crab.

The phylum Cnidaria is divided into four classes, based largely on dominant body type. Members of the class Hydrozoa (hydroids, fire corals and siphonophores) exhibit both body types, with the polyp phase typically dominant. The classes Scyphozoa (jellyfishes) and Cubozoa (box jellyfishes) have a dominant medusa phase and a reduced or absent polyp phase. Members of the class Anthozoa (stony corals, soft corals, sea anemones and zoanthids) form only polyps. Many zoologists consider anthozoans to be the most advanced cnidarians, in part because they can reproduce both sexually and asexually without a medusa stage.

Cnidarians range in size from almost microscopic polyps to magnificent jellyfish more than 6 ft. across. Over 300 species representing all four classes inhabit Hawaiian waters.

HYDROIDS and SIPHONOPHORES
PHYLUM CNIDARIA. CLASS HYDROZOA

Hydroids and siphonophores belong to the cnidarian class Hydrozoa. Members of this class tend to be most conspicuous in the polyp form and most are colonial. Many can sting humans. There are five orders of hydrozoans, of which the hydroids (order Hydroida) are the largest and most prominent.

The typical hydroid is a polyp colony on a flexible featherlike or fernlike skeleton consisting of a main stem and numerous side branches. Hydroid colonies range in size from fuzzy coatings on hard substrates to branching colonies 10 in. or more in height. The larger Hawaiian hydroids usually occur either on pilings or other surfaces in harbors, or on rocky reefs subject to strong currents or surge. Some hydroids live in the open while others prefer crevices, overhangs and caves. Among the most common is the tiny *Rhizogeton* sp. that lines the dark meandering channels created by snapping shrimps in corals of the genera *Porites* and *Montipora*. Divers or snorkelers who brush against hydroids with bare skin may feel a burning sting. Dermatitis (itching) is another possible consequence of contact with these animals.

The medusa stages of hydroids often resemble transparent jellyfishes. Most are an inch or less across. Some local species appear to rest on the bottom by day, rising into the water column at night, sometimes in considerable numbers. Hawai`i has about 25 shallow-water hydroid species, many with wide distributions. Rafting (attachment to floating objects) is believed to be a common means of dispersal.

Other members of the order Hydroida are small floating polyp colonies (sometimes called chondrophores) and "fire corals," hydroids of the genus *Millepora* which build rigid calcareous skeletons resembling those of true stony corals and deliver a burning sting. Although common on many tropical reefs, fire corals (also called hydrocorals) do not occur in Hawai`i.

The hydrozoan most often encountered by beachgoers in Hawai`i, the Portuguese Man-of-War, is a member of the specialized order Siphonophora. Siphonophores are discussed in more detail below.

CHRISTMAS TREE HYDROID

Pennaria disticha Goldfuss, 1820
Order Hydroida. Family Halocordylidae
• This is the most common large hydroid in Hawai`i, and one of the few hydroids divers are likely to notice. The colony, consisting of a dark main stem with alternating branches, often tapers like a Christmas tree; the polyps (tiny, white, cotton-like tufts on the upper sides of the branches) resemble ornaments. These hydroids can sting; the burning may persist for some time in sensitive individuals. The species grows to its largest size in protected locations, such as Kāne`ohe Bay, O`ahu, where it forms fernlike growths. Smaller hydroids, algae and other organisms may colonize its stems. The species name means "two rows." *Pennaria tiarella* and *Halocordyle disticha* are synonyms. To about 12 in. All warm seas (probably spread in part by shipping). Photo: Lāna`i Lookout, O`ahu. 20 ft.

GREEN HYDROID

Dynamena sp.
Order Hydroida
Family Sertulariidae
• This greenish hydroid grows in the open on reef slopes exposed to current. Portions of the back wall of Molokini Islet, off Maui, are covered with it. Although identified in some publications as *Dynamena cornicina*, that species was described from the Atlantic and is probably not the same as the Hawaiian one (which also occurs elsewhere in the Pacific). *Dynamena mollucana* Pictet, 1893, is close and may indeed be the same species. Branches are 1-2 in. high. Photo: Molokini Islet, Maui. 40 ft.

DIAPHANOUS HYDROID

Sertularella diaphana (Allman, 1885)
Order Hydroida
Family Sertulariidae

• These hydroids commonly occur on the ceilings of arches and caves where there is good water movement. The finely branched white colonies contrast beautifully with the colorful corals and sponges that share this habitat. Colonies almost 8 in. high have been recorded, but the species may not attain this size in Hawai`i. Those pictured here are about 1 in. It occurs in warm seas worldwide. Photo: Palea Point, O`ahu. 25 ft.

FEATHER HYDROID

Gymnangium hians (Busk, 1852)
Order Hydroida
Family Aglaopheniidae

• Caves, arches and overhangs where there is constant surge are the preferred habitat of this hydroid. It often occurs together with *Sertularella diaphana* (above). This grayish to light brown species has also been dredged from 163 fathoms north of Laysan Island. Tightly spaced branches tapering only slightly toward the tip give it a distinctly featherlike appearance. Colonies are 1-2 in. high. Indo-Pacific. Photo: Palea Point, O`ahu. 30 ft.

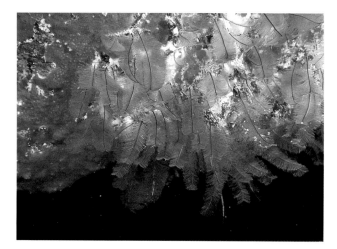

PHILIPPINE HYDROID

Macrorhynchia philippina
Kirchenpauer, 1872
Order Hydroida
Family Aglaopheniidae

• Although not abundant in Hawai`i, this is one of the most common hydroids on Indo-Pacific coral reefs. Each main stem has numerous side branches of equal length. The dark central stems contrast nicely with the white polyps. Known for its powerful sting, this species grows openly on surge-swept rocky reefs. It was described from Philippine specimens but occurs in tropical and subtropical seas worldwide. To about 6 in. high. Photo: Palea Point, O`ahu. 30 ft.

BLACK HYDROID

Lytocarpia niger (Nutting, 1905)

Order Hydroida. Family Aglaopheniidae

• The first specimens of this black to dark brown hydroid were dredged in Hawaiian waters off Laysan Island in 1902. It has since been found throughout much of the Indo-Pacific. Locally it colonizes hollows and crevices along surge-swept rocky walls, sometimes in dense patches. Each branch has numerous side branches. To about 2 in. Indo-Pacific. Photo: Lehua Rock, Ni`ihau. 40 ft.

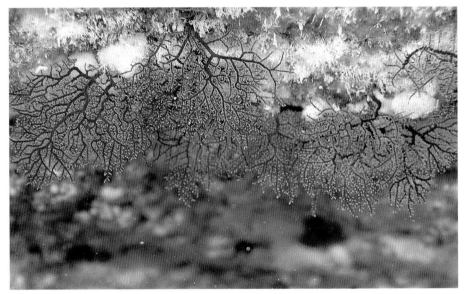

SEA FAN HYDROID

Solanderia secunda (Inaba, 1892)

Order Hydroida. Family Solanderiidae

• These uncommon hydroids form stiff branching colonies that resemble small pink sea fans. They grow in crevices, under overhangs and in cave entrances, usually upside down on the ceiling. They have also been dredged from depths of about 400 ft. To about 2 in. Indo-Pacific. Photo: Lāna`i Lookout, O`ahu. 15 ft.

BY-THE-WIND SAILOR

Velella velella Linnaeus, 1758
Order Hydroida. Family Porpitidae
• These disklike animals float on the surface and are usually encountered when blown ashore after a storm. Some authorities consider them colonies of specialized polyps, others say they are single large inverted polyps. The entire organism is dark purplish blue. The disk is oval and the short tentacles hanging from it are harmless to humans. These animals roam the seas with the aid of a short, transparent triangular sail set on a diagonal, enabling them to tack like a sailboat. The angle of the sail differs among individuals. (Some authors state that most *Velella* above the equator have sails set at one angle, while most below take the opposite angle.) In some parts of the world, By-the-Wind Sailors may blow ashore in tremendous numbers; in Hawai`i they are uncommon. To about 4 in. long and 3 in. wide, although usually much smaller in Hawai`i. All warm seas. Photo: Keoki Stender, Kahuku, O`ahu.

BLUE BUTTON

Porpita pacifica Lesson, 1826
Order Hydroida
Family Porpitidae
• This is a flat, circular, bright blue animal with a floating fringe of long feathery tentacles and no sail. The tentacles are said to be sticky, to break off on contact and to be capable of delivering a painful sting. These animals float on the surface of the open ocean and occasionally wash ashore during periods of strong winds. To 1 1/2 in., but in Hawai`i usually less than 1 in. across. All warm seas. Photo: Mike Severns. Molokini Islet, Maui.

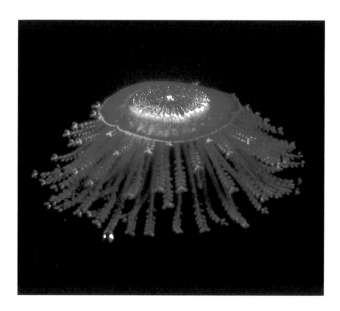

Siphonophores. Order Siphonophora

Siphonophores are pelagic colonies of specialized polyps and medusae equipped with gas- or oil-filled floats or large swimming bells. Nearly all live beneath the surface, rising and sinking in the water column by swimming, changing the amount of gas in their floats, or both. The best known siphonophore, the Portuguese Man-of-War *(Physalia)*, is unique in that it remains at the surface. Siphonophore colonies are formed from one original polyp through the process of budding. Hatched from an egg, the original polyp forms the float. Secondary polyps and medusae bud from the original polyp, each specializing in a certain function. Fishing polyps produce long, stinger-laden tentacles to catch prey; feeding polyps with shorter tentacles digest the prey; reproductive polyps create medusae that produce eggs and sperm. These medusae are not free-swimming but remain attached to the colony and often provide propulsion. The colony is buoyed up by the float and all its polyps and medusae share an interconnecting digestive tube.

PORTUGUESE MAN-OF-WAR • **pa`imalau**

Physalia physalis (Linnaeus, 1758)
Order Siphonophora. Family Physaliidae
• These open-ocean animals are not true jellyfishes (scyphozoans), as many people assume, but floating colonies of hydrozoan polyps and medusae (see above). Each colony is kept afloat by a balloon filled mostly with carbon monoxide and raised into a crest at one end to catch the wind. Sailing the open ocean, the Portuguese Man-of-War trails one or more long tentacles that paralyze or kill small fishes or organisms they happen to touch. The tentacle then contracts, drawing the prey up to be consumed. All polyps in the colony share in the catch by means of an interconnecting digestive tube. Strong winds often wash these animals up on Island beaches by the hundreds. Dark bluish purple or violet, they stand out on the sand but are almost invisible in the ocean. Humans contacting one while swimming experience a fiery sting but seldom see what hit them. Sixteenth century British sailors likened them to enemy battleships (men-of-war). Even beached animals can sting. The pain usually subsides in about 20 minutes and requires little or no treatment. Remove any adhering tentacles and flood the site with fresh or salt water. Vinegar, alcohol, urine or other liquids may do more harm than good. Ice packs may lessen the pain.
    Despite their stinging ability, *Physalia* are preyed upon by the pelagic nudibranch *Glaucus atlanticus* (p. 179) and little pelagic snails, *Janthina* spp. (p. 104) that float on rafts of bubbles. At sea they may be accompanied by small, blue-mottled juveniles of the fish *Nomeus gronovii*. (Although not completely immune, these commensals can tolerate up to ten times the amount of venom that kills other fishes.) Beached *Physalia* are eaten by the Mole Crab, *Hippa pacifica* (p. 263) and ghost crabs, *Ocypode* spp. (p. 288).
    Portuguese Men-of-War occur in all warm seas and appear to have two growth forms. In the Indo-Pacific they usually have a float 1-2 in. long and a single long fishing tentacle. In the Atlantic they usually have floats up to 1 ft. long and multiple fishing tentacles as much as 30 ft. long. The large form occurs only rarely in the Indo-Pacific. Some authorities consider the two forms separate species and use the name *P. utriculus* for the small form. Photo: David B. Fleetham. Maui.

JELLYFISHES

PHYLUM CNIDARIA. CLASS SCYPHOZOA and CUBOZOA

Pelagic scyphozoan jellyfishes such as this *Thysanostoma* are uncommon in Hawaiian waters. The jellyfish is accompanied by juvenile jacks. Its stinging tentacles protect them from would-be predators. Midway Atoll. (Mike Severns)

Free-swimming cnidarians with a dominant medusa stage are commonly called jellyfishes. Two classes exist, Scyphozoa and Cubozoa. Because the word "jellyfish" is often used for other gelatinous or floating animals such as hydroid medusae and Portuguese Men-of-War (class Hydrozoa), or even comb jellies (phylum Ctenophora), scientists usually use the more precise terms "scyphozoans" and "cubozoans." (The root word "scyphus" means "cup" in Latin; its "c" is silent.)

Scyphozoans, or true jellyfishes, are bell- or saucer-shape animals that usually swim on the surface or in midwater; cubozoans—known as box jellyfishes—are similar but with a four-sided bell. Both have four-part radial symmetry, and like other cnidarians are composed of two tissue layers separated by a jellylike mesoglea. Thicker and more substantial than that of other cnidarians, the mesoglea of a jellyfish is responsible for its firmness, shape, buoyancy and common name.

True jellyfishes may have either tentacles or oral arms; many have both. Tentacles are slender, hang from the margin of the bell, and can be long or short. They usually bear stinging nematocysts. Oral arms are thicker than tentacles. Clustering under the bell around the central mouth, if present, they can be short and clubbed, branched and bushy, or long. They often sting. (Any part of a jellyfish, including the bell, can potentially sting.) Box jellyfishes lack oral arms and always have one or more stinging tentacles hanging from each of the four lower corners of the bell.

Most jellyfishes can swim by pulsation of the bell; some, especially box jellyfishes, can move surprisingly quickly. Not all jellyfishes swim, however. The "upside down" jellyfishes of the genus *Cassiopea* rest on the bottoms of shallow, sunlit lagoons with bell side down, oral arms up. An entire order of stalked jellies (Stauromedusae) resemble polyps and spend their adult lives attached to algae or other surfaces by the bell.

Jellyfishes feed in several ways. Cubozoans and scyphozoans with long stinging tentacles "fish" for planktonic animals, crustaceans or fishes. These are the jellyfishes most likely to be dangerous to humans. Others trap plankton in a mucus coating on the oral arms. A few depend largely on the photosynthetic activity of symbiotic algae (zooxanthellae) in their tissues.

Most jellyfishes spend part of their life cycles as attached polyps. These polyps (scyphistomae) are usually small and inconspicuous, but some pack a powerful sting in their own right. In Hawai`i, tiny scyphozoan polyps visible to divers sometimes occur on encrusting sponges under ledges and in caves. Scyphozoan polyps produce tiny immature medusae (ephyrae) that eventually develop into adult bell- or saucer-shape jellyfishes. Polyps of at least one cubozoan transform directly into the four-sided adult form. Most jellyfishes are either male or female and produce sperm and eggs, respectively. Fertilized eggs hatch into larvae that eventually settle to become attached polyps, and the reproductive cycle begins anew. The adult life of most species probably lasts one year or less.

Adult scyphozoans range in size from tiny species a fraction of an inch in diameter to the enormous Lion's Mane Jellyfish or Sea Blubber *(Cyanea capillata)*, which can exceed 6 ft. in diameter, weigh over a ton and trail tentacles 130 ft. long! This primarily cold-water animal gained fame as the culprit in a well-known Sherlock Holmes mystery. About 200 species of scyphozoan jellyfishes have been described. Box jellyfishes are generally modest in size, typically from 1-2 in. diameter, but sometimes as long as 10 in. About 15 species are known.

Fifteen scyphozoan jellyfishes are recorded from Hawaiian waters, including one tiny "stalked" jelly that grows on seaweed in shallow water. Many of these Hawaiian species occur only in quiet harbors, bays and lagoons into which snorkelers and divers rarely venture. Occasionally a pelagic species drifts into shoreline waters. Box jellyfishes are another matter, sometimes washing ashore by the hundreds. They deliver a painful sting and are a regular, though short-lived, nuisance on some of O`ahu's south-facing beaches. As many as 800 swimmers have sought treatment for box jelly stings in one day at Waikīkī Beach, although such numbers are unusual. There are probably four species of box jellyfishes in Hawai`i; two are tiny and currently unidentified. The Hawaiian word for jellyfishes of any kind is **pololia**.

CLASS SCYPHOZOA. ORDER SEMAEOSTOMEAE

Jellyfishes of this order have saucer- or bowl-shape bells with scalloped margins. Slender tentacles hang from the edge of the bell and four oral arms surround a central mouth. The reproductive organs, often colored, usually show through the translucent bell.

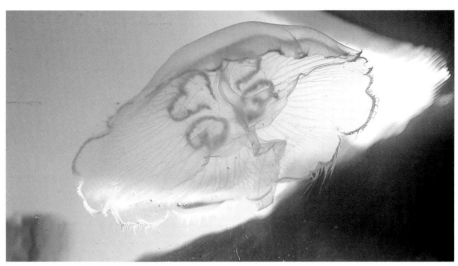

MOON JELLYFISH
Aurelia aurita (Linnaeus, 1758)
Order Semaeostomeae. Family Ulmaridae
• These beautiful, saucer-shape jellies are translucent and almost colorless. Short tentacles hang from the scalloped margin and four frilly oral arms surround the mouth. The reproductive organs, arranged in a four-part pattern, are clearly visible through the top of the bell. These animals pulsate gently and gracefully, primarily to keep themselves afloat, and feed on plankton that accumulates on their mucus-coated surface and other body parts. They are harmless to humans. In Hawai`i they occur in the quiet waters of lagoons and harbors (including brackish areas) attaining a maximum diameter of about 12 in. The species can be plentiful in open coastal waters in other parts of the world. Photo: Keoki Stender. Ala Wai Boat Harbor, Honolulu.

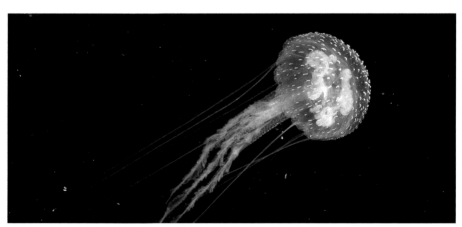

LUMINESCENT JELLYFISH
Pelagia noctiluca (Forsskål, 1775)
Order Semaeostomeae. Family Pelagiidae
• Although rarely seen in Hawai`i, this small pelagic jellyfish is well-known in warm and temperate waters around the world. The bell is covered with nematocyst-laden "warts"; eight long slender tentacles (sometimes retracted) trail from its margin. Frilly oral arms hang from the perimeter of the mouth. The animals vary from translucent to pinkish or purplish (in Australia they are called "mauve stingers) and may also be yellow or brown. At night they can be luminescent. Although painful, the sting is not considered dangerous. Large swarms of these jellyfish in the Mediterranean Sea occasionally cause concern at bathing beaches, sometimes appearing in such numbers that they also interfere with fishing. The species name means "night shiner." Typical length is about 4 in., excluding the tentacles (which can be twice that length or more). Photo: David B. Fleetham. Lāna`i. 30 ft.

Class Scyphozoa. Order Rhizostomeae

Jellyfishes of this large order have prominent oral arms and no tentacles. The oral arms can be long, bushy or clublike (and may sting). Rhizostomes have no central mouth, ingesting food through multiple secondary mouths situated along the oral arms. They are primarily tropical and subtropical animals.

UPSIDE-DOWN JELLYFISH

Cassiopea medusa Light, 1914
Order Rhizostomeae. Family
Cassiopeidae

• These unusual jellyfish rest on the bottoms of shallow lagoons bell side down, oral arms up. The bushy, many-branched arms often obscure the bell, which pulses gently and continually to convey food upward. Symbiotic algae (zooxanthellae) living in the arms and other tissues contribute sugars to the food supply and impart a brownish or reddish coloration. These animals can sting mildly, and if disturbed they release multitudes of free-floating nematocysts into the water. They probably arrived in Hawai`i via ships' ballast water and now occur in harbors and lagoons on O`ahu. The lagoon at Hilton Hawaiian Village (Waikīkī) is full of them. A similar species, *C. mertensi*, has been reported from Kāne`ohe Bay, O`ahu. Jellyfishes of this type are easy to maintain in captivity. Feed them brine shrimp and provide plenty of light. They may even reproduce, forming polyps on the side of the tank. Photo: Lagoon, Magic Island, O`ahu. 5 ft.

CROWNED JELLYFISH

Cephea cephea Forsskål, 1775
Order Rhizostomeae
Family Cepheidae

• Easy to recognize, this animal has a flat or concave upper side crowned with a small central dome often bearing conical projections. Many slender filaments may hang from the branched oral arms. This is a pelagic species that occasionally drifts inshore. The author has handled it with no ill effects. Typical specimens are about 6 in. across. Indo-Pacific. Photo: Kahalu`u Beach Park, Kailua-Kona, Hawai`i. 3 ft.

LAGOON JELLYFISH

Mastigias sp.
Order Rhizostomeae. Family Mastigiidae
• This jellyfish occurs in Kāne`ohe Bay, Honolulu Harbor and Pearl Harbor O`ahu, and perhaps elsewhere. It resembles the Indo-Pacific species *Mastigias papua* and has a mild sting. Like *Phyllorhiza punctata* (below), which it also resembles, it is likely an introduction brought in by ships via ballast water or as attached polyps. Large specimens attain a bell diameter of 5 in. Photo: Thomas Kelly, courtesy of the Waikīkī Aquarium.

WHITE-SPOTTED JELLYFISH

Phyllorhiza punctata von Lendenfeld, 1884
Order Rhizostomeae
Family Mastigiidae
• The bell of this animal varies from brown to blue and usually bears numerous opaque round white spots. From each of the eight short oral arms hangs a long slender appendage slightly swollen at the tip. These impressive jellyfishes can be common in Honolulu Harbor, Pearl Harbor and Kāne`ohe Bay, O`ahu, mainly during the winter months. The species probably arrived by ship, either as immature medusae in ballast water or as attached polyps. It occurs worldwide in warm waters and is often accompanied by juvenile fishes (jacks). Large specimens attain a diameter of about 20 in. Photo courtesy of the Waikīkī Aquarium.

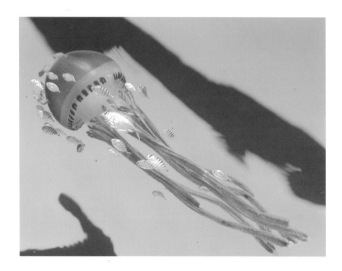

BOX JELLYFISHES. CLASS CUBOZOA

Box jellyfishes have bells with four flattened sides and a long tentacle (or bundle of tentacles) attached to each of the four lower corners. They have well-developed eyes (with lens and "retina"), and some species can swim vigorously up to 2 miles per hour. Cubozoans generally prefer calm environments but sometimes occur a mile or more offshore. Many can deliver a potent sting. The famous Sea Wasp (*Chironex fleckeri*) of northern Australia can kill an adult human in as little as 3 minutes. Hawaiian box jellies are not as potent, but they deserve respect nonetheless.

Craig Thomas and Susan Scott, in their book **All Stings Considered**, offer the following first aid advice: 1) Flood the affected skin with vinegar to disable any undischarged nematocysts. (Do not use sand, fresh water or anything else.) 2) If the eyes are stung, irrigate with fresh water for at least 15 minutes. Do not use vinegar in the eyes. 3) Remove any vinegar-soaked tentacles with a stick or other tool. 4) If the victim has breathing difficulties, weakness, palpitations, eye problems or cramps seek emergency medical care. 5) Apply ice packs if necessary for pain relief. (Most stings subside by themselves within an hour.)

WINGED BOX JELLYFISH
Carybdea alata Reynaud 1830
Class Cubozoa. Family Carybdeidae
• Colorless and transparent, these cubozoans have elongated bells up to about 3 in. high and 2 in. wide. Outside Hawai`i they may be 10 in. high. Four tentacles up to 2 ft. long hang from the edge of the bell, one from each corner. In the water these animals resemble floating plastic bags and are difficult to see. Surprisingly strong swimmers, they move towards light at night and sometimes plague south-facing O`ahu beaches seven to ten days after a full moon, possibly in relation to a breeding cycle. Hundreds or thousands can wash up during a single night. They occur most often at the surface but can also swim in midwater and can be found a mile or more offshore. Contact with their tentacles results in an immediate and intense burning sensation equal to or exceeding that of the Portuguese Man-of-War (a colonial hydroid, not a jellyfish). The pain usually subsides by itself in less than an hour but can linger as long as eight. If weakness, cramps or breathing difficulties occur the victim should seek emergency medical treatment. The species name means "winged," possibly in reference to the speed at which they swim. They occur in warm seas around the world. Photo: David B. Fleetham. Molokini Islet, Maui. 60 ft.

Top picture on next page ➤

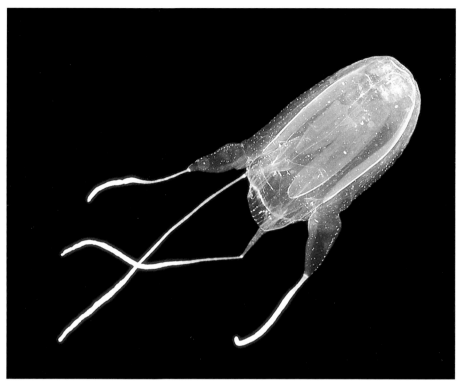

◄ Winged Box Jellyfish

RASTON'S BOX JELLYFISH

Carybdea rastoni Haacke, 1886
Class Cubozoa.
Family Carybdeidae

• The transparent, cubelike bells of these box jellies are about 1 in. on a side; a tentacle hangs from each corner. Contact with the tentacles results in a burning sting; delayed reactions (usually itching) can occur one to four weeks later. Generally less common in Hawai`i than *C. alata* (above), these are usually seen only at night; some observers suggest that they rest near the bottom during the day. (Two smaller box jellyfishes occur in Hawai`i, one with red tentacles, the other with banded tentacles. The bells of these tiny creatures are usually less than 1/4 in. across. One of these animals is visible in the lower right corner. Indo-Pacific. Photo: Waikīkī, O`ahu. Surface of Natatorium.

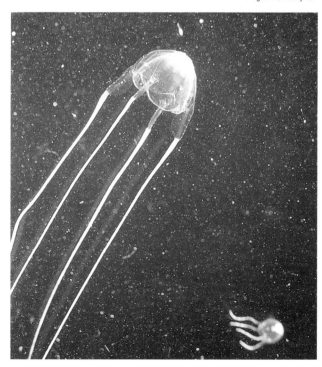

Blue Octocoral *(Anthelia edmondsoni)*, a soft coral endemic to Hawai`i. Lāna`i Lookout, O`ahu. 15 ft.

Soft corals and gorgonians—known collectively as octocorals—are colonies of eight-tentacled polyps with an internal skeleton. (Stony corals usually have an external skeleton and tentacles in multiples of six.) Octocoral tentacles are usually pinnate, meaning they have side branches like a feather. The polyps are typically connected by a mat of common tissue through which run channels connecting their digestive cavities. Octocoral colonies secrete skeletons of a flexible horny material or of calcareous spicules, the latter sometimes fused into a rigid structure. In all cases, the polyps completely enclose the skeleton.

Octocorals tend to have fewer stinging cells than other cnidarians; some appear to feed primarily on planktonic plant cells (phytoplankton) or other small particles. Many protect themselves by producing toxic chemicals, some of which have medical potential.

Octocorals are common shallow-water animals in many parts of the tropics. The pink, red and gold precious corals harvested from deep water also belong to this group. Lesser known octocorals include the sea pens, sea pansies, blue coral and organ-pipe coral (none of which occur on Hawai`i's shallow reefs).

Of six orders of octocorals, only three are recorded from shallow Hawaiian waters: the soft corals (order Alcyonacea), the gorgonians (order Gorgonacea) and an order with no common name, Telestacea. Each is represented by only one or two species.

SNOWFLAKE CORAL

Carijoa riisei (Duchassaing & Michelotti, 1860)
Order Telestacea. Family Telestidae
• Corals of the order Telastacea secrete internal skeletons composed of calcareous spicules. This species forms flexible tapering stems that may or may not have branches. The white polyps open along the sides and at the tip, generally only at night or when a current is running. With polyps closed, the naked branches are red or orange. This coral forms dense colonies in cavities along vertical walls or on the ceilings of caves and overhangs where current is strong, or under docks in harbors where plankton is plentiful. It is abundant on wrecks. Unknown in Hawai`i prior to 1974, it was probably introduced. It now occurs in all the main Hawaiian Islands and at Midway. It is native to the Caribbean and tropical Western Atlantic. Branches attain 4-6 in. Photo: Mākua, O`ahu. 20 ft. (polyps closed). (See also p. 17.)

LEATHER CORAL

Sinularia abrupta
Tixier-Durivault, 1970
Order Alcyonacea
Family Alcyoniidae
• This soft coral forms short, tough flexible lobes on a flat encrusting base. Growing on rocky substrate, it is most common at depths of 15-30 ft. in clear turbulent water. (It also occurs to depths of over 3,000 ft.) Individual colonies may be as much as 3 ft. across but are usually considerably smaller. Lobes attain a height of about 2 in. [A smaller species of *Sinularia* occurring in Hawai`i, *S. molokaiensis*, was described from specimens overgrowing Finger Coral *(Porites compressa)* off the southeast side of Moloka`i.] Western and Central Pacific. Photo: Mōkapu Rock, Moloka`i. 20 ft.

33

BLUE OCTOCORAL

Anthelia edmondsoni (Verrill, 1928)
Order Alcyonacea. Family Xeniidae
• Light blue to purplish patches of these small polyps are common in many locations throughout the Islands. They grow on both hard and soft surfaces, sometimes in very shallow water. The polyps, which do not form recognizable skeletons and cannot completely retract, are connected to each other by tissue at the base. Polyps harbor symbiotic algae (zooxanthellae) and are preyed upon by the nocturnal nudibranch *Tritonia hawaiiensis*. Both the coral and the nudibranch occur only in Hawai`i. Colonies range from 3-12 in. across; in exposed habitats adjacent colonies may form large continuous patches. Individual polyps fully extended are about 1/4 in. across. Photo: Pūpūkea, O`ahu. 30 ft. (detail, showing eight pinnate tentacles) (See also p. 32)

BICOLOR GORGONIAN

Acabaria bicolor (Nutting, 1908)
Order Gorgonacea
Family Melithaeidae
• Gorgonians include the sea fans, sea whips, precious corals and others. Some are calcareous; others have flexible horny skeletons. This tiny species is the only native shallow-water gorgonian recorded from Hawai`i. (Over 90 species occur in deeper water, including the commercially important precious corals.) It typically grows in rocky crevices in surgy or current-swept locations as shallow as 6 ft. but usually deeper—to 1,400 ft. Most commonly bright red or yellow, it may also be white or pink. Large colonies are about 2 in. across and 1 1/2 in. high. Known only from Hawai`i. Photo: Palea Point, O`ahu. 20 ft.

SEA ANEMONES

PHYLUM CNIDARIA. CLASS ANTHOZOA.
SUBCLASS HEXACORALLIA. ORDER ACTINIARIA.

Pleasing Anemone *(Telmatactis decora)* Palea Point, O`ahu, 40 ft.

Sea anemones are solitary anthozoan polyps that do not secrete a skeleton. They range in size from a fraction of an inch to 2 ft. or more in diameter and their tentacles usually occur in numbers divisible by six. An anemone's cylindrical body (column) terminates at the upper end in an oral disk containing a central mouth surrounded by rings of tentacles. Its lower extremity usually attaches to solid substrate but is sometimes adapted for anchoring in sand or mud. Although most anemones remain stationary, these animals sometimes creep about. A few can swim or somersault; one genus is pelagic.

When disturbed or exposed at low tide, anemones close by contracting muscles that pull the upper column tightly over and around the tentacles. In many species the column bears wart-like protuberances (verrucae) to which sand, gravel or bits of shell may adhere for protection and camouflage. The inner cavity of an anemone is partially divided into pie-shape sections by partitions called mesenteries, which possess stinging nematocysts to help kill ingested prey. The mesenteries greatly increase the inner food-absorbing surface area, enabling anemones to attain larger sizes than might otherwise be possible.

Most anemones feed on small bottom-dwelling or planktonic organisms; larger species kill and devour fishes and crustaceans. Symbiotic algae (zooxanthellae) living in the tentacles of some anemones provide additional nutrition by "leaking" photosynthetic sugars to their host. In many parts of the tropics the stinging tentacles of anemones provide havens for commensal shrimps, crabs and fishes that are either resistant to the sting or agile enough to avoid it. This phenomenon occurs in Hawai`i only to a limited extent.

Like other members of the class Anthozoa, anemones have no medusa stage. They may be hermaphroditic (both sexes in one individual) or have separate sexes. Individuals release their sperm and eggs into the water through the mouth. Anemones can also reproduce by splitting in half (fission), or by detaching bits of tissue, usually from the foot, that regenerate into complete animals (laceration). Some eject fully formed "babies" from their mouths.

Often compared with flowers, sea anemones are probably named after the European wood anemone *(Anemone nemorosa)*, a poisonous, white-flowered plant that causes blistering of the skin. Many sea anemone species thrive in captivity. One collected by a Scottish baronet in 1828 lived in its glass jar until 1887, outliving its owner and several heirs! (Water was changed three times a week by a loyal gardener.) Species of the genus *Aiptasia* often multiply so quickly in aquariums that they become pests. Unfortunately many others, such as the giants that host anemonefishes on Indo-Pacific reefs, do poorly. Even in well-maintained "living reef" aquariums, they usually bleach and waste away. In the wild, giant anemones probably live over a century, reproducing rarely. Collecting them is discouraged, and importation of these animals into Hawai`i is illegal.

In ancient Hawai`i anemones were known as `**ōkole**, a word also meaning "anus" (no doubt referring to the appearance of the animal when closed). Sometimes they were called `**ōkole emiemi**, or "shrinking anus." Despite the name, Hawaiians relished some anemones as food, removing the tentacles and eating them raw or cooked in **ki** (or **ti**) leaves. Sometimes they roasted anemones over a fire and ate them whole. The center was said to be particularly sweet. (Note: eating anemones without adequate knowledge and preparation can be unpleasant or even fatal. Some can produce severe reactions in humans. Care should always be taken when handling these animals.)

Compared with other tropical locations, Hawai`i is home to few anemone species. Only about 20 are known, most small and cryptic. Nine are pictured here. Tube anemones (order Ceriantharia), more closely related to black corals than to true anemones, are discussed in their own chapter (p. 66).

SWIMMING ANEMONE

Boloceroides mcmurrichi
(Kwietniewski, 1898)
Family Boloceroidididae
• These anemones have a large moplike crown of long reddish brown to whitish tentacles, often ringed with lighter bands. The tentacles are sticky and easily shed. The species lives on sand or mud bottoms or on seaweed and is mimicked and preyed upon by the nudibranch *Berghia major* (p. 180). To escape the nudibranch, it swims by paddling with its tentacles. These anemones occur from the shallows to depths of about 50 ft. and are at times abundant. They can reproduce by detaching bits of tentacle that regenerate into miniature anemones. Large specimens have a crown diameter of about 6 in., but most are much smaller. Indo-Pacific. Photo: Gustav Paulay, Guam.

PROLIFIC ANEMONE

Triactis producta Klunziger, 1877
Family Aliciidae

• This is the anemone carried by the Pom-Pom Crab, *Lybia edmondsoni* (p. 281). The crab, however, carries only juveniles. Larger specimens occasionally occur on the reef in shaded crevices, often in dense assemblages. They have up to 60 long slender tentacles atop a tall column that bears a ring of branching outgrowths and bulbous swellings (often the most visible part of the animal, but not present on juveniles). The overall color is usually light brown, the bulbs greenish. Polyps attain a height of about 1/2 in. in Hawai`i but may grow to 3 in. elsewhere and pack a powerful sting. Indo-Pacific. Photo: Scott Johnson. Koko Head, O`ahu.

DUSKY ANEMONE

Anthopleura nigrescens (Verrill, 1928)
Family Actiniidae

• These small anemones are abundant under stones and in holes and cracks on shallow reef flats, often above the low tide mark. The columns are dark brown, the tentacles grayish white. Vertical rows of sticky warts (verrucae) on the column collect bits of shell and gravel, completely encasing it in localities where wave action is strong. When exposed at low tide the anemones close, resembling dark slimy slugs; finger pressure may cause them to squirt a small jet of water. The species harbors no zooxanthellae and if kept in an aquarium does not need strong light. It will reproduce in captivity by dividing longitudinally. The Teddy-Bear Crab, *Polydectus cupulifer* (p.282), sometimes collects juveniles to hold in its claws. An eolid nudibranch, *Herviella mietta*, mimics and preys upon this species. Large individuals attain almost 1 in. in diameter and height. In old Hawai`i they were gathered and eaten. The species is reported only from Hawai`i and India but is possibly more widespread. Photo: Kāne`ohe Bay, O`ahu.

SESERE'S ANEMONE

Actiniogeton sesere (Haddon & Shackleton, 1893)
Family Actiniidae

• These anemones resemble zoanthids of the genus *Protopalythoa* and often cluster in the same habitat—the low intertidal of well protected reef flats—their columns typically buried in soft substrate. They also occur along more exposed shores. The disk and tentacles vary from bright green to brownish green, often with flaky white marks. The grayish columns bear small white warts (verrucae), each with a bright green tip. These warts are most visible as the anemones close, and they help to differentiate them from zoanthids. Described from the Torres Straits, off northern Australia, "this species is named after Sesere, the legendary hunter of the dugong, who lived on the neighboring island of Badu." To almost 1 1/2 in. across and 1 in. high. Known from Australia, Papua New Guinea and Hawai`i, and probably more widespread. Photo: Wailupe, O`ahu.

MANN'S ANEMONE

Cladactella manni (Verrill, 1899)
Family Actiniidae

• This is Hawai`i's largest intertidal anemone. The column may be dark rose, dark maroon, dark brownish green or dark copper green and is completely covered with small rounded bumps (vesicles); the tentacles are dark rose and sticky to the touch. These uncommon animals occur singly along turbulent rocky shores in crevices or pockets and under ledges constantly washed by waves. Because of their habitat they are easily overlooked. In captivity they do poorly. To about 4 in. across and 2 in. high. Known only from Hawai`i. Photo: Lāna`i Lookout, O`ahu. Surge pool.

PLEASING ANEMONE

Telmatactis decora (Hemprich & Ehrenberg, 1834)

Family Isophellidae

• This anemone lives in crevices or under stones from the intertidal zone in well-protected areas to depths of at least 100 ft. It expands its long tubular body after dark to expose an attractive crown of tentacles. If disturbed, it rapidly retracts into its crevice. The tentacles may be brown, greenish, orange or rosy red, often flecked or ringed with white. They taper gently but may swell out in a little ball at the tip. The tentacles nearest the mouth are longest, those at the edges much shorter. The Teddy-Bear Crab, *Polydectus cupulifer* (p. 282) usually carries a juvenile of this species in each claw. Large specimens may be 3 in. long and 1 1/2 in. across at the crown. Indo-Pacific. Photo: Palea Point, O`ahu. (See also p. 35.)

SAND ANEMONE

Heteractis malu (Haddon & Shackleton, 1893)

Family Stichodactylidae

• These are Hawai`i's largest anemones, found on shallow sand flats in Kāne`ohe Bay, O`ahu, deeper sand bottom at Mā`alaea Bay, Maui, and on protected rocky reef flats at Lā`ie and Kahuku, O`ahu. Those living in sand remain buried with only their pale crown of tentacles exposed. If disturbed they contract and disappear. The numerous short tentacles harbor symbiotic algae (zooxanthellae). The sting is not powerful enough to affect humans. Juveniles of the Hawaiian Domino Damselfish *(Dascyllus albisella)* and the Ambon Shrimp *Thor amboinensis* (p. 233) occasionally shelter in or about the tentacles. Outside Hawai`i this species hosts the anemonefish *Amphiprion clarkii.* These anemones sometimes attract the interest of local aquarists, who cannot legally obtain the more colorful Indo-Pacific and Caribbean species. Unfortunately, they seldom survive long in captivity. For that reason, and because their distribution in Hawai`i is limited, it is best not to collect them. The species has been known under a number of names, including *Antheopsis papillosa* and *Marcanthea cookei.* To almost 6 in. across. Known from the eastern Indian Ocean to Japan and Hawai`i. Photo: Mike Severns. Mā`alaea Bay, Maui. 40 ft.

HERMIT CRAB ANEMONE

Calliactis polypus (Forsskål, 1775)
Family Hormathiidae
- Because of its association with the common Jeweled Anemone Crab *Dardanus gemmatus* (p. 258), this is probably the most frequently seen anemone in Hawaiian waters. It can also be one of the most colorful; the tentacles, although usually pale or grayish brown, are sometimes yellow, pink or orange, especially near the tips. The animal's broad base, often pinkish, is marked with longitudinal lines terminating in a row of whitish bumps. Copious pink threads (acontia) are expelled through these bumps when the anemone is disturbed. Laden with nematocysts, the threads entangle and sting potential predators. A small, undescribed white anemone (*Anthothoe* sp.) occurs with the same crab, usually on the upper edge of the shell near the crab's colorful eyes. An amphipod and a flatworm often live on the crab too, and in deeper water a small stalked barnacle, *Koleolepas tinkeri*, may attach under the anemone's foot. Widely distributed, this anemone occurs from East Africa and the Red Sea to the west coast of the Americas, almost always in association with hermit crabs of various species. It feeds, at least in part, on small molluscs (whose shells it subsequently disgorges). Large specimens fully expanded are about 3 in. high. Photo: Mākua, Oʻahu. 30 ft.

Glass Anemone (a)

GLASS ANEMONE

Aiptasia pulchella Carlgren, 1943
Family Aiptisiidae
• This is the only anemone truly abundant in the Islands. Often in groups, it lives in crevices or under stones in tide pools and on shallow reef flats. It is usually light greenish brown to dark brown (due to zooxanthellae in the tissues) but specimens from dark locations are transparent. The column is often lightly marked with parallel longitudinal lines. Flecks of white may be present near the tentacles or may cover the entire animal in small specimens. If disturbed, these anemones eject white stinging threads (acontia). In well-protected environments, such as large stagnant tide pools or aquariums, they grow in the open and can multiply prolificly, detaching tiny bits of tissue from the foot that soon become miniature anemones. Eventually these will cover every available surface. They also give live birth, ejecting fully formed young from the mouth. Fully expanded in an aquarium, these anemones are attractive animals; on the reef they usually remain hidden in crevices with only the tentacles protruding. Large specimens attain a height of 3 in. or more, with a similar crown diameter; most are far smaller. The species name means "beautiful." Western, Central and Eastern Pacific. Photos: a) Lāna`i Lookout, O`ahu. Tide pool (completely exposed); b) Kāne`ohe Bay, O`ahu, 3 ft (typical appearance on the reef).

Glass Anemone (b)

ZOANTHIDS (COLONIAL ANEMONES)

PHYLUM CNIDARIA. CLASS ANTHOZOA.
SUBCLASS HEXACORALLIA. ORDER ZOANTHINARIA

The common Blue-Gray Zoanthid *(Palythoa caesia)* forms extensive mats in the surge zone. Palea Point, O`ahu. 10 ft.

Zoanthids are small, anemone-like polyps often called colonial anemones. They may have any number of tentacles greater than 12. Polyps are typically connected to their neighbors by common tissue containing channels that link their digestive cavities. There is no hard skeleton. Zoanthids generally occur in shallow water. Some overgrow exposed rock; others live partly buried in soft substrate. Like anemones, they capture planktonic organisms with stinging tentacles and may also gain nourishment from sugars produced by symbiotic algae (zooxanthellae). As members of the class Anthozoa, zoanthids have no medusa stage. They reproduce both by budding and by the direct release of eggs and sperm into the water.

Zoanthid nematocysts are too weak to sting humans or deter predators. Their bodies and connective tissues, however, are often tough and leathery, and in many species are filled with embedded sand grains. Like anemones, zoanthids can close tightly when disturbed. As further protection, some contain a potent poison.

Aware of these toxic properties, Hawaiian warriors in the Hāna district of Maui often smeared their spear tips with zoanthid mucus before battle. Wounds from such spears were usually fatal. These animals, called **limu-make-o-Hāna** ("the deadly seaweed of Hāna") lived in tide pools at Mūʻolea. In December 1961, despite repeated warnings from local people that the area was **kapu** (forbidden), researchers from the University of Hawaiʻi investigated the Mūʻolea pools to look for the "deadly seaweed". While collecting zoanthids, a student assistant absorbed a tiny amount of toxin through abrasions in his hand. He became dizzy and nauseous, suffered headache, muscle cramps and swelling, and was hospitalized for two days; symptoms persisted for a week. The scientists, meanwhile, returned to O`ahu to find their Coconut Island laboratory destroyed by fire. The unexplained blaze had occurred the very day they visited the tide pools.

This strange story may have a happy ending. Palytoxin, as the poison is now known, has been found to possess anticancer properties. The zoanthids containing it were scientifically described and named *Palythoa toxica*. (This species' status is now in question.) Scientists believe the toxin to be a product not of the zoanthid itself, but of symbiotic algae or bacteria living in its tissues. Zoanthids from other parts of the world contain similar compounds.

The appearance and form of zoanthids varies remarkably. Recent studies from Australia's Great Barrier Reef show that many named "species" are genetically indistinguishable. Furthermore, genetically distinct zoanthids may appear almost identical even when dissected. Although eight zoanthid species have been recorded from Hawaiian waters, the number will probably be reduced to four or five after genetic analysis. Until that time, precise identification of many Hawaiian zoanthids is not possible.

BLUE-GRAY ZOANTHID

Palythoa caesia Dana, 1848
• This common species forms tough mats on rocks and dead coral from the shallows to depths of about 25 ft. in areas of good water movement. The spaces between polyps are completely filled with connective tissue and, on reef flats, sand; when the polyps close, the colony appears smooth and almost featureless. The overall color is typically light bluish gray, but pastel pinks and greens and various shades of brown are not unusual. This zoanthid is often identified as *P. tuberculosa*, a species described in 1791 from an unnamed locality. It is difficult to use that name with confidence, given what is now known about zoanthid identification. *Palythoa caesia*, described from Fiji in the last century, provides a better reference for the present. The species name means "blue-gray". Polyps are about 3/10 in. across. Indo-Pacific. Photos: a) Magic Island, O`ahu. 10 ft. (polyps closed). b) Palea Point, O`ahu, 15 ft. (polyps open). (See also p. 42.)

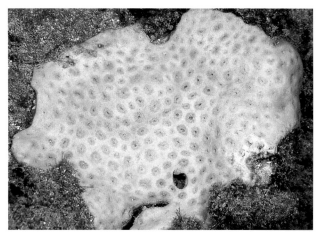

Blue-Gray Zoanthid (a)

Blue-Gray Zoanthid (b)

43

Protopalythoa "A"

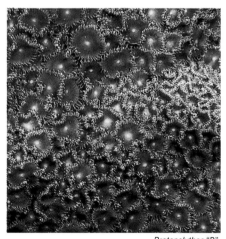

Protopalythoa "B"

Protopalythoa "C"

Protopalythoa spp.
• The three colonies illustrated here all belong to the genus *Protopalythoa*. Whether they represent the same or different species can only be determined by genetic analysis. Colony "A" was photographed at Mākaha, O`ahu, at a depth of about 25 ft. Colony "B" is common at Hanauma Bay, O`ahu, in several feet of water. Colony "C" with its green crown comes from Kāne`ohe Bay, O`ahu, where it grows close to the low tide mark. The Hawaiian zoanthids previously identified as *Palythoa vestitus*, *Palythoa psammophilia* and *Palythoa toxica* all belong to the genus *Protopalythoa* and may be forms of the same species. Because some are toxic, avoid handling any of these zoanthids.

Zoanthus "A"

Zoanthus "B"

Zoanthus "C"

Zoanthus spp.
• The colonies illustrated belong to the genus *Zoanthus*, in which the polyps (about 1/4 in. across) are connected to one another only at the base. All occur to depths of about 30 ft. Although two Hawaiian species, *Z. pacificus* Walsh & Bowers, 1971 and *Z. kealakekuaensis* Walsh & Bowers, 1971, have been proposed, they may in fact be the same. Colonies at Kealakekua Bay on the Kona coast of the Big Island are recorded as toxic. For this reason, avoid handling any of these zoanthids. Colony "A" from Pūpūkea, O`ahu, shows polyps with a green disk and buff tentacles both open and closed. Colony "B" from Magic Island, O`ahu shows a common color variation with green tentacles and pinkish disk. Colony "C" is from the low tide mark at Kewalo, O`ahu.

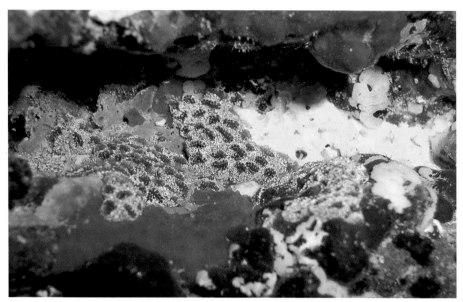

Unidentified zoanthids
• These sand-encrusted zoanthids occur in crevices along exposed vertical walls exposed to strong surge at depths of 15 ft. or less, usually growing near sponges. If truly sponge associated they could be *Epizoanthus* or *Parazoanthus*. Photo: Lāna`i Lookout, O`ahu. 10 ft.

STONY CORALS

PHYLUM CNIDARIA. CLASS ANTHOZOA
SUBCLASS HEXACORALLIA. ORDER SCLERACTINIA

Stony corals are anemone-like polyps that secrete calcium carbonate skeletons. They are also called "scleractinian" corals, from *skleros*, the Greek word for "tough" or "hard." Almost all have weakly stinging tentacles whose number is a multiple of six. Most stony corals are colonial. Their individual cuplike skeletons, called calyces (singular, calyx), join to form solid structures of many shapes and sizes, including tables, plates, branching fingers, bushes, lobes and mounds. Large colonies, often called "heads" or "bommies," may be 1,000 years old and attain a height of 25 ft. The accumulated skeletons of coral polyps, cemented together by limestone-secreting algae, form coral limestone—the underlying basis of living coral reefs. The great atolls and barrier reefs of the Pacific Ocean consist of coral limestone deposits up to 5,000 ft. thick. They are the largest structures on Earth made by living organisms,

Although stony corals occur in all oceans, they are most abundant in shallow, sunlit, tropical seas. Warm-water corals usually harbor symbiotic algae in their tissues that provide oxygen and sugars to the coral through the process of photosynthesis. These algae (zooxanthellae) may contribute as much as 98 percent of a colony's nutrients (depending on coral species), using the polyps' waste products in the process.

Symbiotic algae also facilitate the colony's secretion of calcium carbonate, making possible skeletal growth of up to 4-6 in. per year in some species. For this reason, corals harboring zooxanthellae are the principal reef-building, or hermatypic, corals. Cold-water corals and those from

At least six species of coral are visible in this deep tide pool at Kapoho Bay, Hawai`i: *Porites lobata*, *P. compressa*, *Pavona varians*, *Montipora capitata* and *Pocillopora meandrina*. (Keoki Stender);

dim environments, such as deep water or caves, usually lack zooxanthellae and are of little consequence in reef formation. (They are ahermatypic).

Living coral colonies are completely covered by a thin double layer of tissue that connects the polyps and allows them to share nutrients. This tissue layer also secretes skeletal material between the calyces, prevents fouling organisms from overgrowing the skeleton, and gives the colony color—usually brown, yellow or olive. Golden brown zooxanthellae contribute most of the tissue color, animal pigments the rest. Pigments in some colonies, especially those in very shallow water, may add almost fluorescent green, magenta and blue tints, probably as protection from ultraviolet light. Stripped of their flesh, the underlying skeletons of most stony corals are pure white. (The so-called precious corals, with their gold, pink or red skeletons, are not stony corals at all but deep-water gorgonians.)

Individual coral polyps are generally small, only 1/25 to 1 in. across. Some solitary corals composed of a single polyp, however, attain lengths of up to 10 in. The inner cavities of all coral polyps are divided into 12 or more pie-shape sections by stony partitions called septa (analogous to the mesenteries in anemones). When a coral's tentacles withdraw, as they often do by day, the intricate shapes of the stony calyces and their inward-pointing septa become visible. When the tentacles extend, the calyces are obscured and the colony may appear fuzzy. Coral identification in the laboratory is performed largely on the basis of microscopic examination of the stony calyces after the flesh has been stripped away. The gross forms of colonies and their colors in life are often too variable to allow accurate identification.

Coral polyps lack a medusa stage, as do all members of the class Anthozoa. Like most other cnidarians, they can reproduce both sexually and asexually and have many reproductive strategies. Colonies grow asexually by the budding of new polyps from existing polyps. The new polyp's skeleton remains attached to that of its parent; thus all polyps in a colony are, in theory,

Reef scene at Kahe Point, O`ahu showing *Pocillopora meandrina, P. eydouxi* and *Porites lobata.* 40 ft.

clones of a single original "colonist." Short-range dispersal of colonies may occur asexually when pieces broken and scattered by storms grow into new ones. Also, polyps of certain species (such as *Pocillopora damicornis*) sometimes detach from their skeletons, if the colony is stressed, and float away to found new colonies in more favorable surroundings.

Long-distance dispersal is most often accomplished sexually. When a colony attains a certain size, its polyps mature and begin to produce eggs and sperm. Some corals retain the eggs and brood their larvae, releasing them fully formed. Others release the eggs and sperm into the water (often packaged together in buoyant yellow or pink spherical bundles), where they mix with those from other colonies and cross-fertilize. This type of coral spawning is often triggered by a precise combination of day length, moon phase, and tide. It enables many colonies to release eggs and sperm at once, maximizing the chances of cross-fertilization. Such mass events typically occur only a few times per year in a given location and are usually predictable. Often more than one species is involved. On Australia's Great Barrier Reef over 130 species of corals spawn simultaneously on one or two nights in the month of November. The resulting flood of floating eggs and sperm colors the sea for miles. Predators turn out in force but cannot possibly consume them all.

After fertilization coral eggs hatch into larvae—called planulae—that drift for a time with the plankton. Most are eaten or perish; a few settle in favorable locations, perhaps hundreds of miles away, to start new colonies. Once established, a young colony still faces challenges. Competition for living space on the reef is often a matter of life or death. Wherever colonies of the same or different species touch, the reef becomes a battleground. At night the combatants may extend long stinging tentacles and digestive filaments called "sweepers" to kill polyps of neighboring colonies. A barren "no man's land" the length of the sweeper tentacles often separates such warring colonies, whose advancing or retreating edges may appear stressed and discolored.

Hawai`i has about 50 species of shallow-water corals in 17 genera. These are low numbers compared with the many hundreds of species, in about 80 genera, occurring around Indonesia and the Philippines. Fossil reefs on O`ahu and other Hawaiian islands indicate that more species of corals existed here in the geological past. Hawaiian waters today are too cool for the vigorous growth of many tropical species. Even so, divers and snorkelers can see lush coral gardens in a number of locations, especially off the Kona coast of the Big Island. It should be noted, however, that much of the "coral" limestone along Hawaiian shores is actually the product of limestone-secreting coralline algae. Such algal growths are often mistaken for true stony corals.

While Hawaiian corals do well in captivity if properly cared for, it is illegal to collect them. (See Appendix B, p. 346.) Captive corals can grow prolifically, even spawning in synchrony with wild corals on the reef. The Waikīkī Aquarium takes advantage of this by holding annual "coral spawning" evenings in June or July for the benefit of members.

General Hawaiian names for coral include `**āko`ako`a**, **ko`a** and **puna kea**. In ancient times stony coral skeletons were used as abrasives for smoothing canoes and rubbing off pig bristles; Mushroom Coral was a favorite. Bleached white coral skeletons were also placed on fishing shrines and used to mark paths across dark lava flows. Eighteen of Hawai`i's most common or conspicuous stony corals are described and illustrated below.

FAMILY POCILLOPORIDAE

Members of this family in Hawai`i form discrete branching colonies, or "heads." The branches are conspicuously "bumpy"—their surface area increased by closely spaced wartlike protuberances called verrucae. The calyces, which are small, occur both on the verrucae and between them.

LACE CORAL

Pocillopora damicornis
(Linnaeus, 1758)
Family Pocilloporidae
• This shallow-water coral forms small bushy heads up to 6 in. across. In protected environments its branches are slender and delicate; in more turbulent areas they are thicker and shorter. Wartlike bumps (verrucae) at the branch tips sometimes grow into new branchlets. Colonies are light brown to brown, depending on depth and available light (deeper colonies are darker). In Hawai`i isolated heads occur primarily on protected reef flats several feet deep. Elsewhere, however, the species may dominate shallow reefs down to about 30 ft., growing in extensive beds and crowding out all other corals. This almost ubiquitous Indo-Pacific and Eastern Pacific species was the first coral named by Linnaeus

(Continued on next page)

Lace Coral, delicate form

(Lace Coral—continued)

(in 1758) and is one of the most thoroughly studied corals in the world. It is often host to the Coral Gall Crab, *Hapalocarcinus marsupialis* (p. 289), which creates and occupies hollow galls on the tips of branches. Indo-Pacific and Eastern Pacific. Photos: a) Kāne`ohe Bay, O`ahu. 6 ft. (delicate form) b) Ali`i Beach Park, O`ahu. 3 ft. (robust form)

Lace Coral, robust form

CAULIFLOWER CORAL

Pocillopora meandrina Dana, 1846
Family Pocilloporidae

• Also called Rose Coral, this common species forms compact branching colonies on hard substrate. It may occur anywhere from splash pools above high tide level to depths of 100 ft. or more. Thriving in high energy environments where few other Hawaiian corals (and no other branching ones) will live, it is typically the first coral to colonize new submarine lava flows. The flattened branches, usually equal in length, often curve into a "C" shape at the tips. Viewed from above, channels appear to meander between the branches, hence the species name. Coral guard crabs, shrimps and fishes of several species commonly inhabit these spaces. Colonies are ordinarily brown, but they can be green or rose-pink in very shallow water. (These pigments may help protect the coral from harmful ultraviolet radiation.) Daytime spawning has been observed during spring high tides at Hanauma Bay, O`ahu. Two endemic Hawaiian corals, *P. ligulata* and *P. molokensis*, are similar but far less common. To about 18 in. Western Pacific and Eastern Indian Ocean. Photo: Hanauma Bay, O`ahu. 10 ft.

ANTLER CORAL
Pocillopora eydouxi Milne-Edwards & Haime, 1860
Family Pocilloporidae

• This is Hawai`i's largest, most impressive *Pocillopora*. Colonies have tubular upright branches, often flattened and forked at the tips, and may be 3 ft. high and as many across (reminding some divers of moose antlers). Smaller heads in shallow-er water often resemble Cauliflower Coral (above), but with wider spaces between the branches. As in all species of *Pocillopora*, closely spaced wartlike bumps cover the branches. Colonies are often isolated or far apart and occur between depths of about 5 and 130 ft. They are typically home to schools of Hawaiian Domino Damselfish (*Dascyllus albisella*), one or more hawkfishes, and numerous crabs; they also serve as temporary refuge for many other fishes. Tiny commensal crabs, *Pseudocryptochirus kahe* (p. 289), sometimes occur in discolored pits at the ends of the branches. The species name (pronounced "ee-dew-eye") honors French naturalist Joseph Fortune Eydoux (1802-1841), who died in Martinique while col-lecting specimens for the National Museum of France. Indo-Pacific. Photo: Molokini Islet, Maui. 50 ft.

FAMILY ACROPORIDAE

Members of the family Acroporidae are the principle coral reef builders just about everywhere in the tropics except Hawai`i. The large and important genus *Acropora* (which includes the staghorn and table corals so conspicuous on most Indo-Pacific reefs) is virtually absent in Hawaiian waters. Fossil corals from upraised reefs on O`ahu, however, indicate that they were once widespread. Today, species of *Acropora* occur principally around French Frigate Shoals in the northwestern chain. A few rare colonies have been found off Kaua`i. These corals prob-ably grow from larvae drifting northward from tiny Johnston Island, where *Acropora* corals are abundant. Unfortunately, Hawaiian waters appear to be too cold to allow them to reproduce.

Far more common in the Islands is the genus *Montipora*, represented by seven species. *Montipora* corals usually form encrusting or platelike colonies. The minute calyces are typi-cally surrounded or separated by high, smooth, rounded projections that give the surface a rough texture. Unusually extensive beds of *M. capitata* and *M. patula* occur in the center of the submerged crater at Molokini Islet, Maui, where they overgrow reef-building corals of the genus *Porites*.

51

RICE CORAL

Rice Coral (a)

Montipora capitata (Dana, 1846)
Family Acroporidae

• This coral grows in a variety of forms. Although typically encrusting, it may form massive colonies on protected shallow reef flats and broad plates, cups and delicate branching pinnacles on sheltered reef slopes. Pinnacles occur in bright light, plates in dim. Both forms can grow in a single colony, with pinnacles above and a skirt of plates below. The color varies from dark brown (with light edges and tips) to beige or cream. Rounded projections arise between the calyces like grains of rice set endwise. In sheltered areas such as Kāne`ohe Bay, O`ahu, impressive colonies are common on reef faces within a few feet of the surface. Elsewhere, colonies prefer depths free from turbulence and surge, down to about 150 ft. This species sometimes colonizes the dead bases of Finger Coral *(Porites compressa)*, eventually overgrowing its living branches. In Kāne`ohe Bay, colonies spawn simultaneously on the third night after the new moon in June, releasing little white bundles of eggs and sperm. *Montipora capitata* is endemic to Hawai`i. (Older publications refer the Hawaiian population to the Indo-Pacific species *M. verrucosa.*) Two additional endemic corals of the genus also occur here: *M. studeri* and *M. dilitata*. Neither is common. Photos: a) Kāne`ohe Bay, O`ahu. 10 ft (plates and pinnacles); b) Olowalu, Maui, 50 ft; (overgrowing Finger Coral); c) Laupāhoehoe, Hawai`i. 50 ft (encrusting plate).

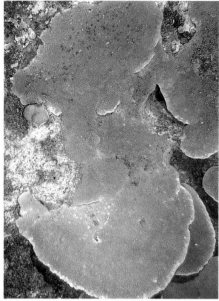

Rice Coral (b)   Rice Coral (c)

BLUE RICE CORAL ➤

Montipora flabellata Studer, 1901
Family Acroporidae
• The vivid fluorescent blue of this species makes it easy to identify, although in heavily shaded areas it may also be brown. The blue pigments appear pinkish under artificial strobe light, making realistic underwater photographs tricky to obtain. It encrusts rocky substrate in areas of strong water movement from the shallows down to about 20 ft. Colonies can be 3 ft. across. Like most species of *Montipora*, the surface is rough and grainy due to small rounded projections between the calyces. Smaller than those of *M. capitata*, the projections are often fused. The snapping shrimp *Alpheus deuteropus* sometimes inhabits this and other corals of the genera *Montipora* and *Porites*, creating dark branching channels on the surface. The species name is from *flabellum*, meaning "fan." Probably endemic. Photo: Hanauma Bay, O`ahu. 15 ft. (The similarly colored *M. turgescens* forms pillars and mounds in Midway Lagoon and perhaps elsewhere in the northwestern chain and might be confused with this species.)

Spreading Coral (a)

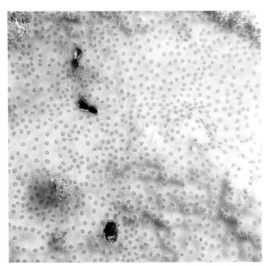

Spreading Coral (b)

SPREADING CORAL

Montipora patula Verrill, 1864
Family Acroporidae

• The Latin word *patula*, meaning "spread out" or "broad," aptly describes this coral, which usually grows both in plates and encrusting colonies (the latter sometimes 6 ft. or more across). The color varies from very light tan to brown, and the relatively small surface projections tend to cluster around the widely spaced calyces. When open, the tiny polyps sometimes appear bright blue or violet. This coral is common in Hawai`i at depths of 10 ft. or more, usually along reef slopes and walls. The limestone reef at "Tunnels" on Kaua`i's north shore appears to consist primarily of this species; only a few live colonies occur there now. Probably endemic. (A similar Indo-Pacific species, *M. verrilli*, is rare in Hawai`i.) Photos: a) "Hale`iwa Trench," O`ahu. 20 ft. (plates); b) Pūpūkea, O`ahu. 30 ft. (detail showing blue polyps)

FAMILY PORITIDAE

Corals of the genus *Porites* are the most important reef builders in Hawai`i. Most have small calyces 1/12 in. or less in diameter crowded together with no space between; their shared walls form polygons. The surfaces of these corals are relatively smooth, without bumps or projections. *Porites* colonies are variable in shape and size, forming rounded or pointed heads, lobes, encrustations, plates or branching fingers. Ringlike "micro-atolls" occur on some shallow reef flats. Because considerable difference in form is possible within a species, field identification may be difficult. Nine species of *Porites* occur in Hawai`i; laboratory examination of the calyces is necessary to identify some of them with certainty. *Porites* colonies are always entirely male or female. Females brood their eggs rather than releasing them to the sea. An eolid nudibranch, *Phestilla lugubris*, preys corals of this genus. In old Hawai`i common species of *Porites* were known as **pōhaku puna**.

Finger Coral (a)

Finger Coral (b)

FINGER CORAL • **pōhaku puna**

Porites compressa Dana, 1846
Family Poritidae

• One of the two most common corals in Hawai`i (the other is *P. lobata*), Finger Coral sometimes forms beds acres in extent. Downtown Honolulu and Waikīkī rest on a fossil reef composed largely of this species. Most abundant on protected leeward shores, it usually grows at depths below 30 ft. to avoid turbulence and surge. In calm locations, such as Kāne`ohe Bay, O`ahu (where it constitutes 85 percent of the reef-building corals), it thrives close to the surface, often forming low rounded heads with fused branches. In deeper water branches tend to be longer and more finger-like, but they may also be short and knobby, resembling other *Porites* corals. Colonies are usually light brown, sometimes yellowish or blue-gray. Fast growing, Finger Coral usually outcompetes other corals in its preferred environment. Once established, its dead under-branches form substrata upon which other corals such as *Pavona varians*, *Montipora capitata* and *Porites rus* can settle and sometimes overgrow the living host. The branches are fragile and easily damaged by storms. Great beds at 40-70 ft. off Kahe Point, O`ahu, were decimated by Hurricane Iwa in 1981. Finger Coral beds provide important habitat for juvenile fishes, and at night they come alive with colorful shrimps and other invertebrates. This species is probably endemic, although a very similar coral occurs in the Persian Gulf, Gulf of Oman and Red Sea. Photos: a) Puakō, Hawai`i. 60 ft. b) Kāne`ohe Bay, O`ahu. 5 ft. (See also pp. 52, 54.)

Lobe Coral (a)

LOBE CORAL • pōhaku puna

Porites lobata Dana, 1846
Family Poritidae

• This is Hawai`i's most common massive coral. Colonies vary greatly in size and shape, depending largely on the amount of wave action. Encrusting patches are common in the turbulent shallows. Huge mounds 10 ft. high or more occur in protected bays, sometimes with secondary lobes and projections of various shapes. "Micro-atolls" (some 20 ft. or more in diameter) occur on some shallow reef flats. Small spires and pinnacles resembling Finger Coral may occur in deeper water. The species grows about 1/3 in. in height a year and a large colony could easily be 400 years old. Colonies are most common at depths of 10-50 ft. but occur to at least 200 ft. As in most *Porites*, the tiny calyces crowd together with no spaces between, their shared walls sharply defined. Color varies from yellowish brown to yellowish green. Colonies often host the snapping shrimp *Alpheus deuteropus*, which creates dark meandering channels in the surface. The tiny xanthid crab *Maldivia triunguiculata* inhabits chambers with surface entrances. A parasitic flatworm (trematode) spends part of its life cycle on this coral, creating conspicuous pinkish nodules on its surface. (Coral-eating butterflyfishes pick at the nodules and ingest the parasite, which continues its life cycle in their bodies.) Hawaiian corals similar to *P. lobata* include *P. evermanni* (p. 57) and three rare species, *P. duerdeni*, *P. pukoensis* and *P. solida*. Only an expert can tell the last three apart. *Porites lobata* has an Indo-Pacific distribution. Photos: a) Hōnaunau, Hawai`i. 15 ft. (large mound); b) detail; c) Keauhou Landing, Hawaii Volcanoes National Park, Hawai`i. 3 ft. (encrusting colonies of several colors riddled with shrimp channels) (See also p. 231.)

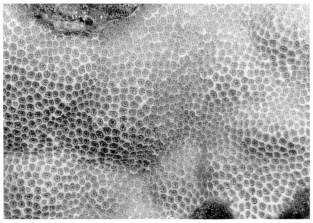

Lobe Coral (b)

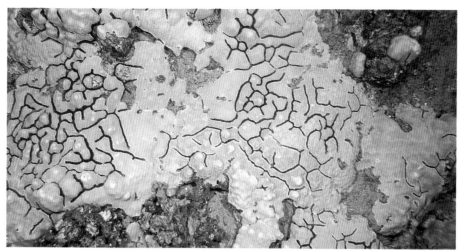

◄ Lobe Coral (c)

EVERMANN'S CORAL • **pōhaku puna**

Porites evermanni Vaughan, 1907
Family Poritidae
• This massive coral varies from chocolate brown to bluish or purplish gray. It never has the yellowish color usually present in *P. lobata.* Heads can be 3 ft. or more across. Colony surfaces are typically crowded with many small rounded swellings or lobes. Often, polyps are not completely retracted, giving the surface a slightly fuzzy appearance. Evermann's Coral has a patchy distribution but may sometimes be common, even dominant, on protected or semi-protected reef flats a few feet deep. (Kahalu`u Beach Park, Kailua-Kona, Hawai`i, and the Koko Head side of Black Point, O`ahu, are good places to see it.) It also occurs to depths of at least 30 ft. The name honors American ichthyologist Barton Warren Evermann (1853-1932), who studied Hawaiian fishes in the early 20th century and later became director of San Francisco's Steinhart Aquarium. Probably endemic. Photos: a) Hekili Point, Maui. 1 ft.; b) Kahalu`u Beach Park, Hawai`i. 3 ft. (surrounded by *P. lobata*)

Evermann's Coral (a)

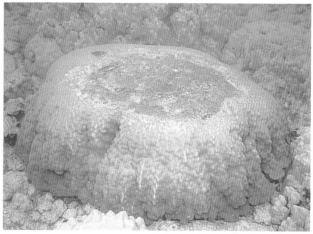

Evermann's Coral (b)

PLATE-AND-PILLAR CORAL

Porites rus (Forsskål, 1775)
Family Poritidae

• This attractive and highly variable coral is common along the Kona coast of the Big Island. It also occurs off west Maui and a few colonies have been found at the "Hale`iwa Trench," off Ali`i Beach Park, O`ahu. It forms thick columns in shallow water and thin stacked plates deeper down. The species may also overgrow colonies of *P. compressa*, forming brackets and plates at the base. Unlike other members of the genus *Porites*, the calyces of *P. rus* are slightly separated. Calyces cluster between irregular raised ridges, especially in the upright column form. Polyps are often visible as tiny yellow spots. Overall color is gray to light brown. Indo-Pacific. Photos: a) Hōnaunau, Hawai`i, 30 ft. (overgrowing *P. compressa*); b) Kealakekua Bay, Hawai`i, 15 ft (pillars); c) Kealakekua Bay, Hawai`i, 90 ft. (plates);

Plate-and-Pillar Coral (a)

Plate-and-Pillar Coral (b)

Plate-and-Pillar Coral (c)

FAMILY AGARICIIDAE

In corals of the family Agariciidae, fine narrow ridges called septo-costae connect adjacent calyces. These are extensions of the septa—the fine partitions radiating inward from the sides of the calyces. In the genus *Leptoseris* (six species in Hawai`i, half from deep water), these ridges are long, fine and often curved. In the genus *Pavona* (at least three local species) they tend to be shorter and straighter. In *Pavona duerdeni* the septo-costae radiating from each calyx form tiny but beautiful star-shape patterns. Corals of the family Agariciidae have extremely fine tentacles. Many feed by trapping suspended particles in mucus rather than by stinging and capturing prey.

SWELLING CORAL

Leptoseris incrustans (Quelch, 1886)
Family Agariciidae
• Hawai`i's most common species of *Leptoseris*, this coral may be identified by the many irregular swellings between the sunken calyces. These swellings are covered with fine wavy ridges (septo-costae) that connect the calyces. The species forms encrusting colonies a few inches across, usually under ledges in deep shade. In deeper water, where light levels are low, colonies may grow in the open. Color varies from greenish to reddish brown. The species name means "encrusting." Indo-Pacific. Photo: Pūpūkea, O`ahu. 30 ft.

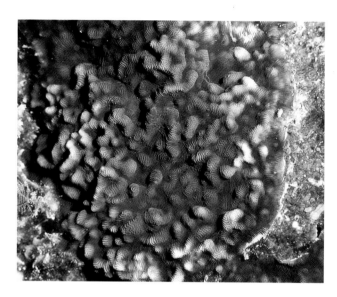

DUERDEN'S CORAL

Pavona duerdeni Vaughan, 1907
Family Agariciidae
• Mature colonies of this coral form lobes flattened on two sides (often resembling thick upright disks, partly buried). Young colonies are encrusting. The species is most common at depths of 20-30 ft. on reefs exposed to moderate wave action. Large lobes are about 12 in. high; entire colonies may attain diameters of 10 ft. or more. The tan surface is finely textured and smooth. The reef off `Anini Beach on the north shore of Kaua`i is largely composed of this coral. The name honors cnidarian specialist J.E. Duerden (1865-1937). Probably endemic. Photo: `Āhihi-Kīna`u, Maui. 20 ft.

Corrugated Coral (a)

Corrugated Coral (b)

CORRUGATED CORAL

Pavona varians Verrill, 1864
Family Agariciidae
• Not uncommon, this coral usually forms encrusting patches that are light yellowish tan. Under shaded overhangs it is brownish gray, greenish or brown. In very protected areas it sometimes forms thin plates. It often overgrows the bases of Finger Coral (*Porites compressa*). Whatever its form or color, the calyces lie in meandering valleys separated by sharp narrow ridges. It occurs from the surface to depths of at least 80 ft. Large colonies are about 8 in. across. Indo-Pacific. Photos: a) `Āhihi-Kīna`u, Maui. 20 ft. b) "Five Pinnacles," Lāna`i. 30 ft. (under overhang)

FAMILY FUNGIIDAE

In this family of free-living corals, polyps are typically large, flattened, solitary and unattached to the substrate. The single calyx has prominent septa and no walls. These corals prefer calm or deep areas where waves won't toss them around. If overturned, however, smaller species are capable of righting themselves. Juveniles (sometimes brightly colored) grow on the reef or directly from the parent, attaching by means of a stalk that eventually breaks. As many as seven species belonging to three genera have been reported from Hawai`i. In addition to the common species below, divers might see *Cycloseris vaughani* (circular with a smooth flat underside living on hard surfaces at depths of about 40 ft. or more) and *Diaseris distorta* (which reproduces by breaking into pie-shape pieces that regenerate into complete disks.)

MUSHROOM CORAL
• `āko`ako`a kohe

Fungia scutaria Lamarck, 1801
Family Fungiidae
• This is Hawai`i's largest and most common mushroom coral. The large free-living polyps occur on shallow reef flats in Kāne`ohe Bay, O`ahu, and also on deeper reefs, usually lodged in crevices or holes. Often several lie together. Immature polyps attached to the reef by thin stalks may occur nearby. The shape is roughly oval, frequently distorted or strongly humped. Sharp ridges (septa) radiate from the central mouth on the rounded upper side. The rough lower side bears many small projections that may help anchor it on the reef. With tentacles retracted, as is usually the case by day, it is tan or brown. The open tentacles of some individuals fluoresce green or magenta in bright sunlight. The species name means "shieldlike," the Hawaiian name, "vagina coral." Large specimens attain about 7 in. Indo-Pacific. Photos: Kāne`ohe Bay, O`ahu. 3 ft. a) adult; b) juveniles growing on dead skeleton of adult.

Mushroom Coral (a)

Mushroom Coral (b)

FAMILY FAVIIDAE

Although poorly represented in Hawai`i, corals of the family Faviidae are important reef builders elsewhere in the tropics. Best known are the brain corals, known locally only from fossils. Faviid corals typically form encrusting or domelike colonies, often with large, easily distinguished calyces. Hawaiian species, however, are generally inconspicuous. The family name comes from *favus*, the Latin word for honeycomb.

OCELLATED CORAL

Cyphastrea ocellina (Dana, 1846)
Family Faviidae
• This species forms small encrusting, clumpy, or even knoblike hemispherical colonies. Encrusting colonies are 2 to 6 in. across, usually at depths of less than 25 ft., and are light brown. The circular calyces (1/16 to 1/8 in. across) have sharply raised edges; they usually crowd together but nevertheless remain distinct. The species name means "little eyes." (The rare *Leptastrea bottae* has larger calyces and is similar in appearance.) Central Pacific. Photo: Hekili Point, Maui. 3 ft.

CRUST CORAL

Leptastrea purpurea (Dana, 1846)
Family Faviidae
• This hardy coral forms encrusting patches in a variety of habitats, from flat, rocky, wave-scoured bottoms near shore to vertical surfaces in deeper or more protected locations (down to 150 ft.). It may overgrow other corals, typically *Porites*, thereby appearing to form fingers or lobes. It is light tan, greenish brown or purplish brown, often with some white within the calyces. The relatively large calyces (usually 1/8 in. or more across) crowd together with walls touching and may vary greatly in size and shape within a single colony. Patches may be up to 6 ft. in diameter; most are much smaller. It is a common Indo-Pacific species. Photos: a) Kepuhi, O`ahu. 10 ft. (on reef flat); b) Molokini Islet, Maui. 25 ft. (on vertical surface)

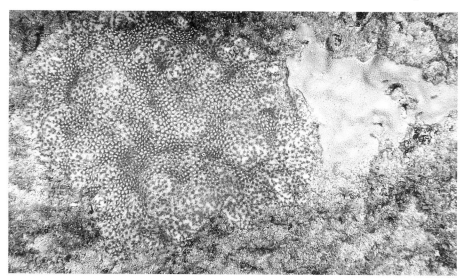

Crust Coral (a)

Crust Coral (b)

Solitary and Colonial Cup Corals (yellow and orange, respectively) with polyps withdrawn. Waimea Bay, O`ahu. 30 ft.

FAMILY DENDROPHYLLIIDAE

Members of this family have long tapering tentacles resembling those of sea anemones. The large, often brightly colored polyps form well-separated tubular skeletons ending in a cuplike calyx. For this reason they are often called "cup corals." Some people call them "flower corals." Hawaiian species host no symbiotic algae. This allows them to colonize dark places, such as the roofs of caves, where algae-dependent corals could not survive. Here they grow amid colorful sponges and other organisms, subsisting on small animals (usually plankton-ic) caught by their stinging tentacles. The sting is not strong enough to be felt by humans. Two genera are common in Hawai`i.

SOLITARY CUP CORAL

Balanophyllia sp.
Family Dendrophylliidae
• These corals are most common in caves and crevices in protected locations, such as O`ahu's "Hale`iwa Trench." They may be yellow or orange, but in dark caves they are dark brown or almost black. The polyps are solitary, sharing no common tissue or skeletal material. (This is not always easy to see in the field.) The calyces tend to be oval in cross-section. The tentacles often extend during the day. The animal pictured could be either of two species reported from Hawaiian waters of less than 150 ft.: *B. hawaiiensis* Vaughan, 1907 and *B.* cf. *affinis* (Semper, 1872). Other solitary cup corals occur in deep water. Photo: "Hale`iwa Trench," O`ahu. 40 ft.

Colonial Cup Coral (a)

COLONIAL CUP CORAL
Colonial Cup Coral (b)

Tubastraea coccinea Lesson, 1831
Family Dendrophylliidae

• In Hawai`i this coral is almost always bright orange; yellow and dark brown colonies may also occur. Colonies (clumps of about 10-20 polyps) typically grow on the roofs of caves, arches and overhangs where current or surge is strong. They occur at all depths attainable by snorkeling and scuba. Tentacles extend most fully at night, when the water is full of plankton, but withdraw completely by day unless current or surge is bringing food. Polyps are colonial, sharing common tissue and a common skeletal base. Calyces are round in cross-section. The eolid nudibranch *Phestilla melanobrachia* (p. 180) both mimics and preys on this coral. Photos: a) Pūpūkea, O`ahu. 30 ft. b) Palea Point, O`ahu. 30 ft. (with red sponge *Clathria (Microciona)* sp.)

TUBE ANEMONES

PHYLUM CNIDARIA. CLASS ANTHOZOA
SUBCLASS HEXACORALLIA. ORDER CERIANTHARIA

Tube anemones are anthozoan polyps that live in soft slimy tubes almost completely buried in sand or mud. Naked, they resemble worms with tentacles at one end. These animals construct their tubes almost entirely with adhesive threads fired from large nematocyst-like capsules called ptychocysts. (Under the microscope, according to one writer, the tube looks like a "chaotically woven rug.") Such tubes, open at both ends, can be exceedingly strong and up to 6 ft. long. When feeding, tube anemones project their beautiful tentacle crowns from the tube opening, which is usually elevated slightly above the bottom. When threatened or at rest they withdraw completely. Their elongate bodies, which may expand to a length of 2 ft., terminate in a narrow pointed foot useful for digging. The animals can "inflate" this foot into a bulb to anchor themselves in the soft substrate.

Unlike ordinary anemones, tube anemones have two sets of tentacles: short oral tentacles immediately surrounding the mouth are paired with longer, food-gathering tentacles along the margin. There may be several hundred tentacles in all. The animals feed primarily on small planktonic organisms but some are capable of killing small fish. They do not harbor zooxanthellae. Little is known of ceriantharian reproduction. Some scientists believe they reproduce only sexually, but aquarists report them expelling fully formed young. Asexual fission and budding have been also observed. About 50 species of tube anemones are known worldwide.

Although only one inconspicuous shallow-water species is officially recorded from Hawaiian waters, at least three larger unidentified tube anemones exist here, some with crown diameters of 12-15 in. Because the larval life of ceriantharians tends to be long, all probably represent widespread Indo-Pacific species. Small to medium-size tube anemones make attractive long-living aquarium animals. A layer of coarse sand at least 4 in. deep should be provided. Specimens of a large Mediterranean species have reportedly survived at the Naples Aquarium for over 100 years!

NIGHT TUBE ANEMONE

Arachnanthus(?) sp.
Family Arachnanthidae(?)
• The tentacles of these tube anemones are pinkish brown and unbanded. The anemones open only at night. When illuminated they close slowly, coiling their tentacles cork-screw fashion. The thin-walled tubes collapse when the animals withdraw. These anemones occur at the "Hale`iwa Trench" off Ali`i Beach Park, O`ahu, at depths of 40 ft. or more on silty sand, and doubtless in other areas as well. At Hale`iwa numerous small tube anemones with banded tentacles and white markings on the oral disk surround the occasional large one. These may be a different species. Crown diameter to about 6 in. Photos: "Hale`iwa Trench," O`ahu. 65 ft.

GHOST TUBE ANEMONE

Isarachnanthus bandanensis
Carlgren, 1924
Family Acontiferidae
• These anemones live on shallow sand bottoms, often at or near the low tide mark, and are abundant just off the beach at Kualoa, O`ahu, and doubtless in other areas. Their pale crowns of faintly banded translucent tentacles extend at night to catch drifting plankton but are almost invisible against the light sand. Like most ceriantharians, they hold their short oral tentacles vertically while the longer marginal tentacles tend to spread more horizontally. Largest observed: about 1 1/2 in. crown diameter. *Isarachnanthus bandanensis* has been reported from sandy, low intertidal habitat on O`ahu where this photo was taken, but the identity of the photographed specimens is not confirmed. The species was probably named for the Banda Islands, Indonesia, and occurs in the Western and Central Pacific. Photo: Kualoa, O`ahu. 2 ft.

DAY TUBE ANEMONE

Cerianthus(?) sp.
Family Cerianthidae(?)
• This large solitary tube anemone lives on sandy bottoms, usually at 100 ft. or more. It opens by day but withdraws instantly when touched. The long tapering tentacles are solid brown. Its thick-walled tube protrudes several inches from the substrate and does not collapse when the animal withdraws. It has been photographed at Molokini Islet, Maui, and Hōnaunau, Hawai`i. Crown diameter is about 6-8 in. Photo: Hōnaunau, Hawai`i. 100 ft.

BLACK CORALS and WIRE CORALS

PHYLUM CNIDARIA. CLASS ANTHOZOA
SUBCLASS HEXACORALLIA. ORDER ANTIPATHARIA

Feathery Black Coral *(Antipathes ulex)*. Mōkapu Rock, Moloka`i. 70 ft.

Black corals and wire corals have flexible horny skeletons much like those of octocoral gorgonians. Their polyps, however, almost always have six unbranched tentacles, placing them closer evolutionarily to the stony corals. Black corals typically grow in branching "bushes" or "trees." Wire corals (also called whip corals) are long, slender and unbranched. Both black corals and wire corals lack symbiotic algae and thrive where light levels are low and currents strong—under arches, in caves and overhangs, and on deep reefs.

Living antipatharians typically appear reddish, brown, yellow-green or even white due to the sheath of flesh covering the skeleton. Only the skeletons are black. Naked antipatharian skeletons (stripped of flesh) bear many small spines or thorns, another characteristic that separates these animals from gorgonians.

The dense dark trunks and branches of large black coral trees have been used since antiquity in Asia and the Middle East for fashioning jewelry, amulets and charms. The hard proteinaceous material takes a high polish and was believed in some cultures to ward off evil and cure disease. (The name *Antipathes* literally means "against disease.") The ancient Hawaiians also used black coral medicinally, grinding it and mixing the powder with other medicinal substances. Detailed recipes still exist for preparations to cure lung trouble and mouth sores. The name used in those days for black coral was `**ēkaha kū moana**.

Black corals have been harvested extensively throughout the Hawaiian Islands since 1958. Unfortunately for the undersea environment, few of these growths remain at sport diving depths. The only marine sanctuary with ample black coral habitat is Molokini Islet, Maui. The sanctuary was created too late to save the large trees that once grew there, although a few moderate- size bushes remain. All other corals enjoy full protection in Hawai`i—black corals and their symbionts need protection too.

Three species of black corals and two wire corals are known from depths of 150 ft. or less in Hawai`i. (Hawai`i's largest black coral, *Antipathes grandis*, is one of these, attaining 6 ft. in height. Uncommon at sport diving depths, it is not illustrated below.) Seven additional black corals and two more wire corals occur in deeper waters about the Islands.

BRANCHING BLACK CORAL

Antipathes dichotoma Pallas, 1766
Family Antipathidae

• This is the black coral most commonly encountered by divers in Hawai`i. In life it is reddish brown. Irregularly spaced smaller branches arise from all sides of the forked main branches; the smallest branches are long and stiff. Colonies generally grow in shaded areas along steep drop-offs, usually at depths of 60 ft. or more, and become more abundant with increasing depth. Bryozoans (up to 30 species) frequently colonize the bases, and Winged Pearl Oysters (p. 185) grow in the branches. Black coral bushes often serve as home territory for the Longnose Hawkfish. Rarely seen Tinker's Butterflyfish may occur in the vicinity as well. Tiny commensal gobies *(Bryaninops tigris)* sometimes live on the branches, scooting up and down and jumping from branch to branch like little monkeys. The species name means "divided in two," perhaps referring to the forking of the main branches. Large trees attain a height of about 4 ft. Indo-Pacific. (The slightly deeper-dwelling *A. grandis* has finer terminal branches and grows to at least 6 ft. It is otherwise similar.) Photo: Lehua Rock, Ni`ihau. 40 ft.

FEATHERY BLACK CORAL

Antipathes ulex Ellis & Solander, 1786
Family Antipathidae
• This coral is easily recognized by examining the smallest branches. Short and flexible, they are nearly equal in length and arranged almost regularly on either side of the parent branches, somewhat like the branches of a feather. The main branches are forked. Underwater its color varies from reddish to whitish, sometimes with a green tinge. The species name means "bushy shrub." To about 4 ft. tall. Central Pacific. Winged Pearl Oysters *(Pteria brunnea)* grow in the branches of the colony pictured here. Photo: Lehua Rock, Ni`ihau. 40 ft. (See also p. 68)

Common Wire Coral (a)

Common Wire Coral (b)

COMMON WIRE CORAL

Cirrhipathes anguina Dana, 1846
Family Antipathidae
• Wire corals (sometimes called whip corals) consist of a single long stem. Rarely, small branches form off damaged areas. This yellow-green to yellow-brown species is often kinky or spiraled, especially near the tip. Polyps occur on two sides of the stem. It attaches to solid substrate, usually on drop-offs at depths of 30 ft. or more, extending the springy stem into the current. Some wire corals harbor small commensal gobies of the genus *Bryaninops* and commensal shrimps of the genus *Pontonides* (two species of each). The gobies clear tissue from a band near the tip on which they lay their eggs. Gobies and shrimps sometimes occur on the same colony, but most wire corals harbor no commensals. The species name means "snakelike." To about 3 ft. long. Indo-Pacific. Photos: a) Hanauma Bay, O`ahu. 40 ft. (entire colony); b) Molokini Islet, Maui. 70 ft. (detail, two colonies intertwined).

RED WIRE CORAL

Stichopathes(?) sp.
Family Antipathidae
• This wire coral has not been identified. It is reddish in life and has polyps on only one side. Less common than *Cirrhipathes anguina* (above), it appears to prefer depths of about 100 ft. or more. It often harbors commensal gobies; shrimps have not been observed on it. Photo: Molokini Islet, Maui. 70 ft.

Two species of tube-dwelling annelid worms: a large Featherduster *(Sabellastarte sanctijosephi)* grows among tangled white tubes of Sea Frost *(Salmacina dysteri)*. Magic Island boat channel, O`ahu. 15 ft.

Legless invertebrates with slender or flattened bodies are commonly called worms. Many diverse animals fit this description. The early taxonomist Linnaeus (1707-1778) tried to place them all in one phylum; today, at least 17 phyla of wormlike marine animals are recognized.

Most marine worms live their lives hidden in sand or mud, under stones, in crevices, or as parasites or symbionts of larger animals. Only a small fraction—principally flatworms, ribbon worms and some annelid worms—are likely to be encountered by divers or snorkelers. Acorn worms might also be added to this list; the animals themselves are not ordinarily visible, but their large fecal piles are conspicuous on some sandy bottoms.

FLATWORMS
PHYLUM PLATYHELMINTHES

Flatworms are the simplest worms and were probably the first animals to develop bilateral symmetry (the left and right sides mirroring each other), an upper and a lower side, and a distinct "head" and "tail." Their bodies are composed of three basic tissue layers, making them ancestral to higher animals such as molluscs, arthropods and vertebrates. Cnidarians, by comparison, have only two tissue layers, and sponges have none.

Flatworms are soft-bodied, breathe through their body surface, have no body cavity, and possess rudimentary nervous and digestive systems. The mouth serves also as the anus. Sensory organs may consist of tiny eye spots or short tentacles.

Many flatworms are symbiotic or parasitic, living their entire adult lives inside other animals. These include the human tapeworms (class Cestoda) and the flukes (class Trematoda) that infest the gills and organs of fishes and other animals. Free-living flatworms (class Turbellaria) are mostly marine, typically occurring under rocks or in cracks and crevices, often in the shallow intertidal zone. The tiny flatworms of the genus *Planaria*, used in biology classes to demonstrate regeneration, live in freshwater streams. (Cut one in half lengthwise and each half will regrow its missing side. Split the head, and in a few days you'll have a two-headed worm.) Most noticeable to a diver or snorkeler, however, are the marine polyclad flatworms (order Polycladida), especially the brightly colored species of the family Pseudocerotidae. ("Polyclad," meaning "many branches," refers to the branched gut of these animals.) Typically about an inch long, they are sometimes confused with the equally colorful but completely unrelated nudibranchs.

Polyclad flatworms and nudibranchs are not hard to tell apart. The worms are flatter, often with frilly or ruffled edges, and typically faster moving (forward motion being accomplished by the beating of microscopic hairs, or cilia, in a bed of slime secreted by the worm on its underside). Flatworms lack the external gills and large sensory tentacles (rhinophores) often found on nudibranchs. In addition, many swim; most nudibranchs cannot.

Most polyclad flatworms are nocturnal or cryptic. The few openly active by day are usually brightly colored and probably toxic or bad tasting. All are carnivores. Many feed on small worms, crustaceans or tunicates; others engulf small snails, extracting the animal and leaving the shell; a few are cannibalistic. Some flatworms immobilize their prey with a sticky glue secreted from special glands near the mouth.

Flatworms are hermaphroditic, possessing both male and female copulatory structures. To mate, many reef-dwelling polyclads simply rear up and stab each other randomly with a hard penile stylet. Tissue damage is repaired in a few hours. Eggs, often deposited in gelatinous strands or masses, either develop directly into tiny flatworms or into free-swimming larvae. Many flatworms can also reproduce asexually by budding or by dividing in half.

Because of their diet, polyclad flatworms are not good candidates for aquariums. Long-term survival is most likely in a well-established tank containing an abundance of small organisms, especially colonial tunicates and bryozoans. Flatworms are sometimes introduced unintentionally in pieces of rock or coral and at least one tiny species (probably of the order Acoela) can multiply uncontrollably in the marine aquarium, appearing as harmless but unsightly brown spots on living corals.

Free-living (turbellarian) flatworms remain a poorly known group with about 3,000 described species worldwide. Only 40-50 are documented from Hawai`i; more remain to be discovered and named. Twelve commonly seen species are illustrated here, all belonging to the order Polycladida.

FEW-EYED FLATWORM

Paraplanocera oligoglena (Schmarda, 1859)

Family Planoceridae

• The translucent body of this worm is spotted and mottled with white and brown; interior organs are visible at its center. Like all polyclads, it seems to glide or flow effortlessly over the substrate. It is found under stones. The species name means "few eyes." To about 2 1/2 in. Indo-Pacific. Photo: L. Newman & A. Flowers, Madang, Papua New Guinea.

HYMAN'S FLATWORM

Pericelis hymanae Poulter, 1974

Family Pericelidae

• These white flatworms are locally common under stones in shallow water at areas such as Black Point, O`ahu. They are thought to be associated with the Brown Purse Shell, *Isognomon perna*, (p. 186) and are named for American zoologist Libbie H. Hyman (1888-1969), a specialist in free-living flatworms and the author of a widely-used multivolume text on invertebrates. To almost 2 in. Known only from Hawai`i. Photo: Nāpili Bay, Maui. 10 ft. (identification not confirmed)

WHITE STRIPE FLATWORM

Family Pseudocerotidae

• One of the more common nocturnal flatworms around O`ahu, this undescribed animal belongs to no established genus and is currently under study. It is light brown, thinly edged in white and speckled with minute white and dark spots. A conspicuous white stripe starts just behind the tentacles and runs about 3/4 of the way down the center of the body. To about 1 in. Known only from Hawai`i. Photo: Hālona Blowhole, O`ahu, 20 ft.

DIVIDED FLATWORM
Pseudoceros dimidiatus von Graff, 1893
Family Pseudocerotidae
• Black, yellow and orange, this conspicuous worm is active by day and almost certainly toxic or distasteful. The "branches" of its color pattern are sometimes replaced with large irregular blotches or may be missing entirely. There is always a yellow double line down the center, however, that may be slender or very broad. The first scientific specimen had only the central yellow double line, thus the species name meaning "divided in half." To about 2 1/2 in.; Hawaiian specimens rarely attain this size. Central and Western Pacific. Photo: "Casa de Emdeko," Kailua-Kona, Hawai`i. 30 ft.

FUCHSIA FLATWORM
Pseudoceros ferrugineus Hyman, 1959
Family Pseudocerotidae
• Rimmed with scarlet and gold, this flatworm is one of the most colorful creatures of Hawai`i's reefs. In deep water, without the aid of artificial light, it appears deep blue; under a flashlight beam it is brilliant fuchsia. Tiny white spots cover the intensely colored central area. When fully expanded this animal is almost circular. Active both by day and night, it is regularly seen in summer off O`ahu's north shore, as well as at many other locations. The bright color is probably a warning that the worm is toxic or distasteful. The species name means "rusty," possibly the color of preserved specimens. To about 2 in., but always smaller in Hawai`i. Central and Western Pacific. Photo: Pūpūkea, O`ahu. 40 ft.

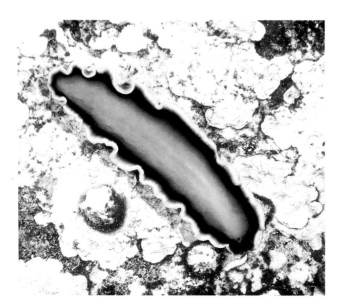

JEBB'S FLATWORM

Pseudoceros jebborum Newman & Cannon, 1994
Family Pseudocerotidae
• This somewhat elongate flatworm is caramel color, ringed with two marginal bands. The inner band is black, the outer cream with a bright yellow rim. It is named for Dr. Matthew Jebb (former Director of the Christensen Research Institute at Madang, Papua New Guinea) and his family. To about 2 3/4 in., but smaller in Hawai`i. Central and Western Pacific. Photo: Honokōhau, Hawai`i. 20 ft.

GOLD RIM FLATWORM

Pseudoceros paralaticlavus Newman & Cannon, 1994
Family Pseudocerotidae
• This flatworm is velvety black with a wide creamy band down the center. There are two marginal bands, the inner white and the outer yellow. It is nocturnal, occurring under rocks by day, and can be found on shallow reef flats. To about 1 1/2 in. Indo-Pacific. Photo: Magic Island, O`ahu. 5 ft

FALSE GOLD RIM FLATWORM

Pseudobiceros sp. 1
Family Pseudocerotidae
• This undescribed flatworm is almost identical in color to the Gold Rim Flatworm (p. 76) but belongs to a different genus. In *Pseudobiceros* the pseudotentacles (simple folds at the front containing sensory organs) are ruffled. In *Pseudoceros* they are plain. *Pseudobiceros* can swim; *Pseudoceros* cannot. The remarkable color similarity of the two species might be explained several ways: one species could be toxic and the other a mimic; both could be toxic, their color patterns converging to reinforce the warning effect; or the color pattern might be advantageous in itself for reasons unrelated to toxicity. Both species are nocturnal. Photo: Hōnaunau, Hawai`i. 30 ft.

HAWAIIAN SPOTTED FLATWORM

Pseudobiceros sp. 2
Family Pseudocerotidae
• This flatworm has a dark brown to purplish brown body covered with dark edged white spots of various sizes that become yellowish and small near the margin. During the day it remains under slabs and stones from the shoreline to depths of at least 50 ft. Emerging only after dark, it sometimes swims in the open with graceful, rippling undulations. The advantage of the striking color pattern is not clear, since the worm is never seen in the open by day. To about 1 1/2 in. Known only from the Hawaiian Islands. Photo: Mākua, O`ahu. 40 ft.

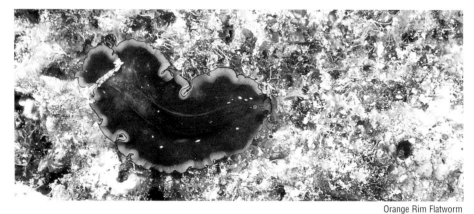

Orange Rim Flatworm

ORANGE RIM FLATWORM ▲

Pseudobiceros sp. 3
Family Pseudocerotidae
• This worm is brown with marginal bands of orange and black. Irregular white spots may be present in the central brown area. To about 1 in. Known only from Hawai`i. Photo: Pūpūkea, O`ahu. 15 ft.

Glorious Flatworm (a)

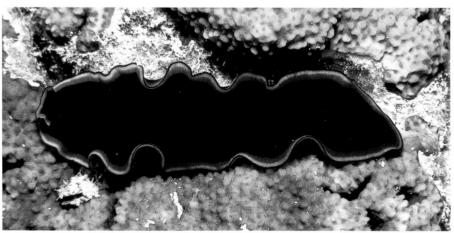

Glorious Flatworm (b)

GLORIOUS FLATWORM

Pseudobiceros gloriosus Newman & Cannon, 1994
Family Pseudocerotidae
• A black body rimmed with narrow bands of color is a common pattern among flatworms of the family Pseudocerotidae. This species has an inner orange band, a middle pink one, and a barely visible outer burgundy band. The underside is burgundy to deep pink. Like many flatworms of its genus, it is most active at night and capable of swimming well off the bottom with undulations of the body. It grows to about 3 1/2 in., but in Hawai`i may not attain this size. Central and Western Pacific. Photos: a) Waimea Bay, O`ahu. 30 ft. (swimming); b) L. Newman & A. Flowers, Great Barrier Reef, Australia, 30 ft.

PLEASING FLATWORM

Pseudobiceros gratus (Kato 1937)
Family Pseudocerotidae
• Many pseudocerotid flatworms have frilly margins, but few are more deeply and beautifully ruffled than this. Its ground color varies from white to tan; a thin black rim accentuates the delicate curly edges. Three (sometimes four) wide dark stripes run lengthwise, pairs joining at the worm's back end. Its front is distinguishable by two pseudotentacles, small folds containing primitive sensory organs (true of many other polyclads as well). This attractive worm remains under rocks by day; it probably emerges at night. The species name means "agreeable" or "pleasing." Photo: Palea Point, O`ahu. 30 ft.

RIBBON WORMS
PHYLUM NEMERTINA

Ribbon worms include about 900 species of long, thin, soft-bodied animals. Carnivores, they are characterized by a long proboscis (usually tipped with barbs) that they shoot from an opening at the tip of the head to impale or entangle their prey. Some ribbon worms from temperate waters attain lengths of 4-5 ft.; one from the north Atlantic is reported to reach 60 ft! Although long, they are highly contractile—a 5 ft. worm can shrink to 6 in. Ribbon worms were probably the first animals to have both a mouth and an anus connected by a complete digestive tract. There are fewer than 10 known species in Hawai`i.

BANDED RIBBON WORM • **ko`e kai**

Baseodiscus cingulatus (Coe, 1906)
Phylum Nemertina. Class Anopla
• The first specimens of this ribbon worm were dredged from about 250 ft., but the animal is also encountered in shallow water and even in tide pools, usually at night. Many narrow, reddish brown bands encircle the slender body, which, like many ribbon worms, can contract or expand. Fully extended, large specimens attain as much as 4 ft. The thicker end, marked with many small spots, is the head. The species name means "girdle." Known only from the Hawaiian Islands. (*Baseodiscus hemprichii*, a related species, is marked with a single reddish brown or black line running the length of its white body.) Photo: Pūpūkea, O`ahu. 20 ft.

Bristle worms, often called polychaete worms, form a class within the large phylum Annelida. Annelida means "little rings," and the approximately 11,500 members of this group (earthworms are the most familiar) have bodies composed of many segments, or rings. Marine bristle worms are characterized by the presence of stiff hairs, or bristles, on the segments. The class name is derived from "poly," meaning "many," and "chaeta," meaning "hair."

Polychaetes are a diverse group far more evolutionarily advanced than flatworms and ribbon worms. Some roam freely, while others are burrowers or live anchored to the reef in calcareous or parchment-like tubes of their own making. A number of polychaetes actively hunt their prey, but most merely filter food from the sand, sediments and water in which they live. Many tube-dwelling polychaetes depend on food gathering tentacles of one sort or another. Polychaetes include the well-known fireworms, feather duster, Christmas-tree and spaghetti worms. Most polychaetes, however, are small, cryptic, and virtually never seen. Nevertheless, they are abundant and important to the ecology of the reef in many ways, serving in particular as food for some of the larger and more conspicuous organisms such as cone shells and butterflyfishes.

Of some 8,000 species worldwide, about 250 marine bristle worms are known from Hawai`i, representing 43 families. Nine are described below.

FIREWORMS. FAMILY AMPHINOMIDAE

Fireworms are free-living carnivores that live under stones and in crevices from shallow reef flats down to at least 600 ft. Their segments bear tufts of sharp toxic bristles on each side that easily penetrate and break off in human skin, causing burning, itching or numbness that may last several hours. A rash may also occur. It is best not to touch these worms. If stung, you might be able to remove some of the bristles with the sticky side of adhesive tape. Two species are shown here. At least six others also occur in Hawai`i. All belong to the family Amphinomidae. They are known in Hawaiian as **'aha huluhulu**, literally "hairy cord."

Lined Fireworm (a) ➤

LINED FIREWORM • `aha huluhulu   Lined Fireworm (b)

Pherecardia striata (Kinberg, 1857)
Family Amphinomidae
• Lined Fireworms are common in Hawai`i. Nosing about the reef in broad daylight, they somewhat resemble underwater centipedes: the white tufts of bristles on each side are reminiscent of legs. Although lacking jaws and teeth, these worms prey on crustaceans, sea stars, molluscs, other worms, algae and carrion; bait placed on the reef at night will soon be covered with them. In the Eastern Pacific they are known to attack Crown-of-Thorns Sea Stars (p. 293), entering through wounds caused by Harlequin Shrimps (p. 229) and killing the stars in about a week. On certain moonlit nights around high tide, sexually mature worms of this species swarm at the surface to spawn. Swollen with eggs and sperm, they appear larger and darker than worms seen on the reef. To about 8 in. Indo-Pacific, especially in areas where pocilloporid corals are common. Photos: a) Pūpūkea, O`ahu. 30 ft; b) Kailua-Kona, Hawai`i. (swimming at the surface)

ORANGE FIREWORM • `aha huluhulu

Eurythoe complanata (Pallas, 1766)
Family Amphinomidae
• This common species is pink, orange or greenish, often with some iridescence. It lives under rocks in shallow water and at moderate depths. Introduced unintentionally into aquariums, it can reproduce prolifically, feeding on live corals, molluscs and other worms. Small traps specifically designed for controlling fireworms in aquariums are available commercially. To about 6 in. Worldwide in tropical seas. Photo: Wai`alae Beach Park, O`ahu.

PARCHMENT WORM

Chaetopterus sp.
Family Chaetopteridae
• These unusual worms inhabit parchment-like tubes open at both ends through which they pump water by means of fan-like paddles. They collect food particles in a bag of mucus, which they periodically roll into a ball and swallow. They never leave their tubes. The worms produce a luminescent slime the function of which is unknown. Aquarists have successfully removed these worms intact from their tubes and placed them in glass tubes of the same size where their fascinating behavior can be observed. The worms appear to survive well in these conditions. The Hawaiian species illustrated here has long been identified as *C. variopedatus* Renier, 1804, but the genus is presently under review and the name may change. This or similar species are common on harbor pilings throughout the warm waters of the world. The tubes are 3-5 in. long and irregularly curved. Photo: Pearl Harbor, O`ahu. 10 ft.

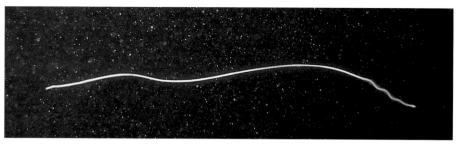

PALOLO WORM

Palola siciliensis (Grube, 1840)
Family Eunicidae
• Palolo Worms bore into coral rock and are seldom seen. To reproduce they detach special segments of their body bearing reproductive organs (epitokes); these swim freely and spawn, leaving the worm itself safe in its hole. In many areas of the South Pacific epitokes swarm by the millions at certain times of the year. Considered a delicacy, they are gathered by the local people who know exactly when the event will occur. In Hawai`i and many other parts of its range, the Palolo Worm does not swarm. Individual epitokes, however, are often seen during the summer months writhing and spiraling through the water at night, as pictured here. The Fijians call these worms "mbalolo," apparently the source of our common name. They may exceed 4 in. length and occur worldwide in warm seas. Photo: Pūpūkea, O`ahu, 20 ft. (identification not confirmed)

FEATHER DUSTER WORM

Sabellastarte sanctijosephi (Gravier, 1906)
Family Sabellidae
• The most conspicuous part of this worm is its large fan, or crown; the body, enclosed in a leathery tube, remains mostly in the interstices of the reef. The worm uses its crown both for respiration and for filtering suspended particles from the water, withdrawing it instantly when it senses a shadow or is touched. Feather Duster Worms prefer turbid water (not surprising considering their feeding habits) and are often abundant in harbors and protected locations such as Kāne`ohe Bay, O`ahu. They occasionally occur in higher energy, clear-water environments to depths of at least 100 ft. These worms do well in captivity but the tubes extend deep into the substrate and must be collected very carefully. In the aquarium it is fascinating to watch small particles of food move down the groove at the center of each tentacle toward the mouth, propelled by tiny hairlike cilia. Particles are sorted by size at the crown base; small edible ones are consumed, while larger grains are discarded or mixed with mucus to extend the tube. These worms occasionally shed their crowns but will grow new ones if the aquarium is well oxygenated. The worm, including the crown, grows to just over 3 in.; its tube is much longer. Indo-Pacific. Photo: "Kewalo Pipe," O`ahu. 40 ft. (See also p. 72)

CHRISTMAS-TREE WORM • **kio**

Spirobranchus giganteus (Grube, 1862)
Family Serpulidae
• The only conspicuous parts of this worm are the two brightly colored spiral fans used for feeding and respiration. These fans (consisting of yellow, blue, tan or orange tentacles) can be instantly withdrawn and the tube's opening sealed with a trap door (operculum) bearing antler-like spines. The Christmas-Tree Worm is one of many segmented worms that live in rigid calcareous tubes of their own making. Its tube is usually buried in living coral. In Hawai`i it is most common on Lobe Coral *(Porites lobata)* but also occurs in other corals. The presence of these worms is thought by some scientists to help protect the coral colony from predation by the Crown-of-Thorns Sea Star. The body grows to 1 3/4 in.; the tiny crown is less than 1/2 in. high or across. This species will survive in aquariums but collecting is illegal in Hawai`i if attached to rock or living coral. It occurs in all tropical seas and in many parts of its range grows to twice the size it does in Hawai`i. Photo: Kahe Point, Oahu. 25 ft. (on Lobe Coral)

SEA FROST

Salmacina dysteri (Huxley, 1855)
Family Serpulidae.
• These masses of white intertwining, threadlike tubes remind some people of the branching patterns of frost on a windowpane. The tubes are formed by tiny polychaete worms (closely related to the Christmas-Tree Worm) whose nearly colorless tentacles can barely be seen protruding from their ends. The tentacles are quickly withdrawn upon too close an approach. This species occurs in all warm seas from tide pools to deeper reefs and often grows thickly on harbor pilings. Photo: Kahe Point, O`ahu. 15 ft. (See also pp. 72, 90)

MEDUSA SPAGHETTI WORM • **kauna`oa**

Loimia medusa (Savigny, 1818)
Family Terebellidae
• Bluish white tentacles strung haphazardly over the rocks and rubble indicate the presence of a spaghetti worm buried nearby. Like most tube-dwelling worms, its body (encased in a tough buried tube covered with bits of shell and gravel) is never seen. Food particles adhering to the sticky strands are either passed to the worm's mouth through grooves extending the length of the tentacle, or the tentacle contracts toward the mouth dragging the particles with it. The soft conspicuous tentacles are possibly protected by a poison. (Spaghetti worms were used medicinally in old Hawai`i and their tentacles contain anticancer compounds.) Nevertheless, the animal is preyed upon by the large cone snail *Conus spiceri*. This worm occurs worldwide in warm seas. Its body can be almost 12 in. long, with tentacles at least twice that length. A similar but less common species, *Lanice conchilega*, also occurs in Hawai`i, its tube often home to a small commensal pea crab *(Aphonodactylus edmondsoni)*. Several smaller species are found on sandy reef flats in Kāne`ohe Bay, O`ahu, and elsewhere. Miniature spaghetti worms sometimes turn up on live rocks in aquariums; it may take a magnifying glass to see them. Photo: Mākua, O`ahu. 100 ft.

SPINY SCALE WORM

Iphione muricata (Savigny, 1818)
Family Polynoidae
• Scale worms are found underneath stones and can curl into a ball when dislodged. This species has 13 pairs of overlapping scales or plates (elytra). It occurs from the shallows to at least 40 ft. To about 1 in. Indo-Pacific. Photo: Hanauma Bay, O`ahu. 40 ft.

ACORN WORMS

PHYLUM HEMICHORDATA. CLASS ENTEROPNEUSTA

Acorn worms are unsegmented, soft-bodied, solitary animals typically inhabiting a U-shape tube deep under sand or mud. They feed on organic matter filtered from the substrate and their castings are common on some sand bottoms. Most have a short rounded proboscis surrounded by a collar with scalloped edges that somewhat resembles the pointed end of an acorn. They are typically from 3 to 18 in. long, but one species exceeds 7 ft. Unlikely as it might seem, acorn worms are evolutionarily advanced. The phylum to which they belong (Hemichordata) is related to the Chordata, which includes tunicates, fishes, birds and mammmals. One indication of this relationship is the presence of gill slits such as those possessed by sharks and rays.

YELLOW ACORN WORM

Ptychodera flava Eschscholtz, 1835
Family Ptychoderidae
• These worms are common just beneath the sand or under stones on some shallow reef flats. Their yellow color is distinctive. In shallow water they attain lengths of about 1 to 8 in. and seldom create recognizable mounds. Much larger yellow worms, up to 18 in. long and 1 in. in diameter, occur on sandy bottoms at depths of 30 ft. or more. They may be the same species. Although the dark fecal mounds of these large worms can be common, the worms themselves are never seen. If one is dug up, the weight of the mud or sand within it often ruptures its thin body walls. Gould's Auger (p. 142) feeds on both the large and small forms. Photos: a) Kahe Point, O`ahu. 30 ft. (casting of large worm); b) Hekili Point, Maui. 2 ft. (small worm, about 1 in.)

BRYOZOANS
PHYLUM BRYOZOA (OR ECTOPROCTA)

A flexible, mosslike bryozoan *(Savignyella lafontii)* and a calcareous, encrusting bryozoan (*Parasmittina*) grow side by side on the back wall of Molokini Islet, Maui. 70 ft.

Bryozoans are tiny colonial animals common in tropical and temperate waters. Some species look like algae, while others resemble small branching or encrusting corals. Most attach to solid surfaces; some live in sand. The individual animals, typically about 1/25 in. across, are called zooids. In most species each zooid secretes an external chitinous or calcareous skeleton, usually in the form of a cup, tube, or hollow rectangle. Individual skeletons join to neighboring ones forming bushy, branching, fanlike or encrusting colonies that may be rigid or flexible.

Bryozoan zooids feed with a circular or horseshoe-shape crown of tentacles called a lophophore. Tiny beating hairs (cilia) on the tentacles create a current that draws suspended food particles into the mouth at the base of the tentacles. A one-way digestive tract ends in an anus opening outside the ring of tentacles. (This represents an advance over the cnidarians, in which the mouth serves also as the anus.) Certain zooids within a colony may be specialized for defense, reproduction, or other functions. Some defensive zooids (called avicularia after *aves*, the Latin word for "bird") resemble little bird heads with pinching beaks.

Bryozoans are significant members of the "fouling community"—animals such as barnacles, tunicates and sponges that thrive in harbors and marinas, often growing on boat bottoms and other bare surfaces. Colonies may be anywhere from 3 ft. (unusual) to about 1/25 in. across. Although bryozoans are common, few divers and snorkelers are aware of them. Small size, hidden habitats (under stones, in crevices, on pilings beneath docks), and resemblance to other growths (algae, tunicates, hydroids and sponges) all contribute to the relative obscurity of this major phylum. Furthermore, the group as a whole lacks a common name, although bryozoans with flexible algae-like skeletons are sometimes called "sea mosses" or "moss animals," and those with branching calcareous skeletons are often called "lace corals." Recently, a potent anticancer substance called bryostatin has been isolated from *Bugula neritina*, a bushy reddish purple bryozoan common in harbors around the world (including Hawai`i). Useful compounds such as these may help put these animals on the map.

Most bryozoans are marine; a few inhabit fresh water. There are more than 5,000 living species, with many more in the fossil record. About 200 are known from Hawaiian waters. No Hawaiian name is recorded for these animals.

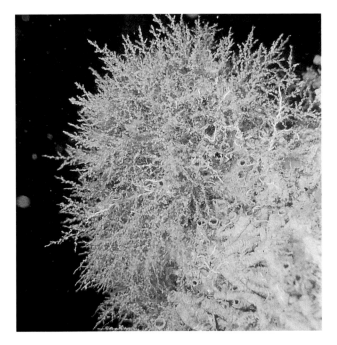

WHITE BUSHY BRYOZOAN

Amathia distans Busk, 1886
Family Vesiculariidae
• This soft, bushy, whitish bryozoan is common in harbors and lagoons throughout the Hawaiian chain. Originally from the tropical Atlantic, it now occurs around the world in warm seas, probably spread by ships from port to port. Photo: Pearl Harbor, O`ahu. 10 ft.

BLUE FAN BRYOZOAN

Bugula stolonifera Ryland, 1960
Family Bugulidae
• This animal produces fanlike branching colonies in harbors and marinas, usually in shaded spots under ledges. The lightly calcified colonies, growing to about 1 1/2 in. across, cannot stand turbulent water. Spread by shipping, this bryozoan now has an Indo-Pacific distribution and is considered an introduced species in Hawai`i. Another introduced bryozoan common in the same habitat, *Bugula neritina*, resembles reddish algae with slightly stiff branches. Photo: Magic Island, O`ahu. 10 ft.

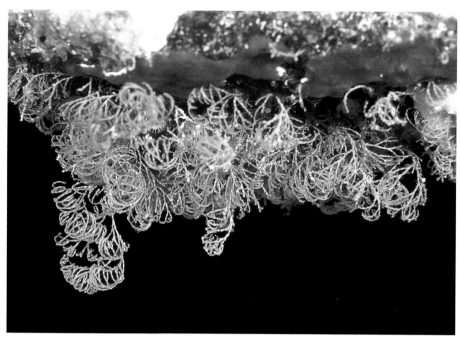

LAFONT'S BRYOZOAN

Savignyella lafontii
(Audouin, 1846)
Family Savignyellidae
• Although best known as a fouling organism in harbors, this "sea moss" also occurs on exposed rocky walls in clear water where it forms curled branching colonies about 1/2 in. high, usually under overhangs or in crevices. In the harbor environment it is typically bright red; in clear water at Molokini Islet, Maui, it is light tan. Occurs worldwide in warm waters. Photo: Molokini Islet, Maui. 70 ft. (See also p. 87.)

YELLOW CRUST BRYOZOAN

Parasmittina(?) sp.
Family Smittinidae
• Bryozoans of this genus form heavily calcified crusts over shell, coral and stone. Some species overgrow flexible objects, such as wire coral stems or seaweeds, forming knobby or even erect formations. Usually yellowish or yellowish white, these crusts may be composed of one or more species (of about a dozen known exclusively from Hawai`i). They are especially common under overhangs along the back wall of Molokini Islet, Maui. It has been suggested that Hawaiian bryozoans of this genus may be the marine equivalent of Darwin's finches. Photo: Molokini Islet, Maui. 70 ft.

ERRATIC BRYOZOAN

Schizoporella errata
(Waters, 1878)
Family Schizoporellidae
• This calcareous bryozoan forms orange-brown encrusting colonies in protected harbors on pilings or other hard surfaces. It occasionally grows in irregular branching formations such as the one pictured here (with the hydroid *Pennaria disticha* and the annelid worm *Salmacina dysteri*). A typical encrusting colony might be 3-4 in. across. The species occurs worldwide in warm temperate and subtropical waters. Photo: Hospital Point Drydock, Pearl Harbor, O`ahu. 15 ft.

LACE BRYOZOAN

Reteporellina denticulata
(Busk, 1884)
Family Sertellidae
• Colonies of this calcareous bryozoan with its fine intercon-nected branches are sometimes mistaken for a delicate coral (and are often erroneously called "lace coral"). The species is common on rocky reefs, usually at depths of 30 ft. or more on or near verti-cal walls. At the shallow end of its range colonies occur in protected crevices; with increasing depth they grow more in the open. The yellowish tan branches break eas-ily when handled. This is one of the largest, most attractive and most easily observed bryozoans in Hawai`i. Colonies attain as much as 3-4 in. diameter. Indo-Pacific. Photo: "Sheraton Caverns," Kaua`i. 40 ft.

"CHEX" BRYOZOAN

Triphylozoon sp.
Family Sertellidae
• This bryozoan forms fragile calcareous ruffles in protected locations, such as inside submarine lava tubes or under coral slabs. The lacy folds are covered with a mesh pattern that has been compared to "Chex" breakfast cereal. Photo: "Red Hill" (Kona coast), Hawai`i. 70 ft.

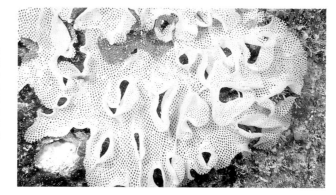

TUNING FORK BRYOZOAN

Crisina radians (Lamarck, 1816)
Family Crisinidae
• This fragile calcareous bryozoan typically grows in fanlike colonies the branches of which often appear to radiate outward from a common center. Branches usually divide in a "tuning fork" pattern. A common species in some areas, it can occur on blades of seaweed as well as on hard surfaces. To about 1/2 in. across. Photo: Waimea Bay, O`ahu. 30 ft.

VIOLET ENCRUSTING BRYOZOAN

Disporella violacea (Canu & Bassler, 1927)
Family Lichenoporidae
• A colony of these small, hard, brightly colored calcareous bryozoans might easily be mistaken for a tiny patch of encrusting coral. They are common in crevices and under overhangs along the rocky reefs of O`ahu's north shore. Typical colonies are about 1/8 to 1/4 in. across and vary from dark blue to violet. Photo: Pūpūkea, O`ahu. 30 ft.

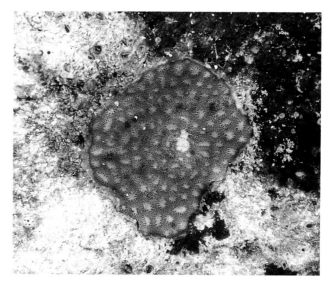

MOLLUSCS
PHYLUM MOLLUSCA

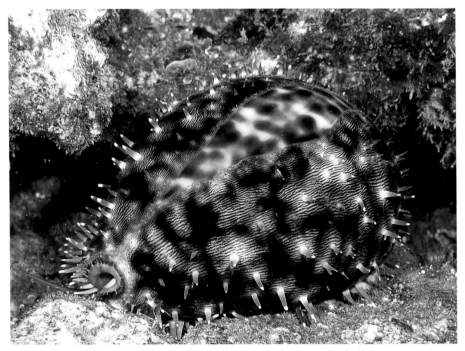

A Tiger Cowry *(Cypraea tigris)* with its mantle out. South shore, Lāna'i. 30 ft.

Molluscs are soft, legless animals that usually secrete calcium carbonate shells. ("Molluscus" means "soft" in Latin.) Clams and oysters, snails, chitons, cuttlefishes, squids and octopuses belong to this group. They range in size from the giant squid (whose 25 ft. body makes it the world's largest invertebrate) to tiny snails about 1/25 of an inch long. Molluscs are thought to have evolved from primitive flatworm-like animals over 500 million years ago. Multiplying and diversifying, they have penetrated all habitats on earth, from the mud and sediment of the sea bottom to the tops of mountains almost 20,000 ft. high. With 85,000 to 110,000 species (estimates vary), molluscs are by far the largest group of animals in the sea. They are second only to the insects (phylum Arthropoda) in total number of species.

Most people are much more familiar with the shells of molluscs than with the animals themselves. Throughout history, mollusc shells have been used for ornaments, tools and even money. Shells may be coiled (snails), conical (limpets), hinged (clams and oysters), segmented (chitons), or tusklike (scaphopods). Some molluscs have internal shells (squids, cuttlefishes, and many slugs) or no shells at all (octopuses and nudibranchs).

Evolutionarily advanced, molluscs have well-defined organs and complex sensory, circulatory, digestive and reproductive systems. A few approach mammals in intelligence. Because of their diversity, classification is complex. Specialists do not even agree on the number of classes in the phylum (five to eight) or their names, much less the number of species in each. Four major classes are described here: the gastropods, containing snails and slugs (class Gastropoda); the bivalves, containing clams and oysters (class Bivalvia or Pelecypoda); the cephalopods, containing octopuses, cuttlefishes, nautiluses and squids (class Cephalopoda); and the chitons (class Polyplacophora or Amphineura). The two largest classes, gastropods and bivalves, together account for 99 percent of all molluscan species and probably constitute a significant part of the earth's biomass.

Molluscs have a number of organs unique to the phylum. Almost all, for example, have a muscular foot, varying in structure and function with the class of mollusc. The names hint at this, almost all containing the root word "pod" meaning "foot." "Gastropod" means "stomach-foot" or "ventral-foot": a snail or slug's flat, wide crawling foot underlies the stomach and viscera. "Pelecypod" means "hatchet-foot": a bivalve's foot is bladelike (useful for digging). "Cephalopod" means "head-foot": an octopus or squid's foot is divided into a crown of arms surrounding the head, used for grasping and clinging.

Other unique molluscan organs include the mantle, a fleshy fold of the body wall that encompasses the body and secretes the shell; and the radula, a tonguelike chitinous feeding organ studded with tiny teeth. Most marine molluscs have gills for respiration, often located in a cavity under the mantle. Terrestrial molluscs (land snails and slugs) lack gills, having mantles modified into a kind of lung.

Most molluscs have separate sexes, but some change sex and others are hermaphroditic. Marine molluscs typically reproduce by releasing eggs and sperm into the water. Some copulate, however, and deposit eggs directly on the substrate. After hatching, trochophore larvae develop into tiny shelled larvae (veligers) that generally drift with the plankton until settling in a suitable habitat to mature.

About 1,650 species of molluscs are known from Hawai`i of which approximately half are marine. Of these, about 20 percent are considered endemic (found nowhere else). Many tiny marine species exist that are as yet unrecorded.

Perched on a sponge, a Gloomy Nudibranch *(Tambja morosa)* stretches out in search of its next meal. Magic Island boat channel, O`ahu. 10 ft.

CHITONS

Chitons are the most primitive of the common seashore molluscs. In a class of their own (variously called Polyplacophora or Amphineura), they have shells composed of eight narrow, usually overlapping plates surrounded by a band of tough flesh called the girdle. Chitons cling tenaciously to rocks with their foot, often in the surge line or just above it, where they creep slowly about scraping algae. If disturbed they clamp down on the rock like living suction cups and are difficult to dislodge. Most grow no longer than a few inches, although the Gum-Boot Chiton *(Cryptochiton stelleri)* of the north Pacific coast of America and Asia is about a foot long. Hawai`i has four recorded species (and probably several unrecorded ones), all small and inconspicuous. In old Hawai`i chitons were sometimes called **pūpū mo`o,** literally "lizard shell."

GREEN CHITON • **kuapa`a**

Acanthochiton viridis (Pease, 1872)
Family Acanthochitonidae
• These animals occur in small depressions on wave-washed limestone reef flats. Greenish overall, their girdles covered with small spines, they are the largest Hawaiian chitons. An old Hawaiian riddle asked "What is the fish that has eight scales?" The answer was **kuapa`a**, or "chiton." (On O`ahu, however, these chitons were called **pūpū-pe`elua**, or "caterpillar shell"). In ancient times chitons were used in a ceremony called **māwaewae** held a few days after the birth of a child. To about 1 1/2 in. Possibly endemic. Photo: Barber's Point, O`ahu.

FLAT CHITON

Ischnochiton petaloides (Gould, 1846)
Family Ischnochitonidae
• These colorful little chitons may be gray, blue, yellow, green, or whitish with orange margins. They occur under rocks in tide pools or reef flats. To almost 1/2 in. Possibly endemic. Photo: Scott Johnson.

SNAILS and SLUGS
CLASS GASTROPODA

Marine slugs are often colorful. Imperial Nudibranchs *(Risbecia imperialis).* Lāna`i Lookout, O`ahu. 20 ft.

With about 30,000 living species, gastropods—snails and slugs—form the largest molluscan class (and for shell collectors the most interesting). Gastropods are the only molluscs to have successfully colonized land, where almost half the known species now live. A few inhabit fresh water; the rest are marine.

Unlike bivalves, gastropods have a distinct head complete with mouth, rasping tongue (radula) and sensory organs. They also have a broad foot used for creeping about or, in some cases, for digging and swimming. Gastropods are sometimes called "univalves" because most have a single spiral or conical shell (as opposed to bivalves, which have two opposing shells). Many adult gastropods, however, have no shell at all.

It might seem logical to split the gastropods into snails and slugs on the basis of shell or no shell. In practice, however, this is not easy. Slugs, for example, often have vestigial shells hidden within their bodies. Instead, zoologists classify marine gastropods on the basis of gill location. Those with gills over their heads (typically hidden under the mantle) are called prosobranchs, whereas those with gills behind the head are called opisthobranchs. "Branch"—pronounced "brank"—means "gill" in Latin; "proso" means "forward" and "opistho" means "behind." Prosobranchs correspond generally to marine snails, opisthobranchs to marine slugs. Members of a third important group, the pulmonates, lack gills entirely. These are the terrestrial snails and slugs, whose mantle cavity has evolved into a kind of lung.

In her landmark volume **Hawaiian Marine Shells** Dr. E. Alison Kay lists 66 prosobranch and 38 opisthobranch families from Hawaiian waters, represented by about 572 known species (119 endemic). Most are snails (prosobranchs). The number of Hawaiian marine gastropods will probably exceed 1,000 as more small species are found. Mention must also be made of the famous Hawaiian land and tree snails, of which 831 species are known (759 endemic). Sadly, most of these colorful little animals are now extinct due to exotic (introduced) predators, overcollecting, habitat destruction and other factors.

SEA SNAILS. SUBCLASS PROSOBRANCHIA

Most sea snails have spiral shells, much like those of land snails. The marine shells, however, are more robust, more varied, and usually more beautiful. Widely collected for centuries, they have made their inhabitants probably the best-known group of marine invertebrates. The spiral shell design is ingenious: a single hard structure protects the soft animal, allows almost unlimited growth, and never needs replacement. Compare this to the hard shell of a growing crustacean such as a crab or shrimp, which must be shed periodically, exposing the animal to danger.

Although the shell has an open end, the retreating snail can seal the opening with a disklike plate called the operculum, giving it protection from all but specialized predators. Some sea snails, limpets for example, have conical rather than spiral shells. These also grow with the animal, however, and the animal can seal off the opening by clamping down on the substrate. Gastropod shells are secreted by the underlying mantle and consist of calcium carbonate (limestone) extracted from sea-water.

Beautiful as it is, the spiral shell has a major drawback: a single opening. Most higher animals breathe and eat at one end (the front or head) and excrete wastes from the rear. Snails, however, must breathe, eat and excrete from the same end. In order to do this their bodies have become twisted 180 degrees, placing both gills and anus near the front. This twisting, called "torsion," is characteristic of all marine snails. Various structural compensations for torsion have arisen, such as the siphon, an organ that enables snails to take in water from a source some distance from the anus. The proboscis, a muscular extension of the mouth, allows many snails to probe or engulf their prey at some distance without emerging from the shell.

The 12,000 known species of living marine snails are divided into three orders: Archeogastropoda, Mesogastropoda and Neogastropoda ("old," "middle" and "new" snails). Their differences, however, are largely internal and meaningful only to a specialist.

The variety of marine snail shells is endless. Fortunately, these animals have been well covered in other publications. Here it will suffice to discuss the major families, presenting one or two examples of species likely to be encountered by snorkelers or divers. The popular cones and cowries, however, are treated here in greater detail.

In old Hawai`i marine snails of all kinds were important as food; their empty shells were used for tools and ornaments. The general name for snails was **pūpū**. (The same word also means "relish" or "appetizer.") Some groups of shells had specific names. Cowries, for example, were **leho**, and augers, `oi`oi.

LIMPETS. FAMILY PATELLIDAE

Limpets, or `opihi in Hawaiian, are snails with conical, caplike shells. They typically cling to surf-swept rocky shores, creeping about slowly to graze on algae; many species have a permanent depression, or "home scar," to which they regularly return. **'Opihi** are highly valued as food by Hawaiians. Harvesters must act quickly both to avoid waves and to dislodge the animals while they are feeding and relaxed. Once disturbed, limpets clamp down firmly. Margaret Titcomb writes: "It was **kapu** for anyone to eat `opihi on shore while a companion was out gathering more. If one broke this **kapu**, the one still collecting would be pounded by the sea. An `opihi gatherer was warned never to turn his back on the sea.... Gathering of `opihi was so dangerous that it was called the fish (creature) of death (**he i`a make**)." Even today `opihi pickers occasionally drown after being swept from the rocks by an unexpected wave. The taking of **'opihi** is regulated by law. See Appendix B, (p. 346).

Hawai'i has four endemic limpets in the family Patellidae (other families of limpets exist as well). The three described below are common on basalt shores of the main islands but absent from the limestone shores of the northwestern chain (where the fourth endemic, the Green-Foot 'Opihi, *Cellana melanostoma*, replaces them). The False 'Opihi, a limpet more closely related to land snails, is included in this section for convenience.

BLACK-FOOT 'OPIHI • 'opihi makaiaūli

Cellana exarata (Reeve, 1854)
Family Patellidae

• The foot of this limpet is dark gray or black. The low ribs of its shell are dark and the troughs between light. The ribs do not extend much beyond the margin of the shell, leaving the edge relatively smooth. Of all 'opihi the Black-Foot lives highest on the rocks and is most easily gathered. Sometimes on still, sunny days it lifts its shell off the rock, probably to keep cool. It is often found in a small depression called a "home scar" to which it always returns after grazing on algae. To about 2 in. across. Endemic. Photo: Wai'ānapanapa State Park, Maui. This species is protected by law. See Appendix B (p. 346).

YELLOW-FOOT 'OPIHI • 'opihi 'ālinalina

Cellana sandwicensis (Pease, 1861)
Family Patellidae

• The Yellow-Foot 'Opihi lives at the low tide mark and just below; it needs constant splash and cannot tolerate drying out as much as the Black-Foot. Often wearing a cap of seaweed, it may be difficult to see. As the name suggests, its foot is yellow. The high ribs extend well beyond the edge of the shell, making it scalloped and sharp. In old Hawai'i the shells of this limpet were useful as scrapers, especially good for shredding coconut meat. To about 2 in. Endemic. Photo: Makapu'u, O'ahu. This species is protected by law. See Appendix B (p. 346).

GIANT `OPIHI • `**opihi kō`ele**

Cellana talcosa (Gould, 1846)
Family Patellidae
• Sometimes called the Kneecap Limpet, this is by far the largest of the Hawaiian `**opihi**. Its shell is smooth and brown and the fine ribs do not extend beyond its edge. The foot is yellow. Always submerged and often covered with seaweed and barnacles, it lives from just below the low tide mark to a depth of about 4 or 5 ft. (although occasionally as deep as 30 ft.), typically between boulders along exposed basalt shores. Rare on Kaua`i and  O`ahu, it is most common from Molokai south. Pearls have been found in it. The large shells, bleached white by the sun, often are found around archeological sites. They were used as scrapers, typically to remove the skin from cooked taro and breadfruit. To about 4 in. Endemic. Photo: Papawai Point, Maui. This species is protected by law. See Appendix B (p. 346).

FALSE `OPIHI • `**opihi-`awa**

Siphonaria normalis Gould, 1846
Family Siphonariidae
• Often abundant in the mid-intertidal zone, this limpet has both gills and lungs and is more closely related to land snails than to true `**opihi**. It can be identified by its prominent, widely separated ribs that have smaller ribs between them. The troughs between the ribs are black. Like all limpets, it subsists on algae scraped from the wave-splashed rocks. The Hawaiian word `**awa** means bitter; the animal was used for medicine or sorcery but not eaten. It grows to about 3/4 in. and has a widespread Indo-Pacific distribution. Photo: Lāi`e Point, O`ahu.

TOPS AND TURBANS. FAMILIES TROCHIDAE AND TURBINIDAE

The tops and turbans are closely related families of herbivorous snails. Top shells are typically conical with a flat base, similar in shape to the spinning toy for which they are named. Their surface is smooth. Turbans are globular and ribbed, named after the spirally wound headdress. Tops have a thin, horny operculum whereas turbans have a thick, calcareous one. The shells of each are composed largely of mother-of-pearl, and species of each family have been extensively harvested in the South Pacific for the manufacture of buttons. (The large top *Tectus niloticus* was once introduced into Hawaiian waters for this purpose, though without success.) Hawai`i is home to nine species of tops and five of turbans.

WOVEN TOP • **hā`upu**

Trochus intextus Kiener, 1850
Family Trochidae
• This is Hawai`i's largest top. It is grayish, greenish or reddish, usually marked with darker vertical marks and dots of red or purple. The interior is pearly. It lives on shallow reef flats in areas of mixed sand and rock; empty shells are often occupied by hermit crabs. For many years the animal was known locally as *T. sandwichensis*; it is still sometimes called the Hawaiian Top. To about 1 1/2 in. Pacific Ocean. Photo: Lāna`i Lookout, O`ahu. Tide pool. (occupied by hermit crab)

HAWAIIAN TURBAN • **'alīlea, pūpū mahina**

Turbo sandwicensis Pease, 1861
Family Turbinidae
• After heavy surf, beaches are sometimes strewn with the round white opercula of this common snail. Beautiful iridescent colors can often be brought out by polishing these "cat's eyes." The Hawaiian name, **pūpū mahina**, means "moon shell," also in reference to the moon-shaped operculum. The snails themselves are common to a depth of about 60 ft. Their globular shells, green or brown with white markings, are heavily ribbed. Earlier books confuse this species with *T. intercostalis*, a widespread Indo-Pacific turban. To about 3 in. Endemic. Photo: Kahe Point, O`ahu. 25 ft.

NERITES AND PERIWINKLES. FAMILIES NERITIDAE AND LITTORINIDAE

Although sharing similar habitat and life style, these two families of snails belong to different orders and are only distantly related. Both are herbivores, inhabiting rocky intertidal and splash zones throughout the tropics and subtropics; both are abundant in Hawai`i. Nerites generally occur from the mid-intertidal to the splash zone at or above the high water mark. Some live in fresh water as adults. (Best known of these in Hawai`i is the limpet-like **hihiwai**, *Neritina granosa*, which was once abundant in streams.) Periwinkles (littorines) typically live high on the rocks above the nerites. Although they retain their gills they are able to breathe air directly. At least nine species of nerites (three from fresh water) and six littorines occur in Hawai`i.

BLACK NERITE • **pipipi**

Nerita picea (Recluz, 1841)
Family Neritidae
• These snails are abundant on rocky shores in the splash zone above the waterline (but below the zone occupied by periwinkles). By day, dozens can often be found clustering under ledges and in crevices well out of the water. At night they graze more actively. They are black with fine spiral lines. The species name means "pitch black." To about 1/2 in. Possibly endemic. Photo: Moku`auia (Island), O`ahu. (*Theodoxus neglectus*, a similar nerite, is pictured below.)

POLISHED NERITE • **kūpe`e**

Nerita polita Linnaeus, 1758
Family Neritidae
• This unusual nerite lives under sand at the high tide line. At night it climbs the rocks to feed on algae. Its thick, smooth gray shell may be variously spotted with white, yellow, pink, orange, red or black. Dr. E. Alison Kay writes: "Known as **kūpe`e** by the Hawaiians, shells of *N. polita* were prized as items of adornment and the animals were used as food. Drilled and made into bracelets and necklaces, the **kūpe`e** was an emblem of mourning for the **ali`i**, or chiefs." To about 1 1/2 in. Indo-Pacific. Photo: Hekili Point, Maui. (The middle snail is *Theodoxus neglectus*.)

DOTTED PERIWINKLE • **pipipi kōlea**

Littoraria pintado (Wood, 1828)
Family Littorinidae
• The largest of three common species in Hawai`i, Dotted Periwinkles are purple-gray, dotted with brown or black. They are abundant in the splash zone on most rocky shores just above the nerites, where waves only occasionally wet them. At low tide or when seas are calm they seal themselves tightly to prevent desiccation, remaining stuck to the rocks with a film of dried mucus. C.H. Edmondson of the Bishop Museum once wrote: "In my laboratory a periwinkle once climbed high up on the wall and stuck there. After almost a year, when taken down and wet with sea water, it promptly became active again." The species name means "painted" or "mottled" in Spanish. East Africa to Clipperton Atoll (Eastern Pacific). To almost 1 in. Formerly placed in the genus *Littorina*. Photo: Lāi`e Point, O`ahu.

WORM SNAILS. FAMILY VERMETIDAE

Worm snails (vermetids) spend their adult lives permanently cemented to rocks or other hard surfaces. Their tubular shells are usually loosely or irregularly coiled. The sharp apertures of some intertidal species can slice open a carelessly placed foot; in old Hawai`i such cuts were considered poisonous. A few species are colonial, creating entire reefs of hard intertwining tubes.

Vermetids feed either by catching suspended food particles in strands of mucus or by creating a current through their mantle cavity and filtering out the suspended food particles in a mucous net. In either case, the mucus and food are periodically ingested and the cycle repeated. There are at least eight species of worm snails in Hawai`i. They had a variety of names in ancient times, including **kauna`oa, ki`o, po`apo`ai** (tightly coiled forms), **pōhaku-pele** and **pohokūpele**.

Variable Worm Snail (a)

Variable Worm Snail (b)

VARIABLE WORM SNAIL • **kauna`oa**

Serpulorbis variabilis Hadfield & Kay, 1972
Family Vermetidae
• Because of its habitat and size this is the worm snail most likely to be noticed by snorkelers, divers and beachgoers in Hawai`i. It occurs in exposed environments such as tide pools, shallow, wave-swept reef flats, and rocky reefs down to about 40 ft. The white to brown shell, entirely coiled or partly straight, is often overgrown with coralline algae or coated with sand grains. This is the only worm snail in Hawai`i that has no operculum. Its coils may attain 1 1/2 in. in diameter. Hawai`i, Marshall Islands, Line Islands. Photos: a) Kona, Hawai`i. (shells on boulder cast up by storm) b) Kewalo, O`ahu. 3 ft. (living animal with strands of mucus)

HORN SNAILS. FAMILY CERITHIIDAE

Horn snails are a large family of tropical herbivores whose long, pointed shells are usually roughened by tubercles and ridges. The shells of a few are sufficiently long and slender to be confused with augers. The canal or notch in the lip of the shell through which the siphon protrudes (siphonal canal) is distinctly turned-back in some species. Many are tiny (about 1/8 in. long) and can be superabundant in tide pools, in anchialine (brackish) pools and on seaweed. At least 16 species inhabit Hawaiian waters, some living in the open, others buried in sand.

PRICKLY HORN • **maka`aha**

Cerithium echinatum Houbrick, 1992
Family Cerithiidae
• These snails are abundant from about 20 to 50 ft. on many Hawaiian reefs (especially off the Wai`anae coast of O`ahu) and occur down to at least 100 ft. Active by day, they crawl in the open on rubble feeding on algae. The empty shells are favored by hermit crabs. To about 2 in. Indo-Pacific. *Cerithium mutatum* is a synonym. Photo: Mākua, O`ahu. 90 ft.

ZEBRA HORN

Cerithium zebrum Kiener, 1841
Family Cerithiidae
• These tiny snails live by the thousands in anchialine ponds along with the red shrimps *Halocaridina rubra* (p. 220). (Anchialine ponds are brackish ponds with indirect underground connections to the sea.) They are also one of the most common snails in tide pools and on shallow reefs (their primary habitat), where they live under stones and in rubble. Varying from white peppered with brown to dark brown, they blend in well and are hard to spot. To about 1/4 in. Indo-Pacific. *Bittium zebrum* is a synonym. Photo: `Āhihi-Kīna`u, Maui. 1 ft.

WENTLETRAPS AND VIOLET SNAILS. FAMILIES EPITONIIDAE AND JANTHINIDAE

Snails of these related families are associated with cnidarians, either as predators or para-sites. Wentletraps always occur with anthozoans, such as corals or anemones, whereas vio-let snails prey on pelagic hydrozoans, such as the Portuguese Man-of-War. Wentletrap shells, also known as "ladder shells," are characterized by prominent riblike varices, the remains of thickened lips formed during earlier stages of growth. Although generally less than 2 in. long, wentletraps rank among the world's most beautifully sculptured shells. Thirteen species occur in Hawai'i, most of them small. One is known to prey on the solitary coral *Fungia scu-taria*, another on the large anemone *Heteractis malu*, and four on the common anemone *Aiptasia pulchella*. Violet snails have extremely thin shells well suited to their pelagic life. Two species wash up regularly on Hawaiian shores. Both wentletraps and violet snails can secrete a purple or pink dye that may be toxic.

HOOKED WENTLETRAP • **pūpū alapaʻi**

Epitonium replicatum (Sowerby, 1844)
Family Epitoniidae
• Although these little snails are seldom seen alive, their shells are not uncommon on some beaches. The inflated whorls bear numerous platelike ribs that are angled and slightly hooked. Little is known of the animal's natural history or feed-ing habits. To about 3/4 in. Indo-Pacific. Photo: Scott Johnson. Pūpūkea, Oʻahu. (*Epitonium perplexum*, another wentletrap sometimes common in tidepools and on shallow reefs, has distinct-ly lower ribs.)

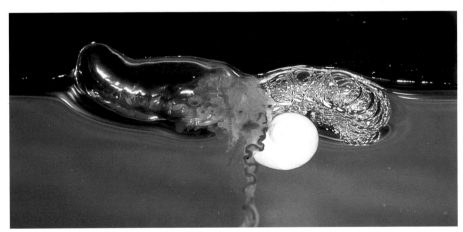

VIOLET SNAIL • **pūpū poni**

Janthina janthina (Linnaeus, 1758)
Family Janthinidae
• Like others of its small family, this pelagic gastropod floats on a raft of bubbles, aperture up. Its shell, violet on the upward-facing side and white to lavender underneath, is among the thinnest of all marine snail shells. It preys on the simi-larly colored pelagic hydrozoans, the Portuguese Man-of-War (*Physalia physalis*) and the By-the-Wind Sailor (*Velella velel-la*). Tiny stalked barnacles of the genus *Lepas* may attach to its shell. Beachcombers on windward shores can find these snails and their prey washed up in numbers after a period of strong tradewinds. Hawaiian, species and genus names all mean "violet." To about 1 1/2 in. All warm seas. Photo: David A. Fleetham.

ECHINODERM SNAILS. FAMILY EULIMIDAE

Most snails in this family are associated with echinoderms, either roaming freely on the surface or permanently attached as a parasite. The typical eulimid is tiny and white, has a high pointed spire, and lives on or near a sea cucumber. Other eulimid snails live on urchins or burrow into the arms of sea stars (in Hawai`i, *Linckia multifora*) but these are less common. At least 21 species occur in Hawai`i, some free-living (i.e., without an echinoderm host).

SEA CUCUMBER SNAIL

Balcis aciculata (Pease, 1861)
Family Eulimidae
• At least six similar snails in the genus *Balcis* are known to live in association with sea cucumbers in Hawai`i. Different species may occur on a single sea cucumber, and up to 15 individuals of a single species *(B. aciculata)* have been found on one host. Most favored as hosts are *Holothuria atra, H. cinerascens, Actinopyga mauritiana, A. obesa* and *Stichopus horrens.* Along some Hawaiian shores up to 60 percent of *H. atra* host eulimids. Photo: Scott Johnson. Diamond Head, O`ahu. (identification not confirmed)

LINCKIA SNAIL

Stilifer linckiae Sarasin & Sarasin, 1887
Family Eulimidae
• These snails dig into the arms of the sea star *Linckia multifora*, forming a large swelling or gall. One to five snails may inhabit a single gall, usually one much larger than the others. These parasites are believed somehow to inhibit autotomy (the shedding of an arm by the star for reproductive purposes). Autotomized arms often fail to survive, which would be fatal to snails living on them. *Thyca crystallina*, another eulimid parasite of *L. multifora*, occurs externally. Stars hosting either of these parasites are uncommon. To about 1/4 in. Indo-Pacific. Photo: Scott Johnson. Nānākuli, O`ahu.

HOOF SHELLS. FAMILY HIPPONICIDAE

Hoof shells attach limpet-like to rocks or other snails. Unlike limpets of the family Patellidae, the apex of the shell is directed backwards showing its spiral origins. Some of these snails secrete a shelly plate between themselves and the substrate and were classified originally as bivalves. These animals remain fixed in one spot and do not move about. Four species are known from Hawai`i.

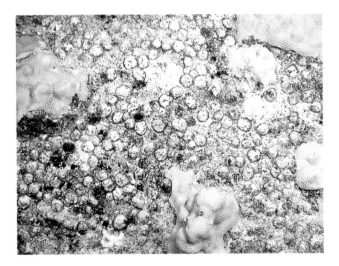

SHINGLY HOOF SHELL

Hipponix imbricatus Gould, 1846
Family Hipponicidae
• These are superabundant on smooth rocky surfaces down to at least 50 ft. along much of the Kona coast of Hawai`i, and in other areas also. Ribs and spiral sculpturing on the shell create a coarse surface upon which algae readily grows. Unable to move about, some members of the family Hipponicidae feed by scraping algae and detritus from the surrounding substrate with their proboscis and radula. The feeding habits of this species are not known. To about 1/2 in. Possibly endemic. Photo: Puakō, Hawai`i. 15 ft.

CONICAL HOOF SHELL

Hipponix australis Lamarck, 1819
Family Hipponicidae
• Despised by collectors, this limpet-like snail attaches to the shell of a larger snail, etching a depression for itself often near the aperture. The permanent scar thus created ruins collectable shells (and has undoubtedly saved the lives of count-less molluscs). Several of these hoof shells may occur on a single host, possibly feeding on its feces. They are especially numerous on shells occupied by hermit crabs. Sometimes a small individual (probably a male) attaches to a larger one (probably a female). To about 1/4 in. Indo-Pacific. *Sabia conica* is a synonym. Photo: Lāna`i Lookout, O`ahu. 5 ft.

STROMBS. FAMILY STROMBIDAE

Stromb shells, often called conchs (pronounced "konks"), are characterized by a long aperture and flaring lip that in some Indo-Pacific species is drawn out into multiple fingers. Such shells (not found in Hawai`i) are called "finger conchs" or "spider conchs." One of the best known shells in America, the large pink Queen Conch of the tropical Atlantic *(Strombus gigas)*, belongs to this family. Living strombs are remarkable in several ways: they can move quickly for a snail by driving their long, bladelike operculum into the sand and vaulting forward in a series of jerky hops or leaps; and they have two large, almost human-looking eyes set on stalks. When the snail is active, one stalked eye protrudes through a special U-shape fold in the lip, the other through the siphonal canal at the forward end of the shell. Seven species of strombs inhabit Hawaiian waters.

Hawaiian Stromb (a)

HAWAIIAN STROMB

Strombus vomer hawaiiensis
Pilsbry, 1917
Family Strombidae
• This animal is rare, live or dead, and usually occurs at 100 ft. or more buried in sand. The shell has a knobby, pointed spire and a greatly flaring outer lip that projects backward into a single finger almost as long as the spire itself. It is cream to light brown, speckled with darker brown. This is an endemic subspecies. Two other subspecies of *S. vomer* occur in the Ryukyu Islands/New Caledonia and in Western Australia, respectively. To almost 4 in. Photos: a) Scott Johnson, Mākaha, O`ahu. (live animal) b) Mākua, O`ahu (shell inhabited by hermit crab)

Hawaiian Stromb (b)

SPOTTED STROMB • **pūpū mamaiki**

Strombus maculatus Sowerby, 1842
Family Strombidae
• Hawai`i's only shallow-water stromb, this snail occurs from the intertidal zone to a depth of about 60 ft., buried in sand by day. The shell, cream with gray-brown mottling, is more often found occupied by a hermit crab than by the snail. To escape predators such as cones, the living animal can jump suddenly backwards, actually clearing the substrate. To about 1 1/2 in. Indo-Pacific. Photo: Scott Johnson.

COWRIES. FAMILY CYPRAEIDAE

Cowries are objects of beauty in any culture. Their lustrous, globular shells have fascinated humankind since the beginning of history, serving as charms, money and adornment. During the 18th century when shell collecting was fashionable in Europe, cowries became even more widely known. The Italians, likening them to little round pigs, named them "porcellana," the origin of our word "porcelain." Rulers of India, the British used the Hindu name *kauri*. Others knew them as Venus shells, the scientific name being taken from Cypraea, the Greek "Venus" (goddess of love).

Today cowries are still the most widely collected sea shells. They are unusual in two ways. First, the outer, final whorl of adult shells wraps around and envelops the inner whorls, thereby hiding the spiral structure so obvious in other snail shells. Second, their mantles expand to cover the shell's entire outer surface, keeping it smooth, glossy and beautiful. The extended mantles of some species—often covered with soft, sometimes branching projections called papillae—are spectacular.

Most cowries are nocturnal, hiding in cracks and crevices during the day. Although temperate and deep-water species exist, the majority occur in shallow tropical waters in rocky areas or coral reefs. Many are herbivores; others feed on sponges or other attached (sessile) animals. In captivity some will even eat dead fish or shrimp. Hawaiian Tiger Cowries have lived at least three years in the aquarium, eating bright green "sea lettuce" (*Ulva*).

Leviathan Cowries *(Cypraea leviathan)*. Lāna`i Lookout, O`ahu. 15 ft.

On the reef, cowries typically remain in one small area, often in breeding pairs. Fertilized internally by males, females lay a mass of eggs on the substrate, sitting on them for up to four weeks until they hatch. Except in a few species that develop directly, eggs hatch into planktonic veligers that eventually settle to the bottom to mature. Juvenile cowries, called bullae, lack the enveloping final whorl and look much like other snails with long, narrow apertures. Gray or brown, sometimes with an encircling dark band, they develop adult coloration only when the rounded adult form with flat base and slitlike toothed aperture is attained. Thereafter, cowries grow very little.

There are about 200 species of cowries worldwide, with at least 35 known from the Hawaiian Islands. Among these are nine endemic species, highly prized by shell collectors worldwide. The list has changed over time but, as currently recognized, the Hawaiian endemic cowries are *Cypraea burgessi*, *C. gaskoini*, *C. granulata*, *C. mauiensis*, *C. ostergaardi*, *C. rashleighana*, *C. semiplota*, *C. sulcidentata* and *C. tesselata*. All are described and pictured on the following pages except *C. burgessi* (most common in the Northwestern Hawaiian Islands) and the rare, deep-dwelling *C. ostergaardi*. The Money Cowry (*C. moneta*), ubiquitous in the Indo-Pacific, is rare in Hawai`i. In olden times white Money Cowry shells were sometimes held over smoke to give them an ivory hue. Cowries so treated were called **leho `uala**, or "sweet potato cowries." Queen Kapi`olani wore a **lei** of such shells in London when she attended Queen Victoria's Jubilee in 1887.

In the Hawaiian language cowries are called **leho**. Traditionally they were eaten and their shells used for scrapers or octopus lures. An older man eying young women would provoke the comment "The octopus notices the little cowries." To escape detection, people said **"Moe a leho"** ("Lie still as a cowry.")

ALISON'S COWRY

Cypraea alisonae Burgess, 1983
Family Cypraeidae
• The back of this slender cowry's shell is whitish covered with irregular brown flecks and blotches. There is a large, sometimes squarish brown blotch near the center. The sides are white with a few brown spots mostly near the base. The slightly translucent mantle is red to orange-brown with widely spaced papillae. The foot has pale spots. When attacked the animal sheds the rear section of the foot to facilitate escape (like a lizard shedding its tail). This is a small, shallow-water cowry found both on reef flats and rocky walls. It eats algae and is reported to do well in aquariums. The name honors Dr. E. Alison Kay, professor of zoology at the University of Hawai`i, longtime student of cowries, and author of the definitive study **Hawaiian Marine Shells**. To about 1 1/2 in. Indo-Pacific. (This cowry was long confused with the Tapering Cowry, *C. teres*; their shells are almost indistinguishable but the living animals of each show subtle and consistent differences. The Indo-Pacific *C. teres* is rare in Hawai`i.)

SNAKEHEAD COWRY • **leho kupa, leho kūpe`e lima**

Cypraea caputserpentis
Linnaeus, 1758
Family Cypraeidae
• One of the most common shallow-water cowries in Hawai`i and the Indo-Pacific, this animal hides in crevices and under stones near the shoreline, emerging at night to graze on algae. The beautiful shell is solid dark brown on the sides, reticulated brown and white on top. There is often a white mantle line where the two edges of the mantle meet. The sides are broad and sloping. The live animal is olive-brown tinged with green and yellow; its mantle is covered with very short papillae that sometimes have reddish tips. The shells of Hawaiian specimens differ slightly from the Indo-Pacific norm. In old Hawai`i they were used for making bracelets (**kūpe`e lima**). To about 1 1/2 in. Photo: Kahe Point, O`ahu. 2 ft.

CHINESE COWRY

Cypraea chinensis Gmelin, 1791
Family Cypraeidae
• This beautiful cowry is rare in Hawai`i. The upper part of its shell, orange-brown and white, resembles that of the common Reticulated Cowry (p. 116) but the sides are cream with violet spots. The grooves between the teeth on the underside are orange. The mantle is red with light and dark marks. In Hawai`i this species generally occurs at depths greater than 40 ft. Elsewhere in its range it inhabits shallow water. To almost 2 in. Photo: Scott Johnson. Pūpūkea, O`ahu. 40 ft.

FRINGED COWRY

Cypraea fimbriata Gmelin, 1791
Family Cypraeidae
• Fully grown, this tiny slender cowry is no more than about 1/2 in. long. The shell is blue-gray speckled with brown spots; a brown double band crosses the center of the back; the tips of the shell are purple. The live animal is completely red except for the pale siphon, the sparse white papillae, and a few white spots. Although one of the commonest cowries in Hawai`i, it is seldom noticed because it is small and nocturnal. It occurs from shallow reef flats to depths of over 200 ft. The species name means "fringed." To about 1/2 in. Indo-Pacific. Photo: Scott Johnson, Pūpūkea, O'ahu. 20 ft.

Gaskoin's Cowry (a)

Gaskoin's Cowry (b)

GASKOIN'S COWRY

Cypraea gaskoini Reeve, 1846
Family Cypraeidae

• The lovely shells of these small cowries are orange or orange-brown with numerous circular pale spots of varying sizes. A band of white at the base contains dark spots. The mantle and foot of the live animal are bright red. A flap of skin at the foot's posterior detaches readily to facilitate escape from predators. This cowry occurs from the shallows to depths of about 100 ft. on cliffs and under ledges, hiding by day and emerging at night to feed almost exclusively on a species of red sponge. The name honors physician and naturalist John Samuel Gaskoin (1790-1858), surgeon to Kings George IV and William IV of England. (The Waxy Cowry, *C. cernica*, is similar but far less common in Hawai`i.) To about 1 in. Considered endemic, but may occur elsewhere in the Pacific. Photos: a) Mākua, O`ahu. 20 ft. (mantle extended); b) Pūpūkea, O`ahu. 60 ft. (inhabited by hermit crab)

Granulated Cowry (a)

GRANULATED COWRY • **leho `ōkala, leho ōpu`upu`u**   Granulated Cowry (b)

Cypraea granulata Pease, 1863
Family Cypraeidae
• This unusual Hawaiian endemic has a pinkish brown shell covered with nodules. The white underside is beautifully grooved. The exposed animal, too, is extraordinary: so numerous and long are its papillae that it resembles a strange sea urchin or a tuft of finely branched red algae. The species occurs from shallow water down to at least 60 ft., usually near overhangs and caves; it is not common. The Hawaiian names mean "rough" or "bumpy." Curiously, subadult shells have a shiny finish like other cowries. To about 1 1/2 in. Endemic. Photos: a) Pūpūkea, O`ahu (fresh dead); b) Mākua, O`ahu. 20 ft. (mantle extended)

Honey Cowry (a)

Honey Cowry (b)

HONEY COWRY

• **leho `ōpule**
Cypraea helvola Linnaeus, 1758
Family Cypraeidae
• This small cowry has a rich brown to purple-red shell marked with an unusual combination of small round white spots and larger dark brown ones. The sides and base are orange-brown. The ends of the shell are tipped with purple. The lovely mantle is thickly covered with short pink papillae and longer branching white ones, somehow resembling fancy lacework. Ranging from the shallows to depths of hundreds of feet, this cowry is most common in Hawai`i at 60 ft. and below. The species name means "pale yellow," possibly because the color of collected shells fades to a honey-like hue. (The rare deep-water Hawaiian endemic *C. ostergaardi* resembles such faded specimens but has a white base.) The Hawaiian name means "variegated." To about 1 1/4 in. Indo-Pacific. Photos: a) Nāpili Bay, Maui. 15 ft. (mantle extended); b) Magic Island, O`ahu. 20 ft.

ISABELLA'S COWRY • **pūleho, leho kūpe`e lima**

Cypraea isabella Linnaeus, 1758
Family Cypraeidae

• Easy to identify, this cowry's elongate shell is light tan or gray-brown and marked lengthwise with fine irregular lines composed of dark dots and streaks. The shell tips of Hawaiian specimens are brown; in other parts of this species' range they are orange. The mantle of the exposed animal is velvety black. The species occurs from the shallows to depths of several hundred feet, usually under stones. In olden times it was used to make bracelets and leis. To about 1 1/4 in. Indo-Pacific. Photo: Lāna`i Lookout, O'ahu. 15 ft.

Leviathan Cowry (a)   Leviathan Cowry (b)

LEVIATHAN COWRY • **leho pāuhu**

Cypraea leviathan Schilder & Schilder, 1938
Family Cypraeidae

• This is the largest of three similar cowry species occurring in Hawai`i. All have creamy brown, cylindrical shells marked with darker red-brown bands; the undersides are white. *Cypraea leviathan* is best identified by its size and the large tufted "shaving brush" papillae (unique to the species) sprouting from the mottled gray and white mantle of the living animal. Shorter spikelike or branched papillae occur as well. In Hawai`i it inhabits caves and crevices, usually 30 ft. deep or less. Similar but slightly smaller cowries are likely to be either *C. propinqua* or *C. carneola*. Differences between the three lie more in size and form of the papillae than in shell coloration. In old Hawai`i shells of the large *C. leviathan* were sometimes used as octopus lures. An octopus captured with such a lure was said to have exceptional medicinal value. The Hawaiian name means "narrow-chested." To about 3 in. Once considered endemic to Hawai`i, it is now recognized as an Indo-Pacific species. Photos: a) Pūpūkea, O`ahu, 30 ft. (mantle extended); b) "Amber's Arches," south shore, Kaua`i. 60 ft. (See also p. 109.)

RETICULATED COWRY

• leho kōlea
Cypraea maculifera Schilder, 1932
Family Cypraeidae
• This cowry's dark brown shell is usually overlaid by so many large, irregular white spots that a netlike effect results, hence the common name, meaning "like a net." The sides at the base are white with blurry dark spots. The mantle of the live animal is dark gray covered with fine, pointed, white papillae that give it a fuzzy appearance. The foot is dark gray. It is common in Hawai`i along rocky coasts where its relatively large size makes it an easy find in holes and under ledges from the shoreline to at least 50 ft. Hawaiian specimens tend to be larger than those found elsewhere. The species name means "dappled" or "spotted." The Hawaiian name refers to the **kōlea**, or Pacific Golden Plover, which has a brown back with golden spots. To about 2 1/2 in. Central Pacific, but has also been found in Guam. Photo: Mākua, O'ahu. 15 ft.

MAUI COWRY

Cypraea mauiensis Burgess, 1967
Family Cypraeidae
• This tiny cowry is found infrequently throughout the main Hawaiian Islands but breeding colonies seem to occur only on the leeward sides of Maui and the Big Island. Because of its shallow, restricted habitat it has been subject to over-collecting and is now quite rare. The shell is cream with sparsely scattered light brown spots. The live animal is yellow. To about 1/2 in. Endemic. Photo: Scott Johnson. Olowalu, Maui.

HUMPBACK COWRY • **leho ahi, leho pa`a, leho pouli**

Cypraea mauritiana Linnaeus, 1758
Family Cypraeidae
• This large handsome cowry is common along Hawai`i's black basalt shores as high as the mid-intertidal zone, especially in areas of heavy wave action. It seldom occurs on limestone but does occur on tuff deposits along O`ahu's southeast shore. The entire shell, including the underside, is brownish black. Irregular light marks dapple the back, which humps almost to a point. This shape and the thick, wide base help the animal withstand the force of pounding surf. The live animal's mantle is dark brown or black, matching the dark rock on which it lives. In old Hawai`i its flesh was eaten and the shell fashioned into scrapers for grating coconut, removing the skin from cooked taro or breadfruit, and preparing **kapa** (bark cloth). Perfect shells with fine spots and a dark reddish hue were called **ipo** (meaning "sweetheart" or "lover") and prized as lures for octopus. The species name comes from *mauros*, Greek for "dark." The Hawaiian names mean "fire" (**ahi**), "solid color" (**pa`a**) and "dark" (**pouli**). To almost 5 1/2 in. Indo-Pacific. Photo: Kahe Point, O`ahu. 15 ft.

POROUS COWRY

Cypraea poraria Linnaeus, 1758
Family Cypraeidae
• This small brown cowry is speckled with white spots, some of which have dark rings. It is easy to confuse with the more common Honey Cowry (p. 114) but has some violet on the base instead of orange-brown. The brown mantle bears numerous stout papillae, some branched. It is a shallow-water cowry occurring most commonly from 12-15 ft. To about 1 in. Indo-Pacific. Photo: Scott Johnson. Mākua, O`ahu.

SCHILDER'S COWRY

Cypraea schilderorum (Iredale, 1939)
Family Cypraeidae
• This species is similar in color to the more common Groove-Tooth Cowry (below). The back of the shell is marked with brown bands on a light background; the sides are thick, solid and without bands. The base is pure white and the teeth are separated by fine grooves. The mantle of the live animal is mottled with black, brown and white, with lighter papillae. Although generally uncommon, in some years this cowry is reported to be almost abundant in Hawai`i. To 1 3/4 in. Central and far Western Pacific. Photo: Pūpūkea, O`ahu. 15 ft.

GROOVE-TOOTH COWRY

Cypraea sulcidentata Gray, 1824
Family Cypraeidae
• This is probably the most abundant of the Hawaiian endemic cowries. The rounded shell is marked with four indistinct brown bands on a lighter background. At the base the sides are creamy brown to deep purple and finely marbled. The underside is light tan. The mantle of the live animal is tan marked with many fine, dark brown longitudinal lines. The many-branched papillae are broad, flat and white. The foot is light tan to white, the tentacles dark gray. This cowry is found under stones and ledges from depths of a few inches to at least 90 ft. The species name, meaning "grooved tooth," derives from the unusually deep grooves between the teeth that extend partly across the base. Schilder's Cowry (above) is similar; it has a pure white underside and fine grooves between the teeth. To about 1 1/2 in. Endemic. Photo: Pūpūkea, O`ahu. 30 ft.

"HALF-SWIMMER" COWRY • **pūleholeho**

Cypraea semiplota Mighels, 1845
Family Cypraeidae

• This small, once-common Hawaiian endemic all but disappeared around the main islands in the 1940s and 1950s and was considered extinct by some. No one could explain the decline. The population has subsequently bloomed and fallen several times but never regained its former numbers. A more stable population may occur in the cooler Northwestern Hawaiian Islands. The back of the shell varies from dark brown to gray-white, well sprinkled with tiny white dots. The underside is white, the grooves between the teeth orange. The mantle, black with numerous papillae, is usually extended by day. The animal is associated with a sponge the same color as the mantle, but it also feeds on algae and can be kept successfully in a home aquarium. It occurs from shallow water to depths of at least 130 ft. The significance of the species name is obscure: *semi* means "half" and *plotos* means "floating" or "swimming." Broad, flattened whitish specimens are sometimes called *C. annae* Roberts, 1869. The Hawaiian name means "dusk" or "twilight." To almost 1 1/2 in. Endemic. Photo: Scott Johnson. Fort Kamehameha Reef, Oʻahu.

MOLE COWRY

Cypraea talpa Linnaeus, 1758
Family Cypraeidae

• This cowry's handsome shell is jet black around the base and on the underside; its back is brown, banded with gold. The mantle is black, densely covered with minute green spots and studded with broad, wartlike papillae. It is found both in sheltered waters and along exposed coasts at depths of at least 20 ft., but it is nowhere common. This cowry attains its largest size in Hawaiʻi. The common name comes from the species name, meaning "mole." To about 3 1/2 in. Indo-Pacific. Photos: a) Kahe Point, Oʻahu. 40 ft. b) Pūpūkea, Oʻahu. 30 ft. (mantle extended)

Mole Cowry (a)

Mole Cowry (b)

RASHLEIGH'S COWRY

Cypraea rashleighana Melvill, 1888
Family Cypraeidae
• Rare in the main Hawaiian Islands, this endemic cowry resembles *C. alisonae* (p. 110) but is smaller, rounder and has more brown spots on the sides. Like *C. semiplota*, it seems to occur in cycles, present some years and absent in others. A more stable population may exist in the cooler northwestern chain. It usually occurs at depths of 50 ft. or more. Melvill named the species for his shell enthusiast friend, Jonathan Rashleigh (1845-1872). To about 1 in. Photo: Scott Johnson. Hale`iwa, O`ahu. 50 ft.

CHECKERED COWRY

Cypraea tessellata
Swainson, 1822
Family Cypraeidae
• This is perhaps the most famous Hawaiian endemic cowry. The best specimens have three large, dark, squarish spots set corner to corner, checkerboard fashion; more often there are only two with fainter squares at the corners. Either way, the pattern is unique. The back is lightly banded. The mantle is smooth and translucent, the shell markings showing through clearly. This cowry occurs from shallow water to depths of at least 200 ft., usually far back under ledges. It is uncommon and much sought after. The species name means "inlaid with small square stones" or "mosaic." To about 2 in. Photo: Mākua, O`ahu. 15 ft.

TIGER COWRY

Cypraea tigris Linnaeus, 1758
Family Cypraeidae

• Hawaiian waters produce the largest Tiger Cowries in the world. Although common throughout the Indo-Pacific, nowhere else do they attain such record sizes, sometimes 6 in. long. Surprisingly, these cowries were not officially recorded from Hawai`i until the 1930s. Dr. C.M. Burgess writes: "The first living specimen...was collected by the late Ted Dranga and Ditlev Thaanum in June, 1936.... The shell was seen in eighteen feet of water. Dranga dived for it and got it. The slippery beauty oozed from Mr. Thaanum's hands and Ted had the pleasure of diving for it a second time." The larger size Tiger Cowries, extraordinarily beautiful, are now difficult to find. Although the pattern is simple, white with dark spots, the varying number and size of the spots can produce shells ranging from very light to almost solid black. It is said that no two are alike. The mantle is mottled dark gray with long, pointed, white-tipped papillae. This cowry is more likely to be seen by day than many others, probably because it is too big to hide effectively. In Hawai`i it usually occurs below about 10 ft. Elsewhere in the Indo-Pacific it ranges also into the intertidal. The species name denotes any of the large cats, many of which (like leopards) are spotted. Photo: a) Molokini Islet, Maui. 30 ft. b) south shore, Lāna`i, 25 ft. (mantle extended) (See also p. 92.)

Tiger Cowry (a)

Tiger Cowry (b)

CALF COWRY

Cypraea vitellus Linnaeus, 1758
Family Cypraeidae
• This large handsome cowry is seldom encountered today in Hawai`i, although it was at one time reportedly common. The shell is brown with round, milk-white spots of varying sizes. The mantle is blotchy gray and white. Its large branching papillae, approaching in size those of the Leviathan Cowry, have been compared in form to a baobab tree. To about 4 in. Indo-Pacific. Photo: Magic Island, O`ahu. 15 ft.

HELMETS. FAMILY CASSIDIDAE

Helmets usually have attractive globular shells with a short spire (the upper whorls and apex); a long aperture leads into a deep notch through which the snail's siphon protrudes. The shell surface may be smooth, grooved or knobby. The animals feed on heart urchins and sea urchins (whose spines appear not to bother them at all). Four species inhabit Hawaiian waters (including an uncommon endemic, the Grooved Helmet, *Phalium umbilicatum*).

HORNED HELMET • **pū puhi**

Cassis cornuta (Linnaeus, 1758)
Family Cassididae
• These are the large, knobbed shells with a flat heavy base sometimes offered for sale at roadside stands. Live specimens are not uncommon along some sandy bottoms at depths of 20 ft. and below, often buried in sand with only the horn tips protruding. Males are said to have relatively few long horns, females numerous short ones. However, a long-horned specimen sitting on eggs has been photographed. The snails feed on urchins (usually the burrowing heart urchins), which they dig out with their long muscular foot. The flat base of the shell acts as a stable "drilling platform." In early Hawai`i the tip of the spire was knocked off and the shell was blown like a trumpet. The largest of its family in Hawai`i, it attains a length of about 15 in. Indo-Pacific. Photo: Hālona Blowhole, O`ahu. 30 ft.

SPIKED HELMET

Casmaria erinaceus kalosmodix (Melvill, 1883)
Family Cassididae
• This helmet is small and delicate compared with the massive species above. A row of tiny spikes along the lip gives it its common name. Living snails are rarely seen, and shells are usually found occupied by hermit crabs. (The snail from the empty shell pictured here had probably been consumed the night before by a cone or triton.) There are three subspecies: one from East Africa to the Western Pacific, a second from Hawai`i and the Central Pacific, and a third from the Eastern Pacific. To about 3 in. Photo: "Sheraton Caverns," Kaua`i. 40 ft.

TRITONS. FAMILIES RANELLIDAE AND PERSONIDAE

Tritons usually have thick, solid, heavily sculptured shells with prominent riblike varices (the remains of the thickened lip created during rest periods in the shell's growth). The inner lip of the aperture is usually toothed. In life the shells are covered with a hairy or fibrous skin-like covering called the periostracum. The animals, often brightly spotted, feed on echinoderms and molluscs, which they subdue with acid secretions. Tritons are related to tuns and helmets. They were called **naunau** in ancient times. About 14 species occur in Hawai`i. Tritons were formerly placed in the family Cymatiidae.

TRITON'S TRUMPET • **pū, `olē**

Charonia tritonis (Linnaeus, 1767)
Family Ranellidae
• The shells of these giant triton snails are relatively smooth, marked with spiral ribs and low varices. Although often obscured by coralline algae, their variegated pattern has been compared to the plumage of pheasants. The large aperture is reddish orange, its inner edge (columella) beautifully striped dark brown and white. The spotted animals feed on a variety of sea stars and urchins including the Cushion Star, the Crown-of-Thorns Star and the Red Pencil Urchin. Sometimes divers find them sitting on a mass of eggs, bright red when freshly laid or deep rust when about to hatch. With a maximum length of almost 20 in., this species' shell is the second largest in the Indo-Pacific. (The largest, *Syrinx aruanus*, an Australian snail, attains 28 in.) Unfortunately, these snails are now uncommon, possibly due to over-collecting. If you find one, resist the urge to take it home. The early Hawaiians used them as blowing shells. The species is named for the Greek sea god, Triton, who controlled the waves by blowing on his shell trumpet. Indo-Pacific. Photo: Kailua Harbor, Hawai`i. 20 ft. (feeding on the remains of a Cushion Star, *Culcita novaeguineae*)

NICOBAR TRITON

Cymatium nicobaricum
(Röding, 1798)
Family Ranellidae
• This predator of other snails has been observed to make meals of 21 different species, including the equally voracious Textile Cone. Though fairly common in Hawai`i it usually hides under dead coral during the day. The white and brown shell is encircled with fine dark lines; the aperture is bright orange. Like many tritons, live shells of this species are covered with a brown, hairy periostracum and the animal inside is brightly spotted. Indo-Pacific. To about 3 in. Photo: Mākua, O`ahu.

DISTORTED TRITON

Distorsio anus (Linnaeus, 1758)
Family Personidae
• One of the more unusual tritons (and certainly the one with the catchiest scientific name), this species' shell has a small, irregular aperture bordered by many teeth. The extensive white underside is covered with knobs and swellings; the top and sides are nearly white, often banded with reddish brown. This snail is uncommon in Hawai`i, usually occurring at 30 ft. or more in pockets of sand and rubble; elsewhere in its range it inhabits shallow water. The periostracum is hairy. To about 3 in. Indo-Pacific. Photo: Scott Johnson. Koko Head, O`ahu.

HIDDEN TRITON

Linatella succinta
(Linnaeus, 1771)
Family Ranellidae
• Spirally banded with dark and light brown, this triton's elegant shell is almost completely hidden in life by an unusual periostracum that resembles a fuzzy clump of algae. The animal lives at depths of 40 ft. or more on hard substrate. To about 2 1/2 in. Indo-Pacific. *Cymatium clandestinum* is a synonym. Photo: Scott Johnson. Kwajalein, Marshall Islands.

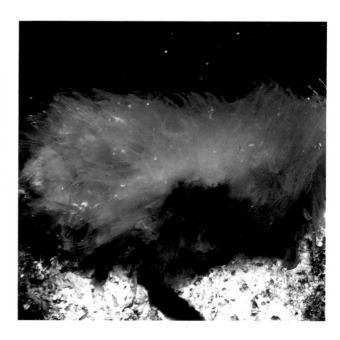

TUNS. FAMILY TONNIDAE

Tuns are large, thin-shelled carnivores related to helmets and tritons. They remain buried in the sand during the day, emerging at night to hunt and feed. Their handsome ribbed shells have a capacious round opening from which extends a muscular foot of surprising size. Empty shells are particularly favored by large hermit crabs. Four species occur in Hawai`i.

PARTRIDGE TUN • **pū`ōni`oni`o**

Tonna perdix (Linnaeus, 1758)
Family Tonnidae
• This is the largest and most common of Hawai`i's tuns, and the only one likely to be found in shallow water. Brown and white mottlings on the shell resemble a partridge's plumage. The snail is encountered most often at night on sand bottoms adjacent to the reef at depths from 2 to at least 30 ft. Fully extended, its plate-size foot appears much too large to fit back in the shell. The animal feeds on sea cucumbers, engulfing them endwise with its enormous proboscis in about 15 seconds. To about 6 or 7 in. Indo-Pacific. Photo: Black Point, O`ahu. 3 ft.

DRUPES AND MUREXES. FAMILIES THAIDIDAE AND MURICIDAE

Drupes and murexes (sometimes called "rock shells") are often combined in the single family Muricidae. Both live openly on rocks or coral, their thick, encrusted shells resembling lumps on the rocky substrate. Murexes are known for their complex (sometimes fantastic) shell sculpture, whereas drupes tend to be heavy, solid and plain. Drupes are common in shallow water along turbulent, rocky shores and are represented in Hawai`i by about 27 species (four endemic). Most of the Islands' eight murex species are from deep water. Drupes and murexes are carnivores, feeding on a variety of organisms including worms, sea urchins and live coral. A special boring organ on the foot enables some to drill neat holes in the shells of other molluscs. Enzymes are injected to digest the victim in its own shell and the liquified tissues ingested. Some drupes and murexes are called "dye shells" because they secrete brightly colored toxic fluids. The famous Tyrian purple of the ancient Greeks and Romans was obtained from a Mediterranean murex. The Latin word *drupa* denotes an overripe, wrinkled olive. The general Hawaiian name for these shells is **makaloa**.

MULBERRY DRUPE • **makaloa**

Drupa morum Röding, 1798
Family Thaididae
• A bright purple aperture gives this species its name. The heavy shell bears blunt projections and is usually thickly encrusted with coralline algae and other organisms. It occurs at snorkeling depths in areas of heavy wave action where it preys on polychaete worms. To about 1 1/2 in. Indo-Pacific. Photos: Näpili Bay, Maui. 10 ft.

SPOTTED DRUPE • **makaloa**

Drupa ricina (Linnaeus, 1758)
Family Thaididae
• These snails have whitish shells with four rows of black tubercles that create a spotted appearance. Growths sometimes obscure the black markings. The outer lip bears five spines. Shells from the intertidal zone have shorter, stouter, spines while deeper-dwelling ones may have long, thin spines and an aperture ringed with orange. The latter, often called "spider drupes" and sometimes regarded as a separate species, *D. arachnoides* (Lamarck, 1816), occur on sponge covered ledges down to at least 20 ft. The living animal is green and white. The species name means "veiled." To about 1 1/4 in. Indo-Pacific. Photo: Hälona Blowhole, O`ahu. 15 ft.

BRILLIANT DRUPE • **makaloa**

Drupa rubusidaeus Röding, 1798
Family Thaididae
• The shell of this animal is cream with a colorful aperture of pinkish purple. Its many strong spines help protect it from shell-crushing predators. It usually occurs on or near ledges at depths of about 30 ft. or less. The species name is from the Latin *rubus* (blackberry), perhaps because the aperture appears stained with blackberry juice. To about 2 1/2 in. Indo-Pacific. Photos: Magic Island, O`ahu. 15 ft.

ARMORED DYE SHELL

Thais armigera (Link, 1807)
Family Thaididae
• These large drupes have thick, spindle-shape shells covered by blunt tubercles. The aperture is yellowish brown. Common along some exposed rocky coasts at depths of 10-40 ft., the shells are typically covered with pink coralline algae. They often occur in pairs. The species name means "bearing arms." To about 3 in. Indo-Pacific. Photo: Hanauma Bay, O`ahu. 5 ft.

GRANULAR DRUPE • **maka`awa**

Morula granulata (Duclos, 1832)
Family Thaididae
• These drupes are covered with spiral rows of low, rounded tubercles. They are abundant on rocky limestone shores with good water movement and adhere to wet rocks at low tide, totally exposed. Not choosy about their food, they drill and consume limpets, worm snails, oysters, barnacles and perhaps other organisms. They also feed on dead molluscs. The Hawaiian name means "sour-face shell" because of the snail's bitter taste. To about 1 in. Indo-Pacific. Photo: Barber's Point, O`ahu.

GRAPE MORULA

Morula uva Röding, 1798
Family Thaididae
• This snail has a white or yellow-orange shell studded with black tubercles. It somewhat resembles the Spotted Drupe (p. 127), but lacks spines on the outer lip. Its aperture is purple. Occurring both in protected and turbulent locations down to about 15 ft., it feeds primarily on worm snails. To about 1 in. Indo-Pacific. Photo: Lāna`i Lookout, O`ahu.

BURNT MUREX

Chicoreus insularum (Pilsbry, 1921)
Family Muricidae
• This is an uncommon snail, usually seen at depths of 60 ft. or more. Its complex shell sculpture, typical of murexes, helps it blend into the reef and makes it awkward to crush or swallow. The pictured specimen is light in color but some are dark brown, hence the common name. It feeds on bivalves and can be maintained in captivity if live clams or oysters are provided. To feed, it climbs onto the bivalve, drills through the shell, injects digestive enzymes, then ingests the softened tissues through its proboscis. The whole process takes several days. It attains about 4 in. Endemic. Photo: Wreck of the YO-257 off Waikīkī, O`ahu. 80 ft.

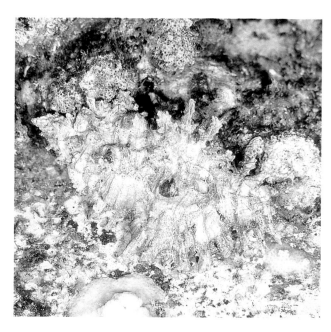

Homolocantha anatomica (Perry, 1811)
Family Muricidae
• The Hawaiian population of this snail was known for many years as *Murex pele* Pilsbry, 1920 and was considered endemic. Some authorities still use that name. The shell may be subtle shades of red, lavender, brown, orange, pink, yellow or white but is usually covered with calcareous growths that blend with the background; its odd shape makes it even more difficult to spot. Uncommon, it occurs at depths of about 50 ft. or more. To about 2 in. Indo-Pacific. Photo: Scott Johnson. Barber's Point, O`ahu.

BASKETS. FAMILY NASSARIIDAE

Baskets, or dog whelks, are moderate-sized to small carnivores and scavengers that live on soft bottoms. Their usually conical shells may be ribbed, smooth, or covered with tubercles. Like many gastropods, they remain buried by day, emerging at night to hunt for food. Seven species inhabit Hawaiian waters.

PIMPLED BASKET

Nassarius papillosus (Linnaeus, 1758)
Family Nassariidae
• Pimpled Baskets are common along the edge of the reef after dark. These scavengers have a keen sense of smell and will quickly home in on a piece of bait left on the sand. They have also been observed attacking damselfish eggs in broad daylight. In the aquarium they usually stay buried but will emerge promptly at feeding time and are capable of wresting food away from hungry hermit crabs. They can also attack and eat small crabs or shrimps. In this photograph the red operculum is clearly visible at the rear of the snail's foot. To about 2 in. Indo-Pacific. Photo: Kea`au Beach Park, O`ahu. 50 ft.

SPINDLES. FAMILY FASCIOLARIIDAE

Like spindles on old-fashioned spinning wheels, spindle shells are typically slender at each end and thick in the middle. The spire is long and pointed, and the aperture is drawn out into a long canal through which the snail's siphon extends. In some spindles this prolongation is pointed and almost equals the spire in length. The ancient Hawaiians called spindle shells **pūpū nuku loa**, literally "long beaked shell." One of the world's largest shells, the so-called Florida Horse Conch (*Pleuroploca gigantea*), is actually a gigantic spindle that attains as much as 23 in. At least 10 species of spindles inhabit Hawaiian waters, including one deep-water endemic. All are carnivores, feeding on a variety of molluscs, worms and other small invertebrates.

WAVY SPINDLE • **pūpū nuku loa**

Fusinus undatus (Sowerby, 1880)
Family Fasciolariidae

• This large spindle lives buried or partially buried in sand. The shell is covered with a thick brown periostracum; the animal is bright salmon red. In the Hawaiian Islands it occurs principally off the southwest shore of Maui and occasionally off west Hawai`i; locally, it is often called the Maui Spindle. Two similar spindles occur in the Islands: the deep-water endemic *F. sandvicensis* and an unnamed species found off O`ahu and Kaua`i. To about 7 in. Indo-Pacific. Photo: Mike Severns. Mā`alaea Bay, Maui.

KNOBBY SPINDLE • **pūpū nuku loa**

Latirus nodatus (Gmelin, 1791)
Family Fasciolariidae

• This snail lives on hard substrate at depths of 8 to 60 ft. and is active by day. The large rounded nodules spiraling up the orange-brown to light yellow shell are distinctive. The aperture is bright pink, and the snail itself bright red (true of most spindles). The commensal Conical Hoof Shell, *Hipponix australis* (p. 106), often attaches to it, blending in well with the knobby exterior. To about 4 in. Photo: Kahe Point, O`ahu. 45 ft.

HARPS. FAMILY HARPIDAE

Harps are among the most beautiful sea shells, their parallel sets of ribs reminiscent of harp strings. These sand-dwellers glide about at night on a large foot searching for small crustaceans, which they trap with the aid of copious mucus. There are at least three species in Hawai`i, none encountered frequently. They look much alike.

LOVE HARP

Harpa amouretta Röding, 1798
Family Harpidae
• This is Hawaii's smallest harp. It is yellow-white to dark straw. Thick brown wavy lines run between the prominent ribs, and the ribs themselves are crossed by fine dark lines. The live animal is beautifully spotted with yellow and brown. In Hawai`i it prefers depths of 60 ft. or more. Elsewhere it occurs on shallow sand flats. To about 2 in. Indo-Pacific. Photo: Scott Johnson. Enewetak, Marshall Islands.

MITERS. FAMILY MITRIDAE

The miters are a large and varied group composed of the family Mitridae and two closely related families, Costellariidae and Volutomitridae. There are over 500 species of miters. The name comes from their typically spindle-shape shells—thick in the center and tapering at each end—thought to resemble a bishop's miter. Some miter shells are smooth, but most have spiral or axial ribs. (Costellarids usually have the axial ribs.) A few are brightly colored. The best diagnostic character for miters in general is a ribbed or folded inner edge of the aperture (columella). All miters are carnivores. They subdue their prey, usually worms or small molluscs, with a venomous sting. Some of the larger and more colorful miters live under the sand, but many others live on hard substrate. At least 49 species of the family Mitridae occur in Hawai`i, varying from less than 1 in. to 7 in. long. Eight are endemic. (Counting all three families, there are at least 94 miters in Hawai`i, 14 of them endemic.) In old Hawai`i, shells of the larger miters were known as `aha`aha.

EPISCOPAL MITER

Mitra mitra (Linnaeus, 1758)
Family Mitridae
• This is the most common large miter in Hawai`i, found from shallow water to depths of several hundred feet. Although dwelling under the sand, it (and similar snails) can be located by the track they leave on the surface. The solid, smooth white shell is marked with a spiral pattern of squarish or irregular orange-red blotches. The largest of all miters, it attains up to 7 in. Indo-Pacific. Photo: Kahe Point, O`ahu. 25 ft. (The similar Papal Miter, *M. papalis*, is pictured on p. 260.)

PUNCTURED MITER

Mitra stictica (Link, 1807)
Family Mitridae
• This miter can be found on shallow sand flats but is more common at depths of 60 ft. or more. The shell is white with closely spaced orange-red blotches. The edge of each whorl is toothed. Rows of tiny punctures encircle the shell, conferring its common and species names. To about 2 1/2 in. Photo: Mākua, O`ahu. 20 ft.

RUSTY MITER

Mitra ferruginea Lamarck, 1811
Family Mitridae
• The spirally ribbed shell of this miter shows its habitat to be hard substrate rather than sand. (Snails that live under the sand are more likely to have smooth shells or axial ribbing that offer little resistance to forward movement.) It is white or cream with rusty markings. To about 3 in. Indo-Pacific. Photo: "Hale`iwa Trench," O`ahu. 40 ft.

CONES. FAMILY CONIDAE

The cones form one of the largest families of marine snails. Their smooth, usually heavy shells are often beautifully patterned and especially popular with collectors. Living cone shells, like those of many other marine snails, are covered by an opaque or translucent skin-like layer called the periostracum that frequently obscures the pattern underneath. Algal and calcareous growths may be present as well. Like many molluscs, cone snails are most active at night. By day they either withdraw completely into their shells (which lie inert on the reef) or bury themselves in sand or rubble. Some cones spend much of their adult lives under sand.

Cone snails are predators. They hunt primarily by smell, waving their siphon to and fro to home in on the marine worms, snails (including other cones), and sleeping fishes that are their typical prey. Many cones sting their victim with a hollow, venomous, harpoon-like tooth at the tip of their proboscis, then consume it at their leisure. Stretching the length of the shell or beyond, the deadly proboscis can penetrate deep into the aperture of another snail, or reach high into a crevice to kill a sleeping fish.

The venom of some cones, especially the fish-eating species, can be dangerous to humans. (Recent research reveals that cone toxins block highly specific neurotransmitters and thus have medical potential.) The Geographic Cone *(Conus geographus)*, not found in Hawai`i, is responsible for at least eight recorded human fatalities. The Textile Cone, which does occur here, can also kill (two recorded fatalities, none in Hawai`i). Many cones, even small ones, are capable of stinging. If you must handle a cone, pick it up at the broad end (although this is no guarantee that you cannot be injured).

Most cone snail stings have the intensity of a bee sting and subside in several hours. Soreness, numbness or both usually occur for some time afterwards; numbness sometimes lasts for weeks. If breathing difficulties, nausea or headache appear seek emergency medical treatment. If possible, bring the cone with you for identification.

Penniform Cone *(Conus pennaceus)* feeding on Gaskoin's Cowry *(Cypraea gaskoini)*.
Campbell Park, O`ahu. (Scott Johnson)

In captivity, cones do well if given live food. Easiest to feed are mollusc and fish eaters. The latter have even been conditioned to accept small frozen fish. Both can be kept alone in very small tanks if provided with plenty of sand in which to bury themselves.

Hawai`i has at least 34 living species of cones, one of which is endemic. Thirteen are illustrated below. An interesting fossil cone, presumed to have been endemic, occurs in almost perfect condition in upraised reefs on O`ahu's Mōkapu Peninsula. It was recently named *Conus kuhiko* after the Hawaiian word for "ancient." In old Hawai`i, cones in general were called **pūpū `alā**; those capable of delivering a bad sting were called **pūpū pōniuniu**, literally "dizzy shell."

Cone eggs, frequently found attached to the underside of stones, often resemble miniature slices of bread laid in rows. The worn spires of cones, naturally perforated by rolling in the surf, are locally called "puka shells." They are popular for stringing into necklaces and bracelets.

ABBREVIATED CONE

Conus abbreviatus Reeve, 1843
Family Conidae
• This tubby little cone is Hawai`i's only living endemic. Shorter and broader than most others, it is creamy or yellowish in life, marked with faint spiral bands and regular rows of widely spaced brown dots. (Empty shells cleaned of the thin, yellowish periostracum have a base color of bluish gray.) The edge of the broad end or shoulder is knobby (coronate). It occurs on shallow reefs and at all sport-diving depths and feeds on bristle worms. Although occasionally found in the Marshall and Line Islands, it is not known to reproduce outside Hawai`i. To about 1 1/4 in. Photo: Ali`i Beach Park, O`ahu. 3 ft.

HEBREW CONE

Conus ebraeus Linnaeus, 1758
Family Conidae
• A heavy white shell marked with three or four rows of dark markings (faintly reminiscent of Hebrew letters) immediately identifies this species. In life, the white shell is covered with a thin yellowish periostracum. A common cone in Hawai`i, it lies exposed on shallow reefs, conspicuous because of its bold pattern. It feeds on bristle worms and attains its greatest size in Hawaiian waters, almost 2 1/2 in. Indo-Pacific. Photo: Pūpūkea, O`ahu. 3 ft.

Conus flavidus Lamarck, 1810
Family Conidae
• This is one of the two most abundant cones on O`ahu's shallow reefs (the other is the Spiteful Cone, p. 138). It varies from almost white to greenish yellow to yellowish brown with white bands. The tip of the narrow end is dark violet, and the aperture is a lighter violet. The shoulder is rounded and smooth, the spire low and flat. (The similar Spiteful Cone has a knobby shoulder.) It preys on bristle worms and attains its largest size in Hawai`i, about 2 1/2 in. Indo-Pacific. Photo: Palea Point, O`ahu. 20 ft.

IMPERIAL CONE

Conus imperialis Linnaeus, 1758
Family Conidae
• The spire at the broad end of this cone is low, almost flat, and its shoulder is ornamented with points like a crown (coronate). The shell is white with numerous encircling lines of brown dots and dashes and two irregular, wide brown bands flecked with white. The tip of the shell and its opening are purple. This cone, like many others, feeds on bristle worms. It attains 3 1/2 in. and has an Indo-Pacific distribution. Photo: Pūpūkea, O`ahu. 50 ft.

Leopard Cone (a)

LEOPARD CONE

Leopard Cone (b)

Conus leopardus (Röding, 1798)
Family Conidae

• The Leopard Cone is Hawai`i's largest. It is found partly buried in sand, usually at depths of 40 ft. or more but often as shallow as 3 ft. The heavy shell, its thick fibrous periostracum usually covered by fuzzy algae and silt, looks much like a stone and is easily overlooked. Carefully turn it over (remembering to handle it briefly and only at the wide end) and the strik-ing pattern of spiral dots may be visible underneath. On smaller specimens with less algae the attractive dotted pattern often shows through the translucent, yellowish brown periostracum. This cone grows to about 9 in. Indo-Pacific. Photos: Kahe Point, O`ahu. 45 ft. a) young specimen showing spots; b) typical appearance.

137

SPITEFUL CONE

Conus lividus Hwass, 1792
Family Conidae

• One of the two most common cones on O`ahu's shallow reefs, this species has a drab olive or yellow-brown shell with a bluish white central band. The knobby shoulder and higher spire distinguish it from the similarly colored and equally common Yellow Cone. Some collectors report that males have a white crown, females a green one. The animal is black, mottled with red. It feeds on bristle worms and attains its largest size in Hawai`i, about 2 1/2 in. Indo-Pacific. Kewalo Park, O`ahu. 3 ft.

MARBLED CONE

Conus marmoreus Linnaeus, 1758
Family Conidae

• This is one of the so-called "tented" cones, whose basic pattern is a dark ground color overlaid with light triangular marks. In this species the tent marks are large and slightly rounded. In Hawaiian specimens they tend to coalesce loosely into two bands, a variant considered by some authors to constitute a distinct subspecies. Identification is easy: in no other Hawaiian tented cone is the shoulder coronate. The Marbled Cone preys on other cones including the Flea Cone, the Soldier Cone and the Abbreviated Cone; it can inflict a nasty sting on humans as well, confirming the old shell-collector's adage, "beware of tented cones." If you must handle it, do so only briefly and by the broad end, remembering that you could still get stung. To about 5 in. Indo-Pacific. Mākua, O`ahu. 20 ft.

SOLDIER CONE

Conus miles Linnaeus, 1758
Family Conidae
• A broad black band at the narrow end of the shell and a slender band near the center make this cone easy to identify. Fine, somewhat wavy lines run lengthwise. The translucent periostracum is green-brown and covered with tiny tufts. This snail preys on bristle worms and occurs in shallow water and at all sport-diving depths. To about 3 in. Indo-Pacific. Photo: Pūpūkea, O`ahu. 50 ft.

PENNIFORM CONE

Conus pennaceus Born, 1780
Family Conidae
• This tented cone has a smooth shoulder, distinguishing it from the Marbled Cone. The spire may be flattened or pointed. The base color varies from reddish to golden brown, and the white or pinkish "tent" marks may be large or small, resulting in patterns of quite different appearance. This cone inhabits shallow water, usually no deeper than 10 ft., where it feeds on other molluscs. It is common in Hawai`i and easily confused with the Textile Cone, which also has small, overlapping tent marks. (The Textile Cone has jagged dark brown lines running lengthwise in the non-tented areas; these lines do not occur on the Penniform Cone.) The Latin word *penna* means "wing," "feather" or "arrow." It attains about 2 1/2 in. Indo-Pacific. Photos: Scott Johnson, Hau`ula, O`ahu. a) large tent pattern, depositing egg capsules; b) small tent pattern. (See also p. 134.)

Penniform Cone (a)

Penniform Cone (b)

139

Conus pulicarius Hwass, 1792
Family Conidae

• This white cone with its dark spots resembles a small Leopard Cone. The spots, however, are irregularly distributed (as if, perhaps, the shell were covered with fleas). The shoulder is somewhat knobby, whereas the Leopard Cone has a smooth shoulder. Juveniles have a translucent yellow periostracum but larger specimens have a thick brown one, much like the Leopard Cone. The Flea Cone lives in sand, reportedly in pairs, where it feeds on bristle worms and heart urchins. The species attains its largest size in Hawai`i, to about 2 1/4 in. Indo-Pacific. Photo: Kahe Point, O`ahu. 50 ft.

OAK CONE

Conus quercinus Lightfoot, 1786
Family Conidae

• This attractive lemon yellow cone is easily identified by its numerous, closely set encircling lines and the small pointed spire at its broad end. On large individuals, however, the color and lines are completely obscured by a thick brown periostracum. It lives in sand from snorkeling to scuba-diving depths and beyond. The smaller, more colorful specimens usually occur in deeper water, the older ones in shallower. Painful stings have been recorded from this species. It attains its largest size in Hawai`i, almost 5 in. Indo-Pacific. Photo: Kahe Point, O`ahu. 50 ft.

STRIATED CONE

Conus striatus Linnaeus, 1758
Family Conidae
• This large cone is dangerous to humans. As with most cones, its proboscis is long and flexible; holding the shell by the wide end provides no guarantee that you can't be stung. Active at night, it stings small sleeping fishes and then engulfs them. The name comes from the closely set fine lines that encircle the shell. It is pinkish white with irregular dark streaks and spots, but the translucent yellow periostracum on live specimens gives them a yellowish brown tint. It occurs in the shallows and throughout sport-diving depths. In captivity these animals will feed on fresh or frozen fish. To 5 in. Indo-Pacific. Photo: Mākua, O`ahu. 20 ft.

TEXTILE CONE

Conus textile Linnaeus, 1758
Family Conidae
• Cones with tentlike markings are said to be dangerous to humans and this species has been responsible for several fatalities (fortunately none known in Hawai`i). Jagged dark brown lines running lengthwise within the golden brown, non-tented areas help separate it from the similar Penniform Cone. It preys on other snails, including cones. After stinging them, it inserts its long proboscis into the shell to consume the entire animal, which it softens with enzymes. Textile Cones are sometimes found under loose pieces of coral, the empty shell of their last meal lying nearby. They do well in captivity if supplied with live snails, which will frantically try to escape the pursuing cone. One Textile Cone lived 9 years in an aquarium, obligingly emptying and cleaning shells for its shell-collecting owner. If you keep a Textile Cone, be sure to give it plenty of sand in which to bury itself. Be extremely careful. This potentially lethal animal, sometimes known as the Cloth-of-Gold Cone`, grows to about 5 in. Indo-Pacific. Photo: Kōloa Landing, Kaua`i. 20 ft.

141

AUGERS. FAMILY TEREBRIDAE

Augers are snails with long, pointed shells reminiscent of the human tools for which they are named. Hawaiians once used the shells as scrapers and stoppers for water gourds, as well as for drills. Augers generally remain just under the surface of the sand, their slender, close-ly spiraled shells offering minimal resistance to forward motion. Even so, tracks appear on the surface as the snails move about. Often in the 5 to 6 in. range, the shells are quite attrac-tive. Augers occur from the wave wash zone of beaches to sandy bottoms hundreds of feet deep. All are predators of specific organisms, usually polychaete worms. Some, like cones, disable their prey with injected venom. Despite their specialized feeding habits, many augers can be maintained in captivity if the aquarium gravel contains a good supply of small worms and other invertebrates. About 31 species of augers occur in Hawaiʻi, but none are seen alive with any frequency except by those who specifically seek them out. In old Hawaiʻi augers in general were known as **pūpū loloa** ("long shell") or **ʻoiʻoi** ("sharp").

GOULD'S AUGER • **pūpū loloa, ʻoiʻoi**

Duplicaria gouldi (Deshayes, 1859)
Family Terebridae
• Gould's Auger is yellowish white with circular bands of pale brown. It is always found in areas inhabited by the Yellow Acorn Worm (*Ptychodera flava*), on which it feeds exclusively. The snails are able to ingest worms twice as long as them-selves, taking about 15 hours to do so. Augers living in the castings of large acorn worms, however, may simply nip bits off the posterior of the worm when it emerges to defecate. (The worm quickly regenerates its lost flesh.) To about 2 1/2 in. Endemic, but rare or absent from the Big Island. Formerly placed in the genus *Terebra*. Photo: Kahe Point, Oʻahu. 30 ft.

WHITE-SPOTTED AUGER • **pūpū loloa, ʻoiʻoi**

Terebra guttata (Röding, 1798)
Family Terebridae
• Light orange with white spots, this is one of the prettiest augers. It lives a buried life, like others of its family, but is occa-sionally found alive on the surface. The one pictured was exposed on a rocky reef some distance from the sand. To about 5 in. Indo-Pacific. Photo: Kahe Point, Oʻahu, 35 ft.

MARLINSPIKE AUGER • **pūpū `olē, `oi`oi**
Terebra maculata (Linnaeus, 1758)
Family Terebridae
• This stout auger is the largest of its family. It is cream with rows of dark, squarish spots that follow the spirals and almost coalesce into lines. Common in some areas, it feeds on a single species of polychaete worm and attains up to 10 in. Indo-Pacific. Photo: Kahe Point, O`ahu. 45 ft. (The similar *Terebra strigilata* is pictured on p. 267.)

SEA SLUGS. SUBCLASS OPISTHOBRANCHIA

Sea slugs and their relatives belong to a fascinating group of molluscs that are in the evolutionary process of abandoning shells in favor of more sophisticated chemical and biological defenses. Collectively known as opisthobranchs (subclass Opisthobranchia), this group includes the sea hares (order Anaspidea), bubble shells and headshield slugs (Cephalaspidea), side-gilled slugs (Notaspidea), sap-sucking slugs (Sacoglossa), and the true sea slugs or nudibranchs (Nudibranchia).

The word opisthobranch means "gills behind," and in general marine slugs and their relatives have gills and anus situated toward the rear of the body. Marine snails, by contrast, have gills and anus at the front. This forward position is due to "torsion," the evolutionary twisting necessary to fit the body into a shell. Because opisthobranchs have undergone detorsion (they have "untwisted"), the gills have moved rearward.

In place of protective shells, opisthobranchs have developed ingenious defenses that include acids, toxins, stings and superb camouflage (discussed in greater detail below). In most other respects, marine slugs and snails resemble each other. Like snails, most slugs have a radula (a toothed tongue or ribbon used to rasp their prey), and some have a pair of chitinous jaws. The radular teeth, also made of chitin, are important for precise identification of these animals.

Opisthobranchs are hermaphroditic: individuals possess both male and female reproductive organs, ensuring that any other animal of the same species will be a suitable mate. After mutual insemination, they lay eggs on the substrate in strands or ribbons. (In the case of nudibranchs these are often coiled; the egg coils are often easier to spot than the animal itself.) Some time after hatching, trochophore larvae develop into shelled larvae (veligers) that swim or drift freely for a time before settling to the substrate to mature. Most of them lose their shells at this time. Chemicals emitted by their prey organisms may stimulate this change, ensuring that veligers can feed and grow to adulthood.

The feeding habits of opisthobranchs are varied. Some are herbivores, and a few capture minute organisms in a "throw net" or oral hood. Most, however, are specialized predators, feed-

143

Eared Sea Hare *(Dolabella auricularia)*. Pūpūkea, O`ahu. Tide pool.

ing on a single species or genus of sponge, bryozoan or hydroid. Some species are seasonal (in synchrony with their food source), while others may vary greatly in abundance from year to year. An example of the latter case is the Imperial Nudibranch *(Risbecia imperialis)*, which was abundant on O`ahu in the 1970s, all but disappeared in the 1980s, and began to be seen again in the mid-1990s.

Most opisthobranchs are unsuited to aquarium life. Special food requirements alone would disqualify most of them. In addition, their toxic secretions or stings could be disastrous to other organisms. Finally, many species (even large ones) have life spans of less than a year; by the time an adult is collected its natural life might be almost over. It is best to leave these animals on the reef where they can reproduce and be enjoyed by others.

There are an estimated 2,000 opisthobranch species worldwide. Dr. E. Alison Kay, in her book **Hawaiian Marine Shells**, lists over 150 from the Islands. A great many more are known to exist, some lacking scientific names. Sixty-four of the most common, colorful or interesting of Hawai`i's opisthobranchs are shown below. The Warty Slug (*Onchidium verruculatum*), an air-breathing relative of common garden slugs (hence not an opisthobranch), inhabits Hawai`i's rocky shores and is included here for convenience.

WARTY SLUG

Onchidium verruculatum (Cuvier, 1830)

Family Onchidiidae

• This slug and other members of its family are more closely related to common garden slugs than to the marine opisthobranchs. They may be undergoing an evolutionary return to the sea. Instead of gills, all have a lunglike sac at the rear of the mantle cavity enabling them to survive out of water. This species inhabits the intertidal zone of rocky shores where it grazes on algae. Its back is covered with light-sensitive warts. The two head tentacles each bear an eye at the tip (not true of opisthobranchs). Glands at the side of the body secrete noxious fluids. To about 1 in. Indo-Pacific. Photo: Fort DeRussy Beach Park, O`ahu. Intertidal.

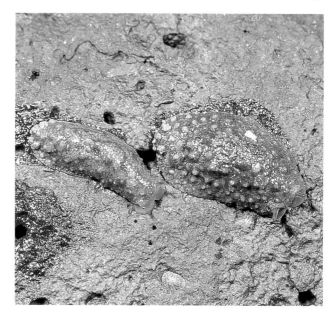

BUBBLE SHELLS AND HEADSHIELD SLUGS. ORDER CEPHALASPIDEA

The cephalaspideans are a diverse group. Together they form the most primitive of the opisthobranch orders. Many live just under the sand, plowing through it with the aid of a wedgelike head shield. The bubble shells have a shell big enough, or almost big enough, to hold the entire animal; headshield slugs retain only internal vestiges of a shell. These animals are sometimes called bullomorphs and tailed slugs, respectively. Some living bubble shell animals are exquisitely beautiful.

CYMBAL BUBBLE SHELL

• **pūpū leholeho**

Haminoea cymbalum (Quoy & Gaimard, 1835)

Family Haminoeidae

• Small but exciting to find, these unusual molluscs have a colorless transparent shell that allows the striking green, white and orange animal to show through. They occasionally "bloom" by the thousands in certain tide pools and shallow reefs. They feed on blue-green algae (*Lyngbya* spp.) and can frequently be found upon it. They also occur down to at least 25 ft. In Hawai`i they are most common from October to December. To about 3/4 in. Indo-Pacific. Photo: Scott Johnson. Pūpūkea, O`ahu. Tide pool.

SWOLLEN BUBBLE SHELL
• pūpū leholeho`ōni`oni`o

Hydatina amplustre (Linnaeus, 1758)
Family Hydatinidae
• The delicate white animal inhabiting this spirally banded pink and white shell is too large to fit entirely inside. Active only at night, it can be found at depths of 1-2 ft. in areas of mixed sand and rock with good water circulation. Hermit crabs inhabiting the distinctive shells are a daytime clue to its presence. The species feeds on small, colorful bristle worms of the family Cirratulidae. Some specialists place it and the two species below in the order Acteonoidea. The species name means "large" or "swollen." To about 3/4 in. Indo-Pacific. Photo: Black Point, O`ahu. 1 ft.

PAPER BUBBLE SHELL • **pūpū leholeho**

Hydatina physis (Linnaeus, 1758)
Family Hydatinidae
• Closely set black spiral lines mark the shell of this animal, which is pinkish, with faintly iridescent blue borders. Fleshy flaps (parapodia) over the back partially cover the shell (true of all bubble shells). It occurs on shallow reef flats, most often from January to May, and is seen at night when it feeds on bristle worms of the family Cirratulidae. The white egg mass is sometimes carried for a time on the side of the shell, which in large specimens may be 1 1/2 in. long. All warm seas. Photo: Scott Johnson. Diamond Head, O`ahu.

WAVY BUBBLE SHELL

Micromelo undata (Brugière, 1792)

Family Hydatinidae

• These animals crawl on algae-covered rocks in shallow water. Their beautiful blue-green mantles, which almost glow in the sunlight, are adorned with white spots and yellowish margins. The shell, attractively patterned with black spiral lines intersected by wavy transverse ones, is too small to hold the animal's entire body. This species feeds on small, colorful bristle worms of the family Cirratulidae, whose toxin they incorporate for their own defense. The species name means "wavy." *Micromelo guamensis* is a synonym. To about 1/2 in. All warm seas. Photo: Magic Island boat channel, O`ahu. 5 ft.

CALYX BUBBLE SHELL

Smaragdinella calyculata (Broderip & Sowerby, 1829)

Family Smaragdinellidae

• These shelled slugs are common on rocky shores around the high tide line, where they feed on algae (often in the company of `**opihi**, or limpets). Their dark green to almost black mantles nearly cover their shells. Mucous secretions keep them from drying out at low tide. To about 1/2 in. Indo-Pacific. Photo: Fort DeRussy Beach Park, O`ahu.

BLUE SWALLOWTAIL SLUG

Chelidonura hirundina (Quoy & Gaimard, 1824)
Family Aglajidae
• Slugs of the family Aglajidae have a small internal shell and a pair of flaps (parapodia) that fold over the center of the body. Often one flap is longer at the rear than the other, forming a "tail." One of the prettiest members of this family, the Blue Swallowtail is also one of the most abundant in Hawai`i. It crawls by day on algae-covered rocks in shallow, well-protected environments and feeds on small flatworms. The animal is velvety black with peacock blue and orange lines; T-shape markings on the head are distinctive. When handled it may release a yellow fluid that temporarily stains the fingers. Both genus and species names mean "swallow." (A similar but undescribed species of *Chelidonura* has bright blue spots instead of lines and is common in the same habitat.) To about 1 in., but usually smaller. All warm seas. Photo: Magic Island boat channel, O`ahu. 4 ft.

PILSBRY'S HEADSHIELD SLUG

Philinopsis pilsbryi (Eliot, 1900)
Family Aglajidae
• This attractively patterned slug lives on sand bottoms in protected areas, emerging after dark. It varies from predominantly white or cream with dark brown markings to the reverse. The markings sometimes form a figure-8 shape. The name honors American malacologist H.A. Pilsbry (1869-1959), a longtime student of Hawaiian molluscs. To about 3 in. Indo-Pacific. Photo: Black Point, O`ahu. Artificial pool.

148

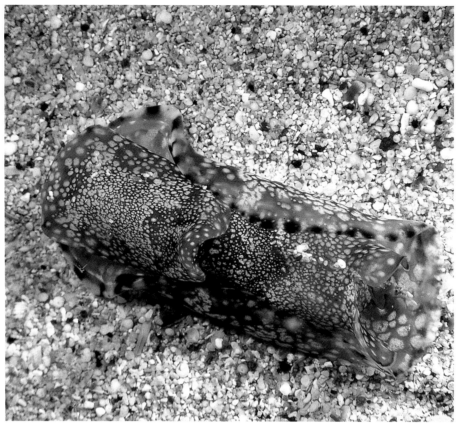

BLUE HEADSHIELD SLUG
Philinopsis speciosa Pease, 1860
Family Aglajidae
• This slug lives just under the sand or on its surface. It varies from light to dark brown, sometimes with pairs of parallel orange-brown lines on the head shield and tinges of blue at the edges of the parapodia. It feeds on other opisthobranchs, engulfing them whole. To about 3 in. Indo-Pacific. Photo: Waimea Bay, O`ahu. 30 ft.

SEA HARES. ORDER ANASPIDEA

Sea hares are plump, fleshy slugs named for the two tentacles, or rhinophores, that sprout like rabbit ears from their small heads. Two additional cephalic tentacles protrude forward. Most species are between 1 and 8 in. long and dull in color. All are herbivores. Many adult sea hares have a vestigial, platelike shell, covered by the mantle and further hidden by large fleshy flaps (parapodia). Some species of sea hares can swim with vigorous movements of their parapodia. The vestigial shell, if present, can often be felt by the fingers.

Like all opisthobranchs, sea hares are hermaphroditic. Many copulate in chains, each acting as a male to the animal in front and as a female to the animal behind. The egg masses of sea hares typically resemble tangled brown spaghetti.

Although sea hares contain poisonous compounds (many of which are of interest to medical researchers), humans can safely handle them if care is taken not to get slime near the eyes. Some of these poisons (notably aplysiatoxin) come from the blue-green algae upon which these animals feed.

In old Hawai`i sea hares were known as **kualakai**. Some were considered food; these were wrapped in **ki** (or **ti**) leaves and baked in an **imu**.

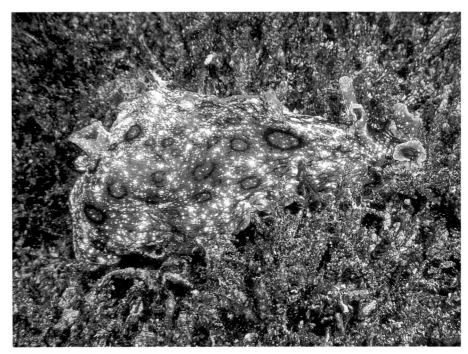

WHITE-SPECKLED SEA HARE • **kualakai**

Aplysia dactylomela Rang, 1828
Family Aplysiidae
• This common sea hare is gray to brown, speckled with small white spots of varying sizes, and marked with a network of black lines that often form large rings. When disturbed it may emit a pinkish purple ink manufactured from compounds in red algae, a principal food. When red algae is unavailable, the ink supply fails. This slug occurs on shallow reef flats, hiding under stones by day, emerging at night to feed on algae. (The animal pictured, however, was crawling in the open at mid-day.) Some aquarists have reported keeping this species for over two years; it requires plenty of green algae. The species name means "dark fingered." To about 8 in., but most are half that size or smaller. Found in warm seas worldwide. Photo: Black Point, O`ahu. 1 ft.

EYED SEA HARE

Aplysia oculifera Adams & Reeve, 1850
Family Aplysiidae
• Rare in Hawai`i, this sea hare is dark green or olive-green speckled with minute white spots. Small black rings, usually with white centers, may be present. (These are the "eyes" for which the slug is named.) It occurs most often in the shallows along turbulent limestone shores. To at least 2 1/2 in. Photo: Barber's Point, O`ahu. 1 ft.

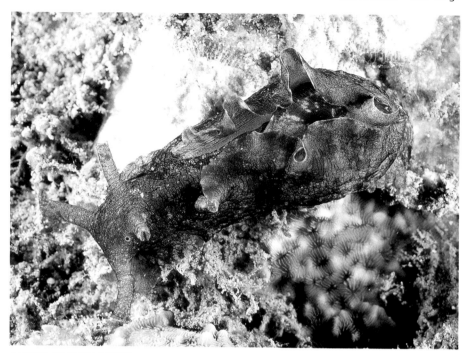

JULIANA'S SEA HARE • **kualakai**

Aplysia juliana Quoy and Gaimard, 1832
Family Aplysiidae
• Most common during the rainy season, this nocturnal sea hare feeds exclusively on *ulva*, or sea lettuce, a green alga which grows best in areas of freshwater seepage. Its color varies from dark greenish brown to white, with blackish blotches. A round sucker on the back of its foot is distinctive. When disturbed, the slug contracts into a ball and may emit a milky fluid. Because it does not eat red algae, source of the purple ink in other sea hares, this species may not be as poisonous. According to Hawaiian scholar Mary Kawena Pukui, this was the sea hare eaten most often in old Hawai`i. Typical local specimens are about 2 in. long and weigh about 2 ounces, although 5 in. specimens have been reported. In Japan and other parts of the world it can weigh up to a pound or more. Indo-Pacific. Photo: Mike Severns. Maui.

SMALL SEA HARE • **kualakai**

Aplysia parvula Mπrch, 1863
Family Aplysiidae
• This tiny sea hare is often found clinging to seaweed in very shallow water, occasionally in chains of three or four mating individuals. It varies from black to light brown and sometimes emits purple ink when disturbed. To about 2 in. Occurs worldwide in tropical and warm temperate seas. Photo: Wai`alae Beach Park, O`ahu. 2 ft.

• **kualakai**

Aplysia pulmonica Gould, 1852
Family Aplysiidae
• This uncommon sea hare is light reddish to greenish brown, covered with white spots and scattered black spots. It sometimes emits pinkish purple ink when disturbed, or swims away using its large, winglike parapodia. This is probably the species known in old Hawai`i as **lepelepe-o-Hina**, which also meant "butterfly." It is encountered most often in Kāne`ohe Bay, O`ahu. Photo: Kāne`ohe Bay, O`ahu. 1 ft.

EARED SEA HARE • **kualakai**

Dolabella auricularia (Lightfoot, 1786)
Family Aplysiidae
• Quite distinctive, this chunky, cone-shape slug looks as if its rear end has been lopped off with a machete. The disklike posterior contains a central orifice. Two flaps (parapodia) converge on its back. The color, dark brownish green with green warts and bumps, blends in well with rocks and algae, rendering the animal almost invisible. It occurs in tide pools and to depths of at least 70 ft., and can occasionally be found in the open during the day. If it has recently fed on red algae it can secrete thick pinkish purple ink. Researchers have recently isolated an anticancer and immunostimulant compound, dolastin 10, from this animal. To about 10 in. or more. The species name means "having ears." Tropical Atlantic and Indo-Pacific. Photo: Kahe Point, O`ahu. 30 ft. (See also p. 144.)

COMMON SEA HARE • **kualakai**

Dolabrifera dolabrifera
(Rang, 1828)
Family Aplysiidae

• These small, shallow-water sea hares cluster on the undersides of rocks in the intertidal zone, emerging at night to feed on algae. Their transparent eggs, laid in zigzags, are often visible nearby. The slugs, which vary from dark brown to gray-green, and may also be reddish or cream, are abundant in some areas, absent from others. Both genus and species names mean "shaped like an ax." To 1 1/2 in. Found in all warm seas. Photo: Pūpūkea, O`ahu. Tide pool.

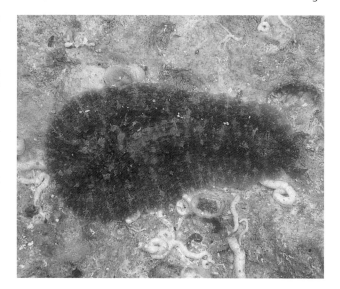

LINED SEA HARE

Stylocheilus longicauda (Quoy & Gaimard, 1824)
Family Aplysiidae

• Easily recognized by the fine, dark longitudinal lines and the numerous little pointed projections (papillae) on the body, these small sea hares may also have black or even blue spots. Amazingly abundant at certain times of the year, they aggregate by the thousands on the sand, travelling in chains up to 30 ft. long—virtual rivers of molluscs. The massed slugs, however, often rarely notice them because they closely resemble clumps and windrows of decaying algae. The massed slugs, however, often spurt purple ink when stepped on inadvertently by waders. They occur from shallow tide pools to depths of at least 100 ft. and feed primarily on the blue-green algae of the genus *Lyngbya*, often called "mermaid's hair"; when the algae disappear so do the short-lived slugs. They will survive for some time in an aquarium, laying tangled strands of eggs all over the glass. Adults lack a shell. To about 1 1/2 in. All warm seas. Photo: Kahe Point, O`ahu. 50 ft.

SAP-SUCKING SLUGS. ORDER SACOGLOSSA

Sacoglossans are a diverse group; some have shells, and some do not. Almost all are herbivores living on or around seaweeds in shallow water. They feed by piercing algal cells with sharp, pointed radular teeth and sucking out the contents. (Teeth are used one at a time until dull or broken, then discarded into a pouch.) Often, all the algal matter is digested except the chloroplasts (bodies within the cell responsible for photosynthesis). These are retained intact in the slug's tissues where they continue to produce sugars that feed the slug. Some of these solar-powered slugs can survive for several months on light alone and, unlike many opisthobranchs, can be maintained easily in captivity.

ORNATE SAP-SUCKING SLUG

Elysia ornata (Swainson, 1840)
Family Elysiidae
• These slugs feed on feathery algae of the genus *Bryopsis*, common on shallow rocky reef flats where there is good water movement. Their overall color (green with yellow spots and streaks, peppered with tiny black spots) blends in well with the algae in which they live. Large, almost winglike parapodia, usually held straight up over the body, have striking wavy orange and black margins that may provide warning coloration. Recent research indicates that these animals manufacture toxic compounds with anticancer properties. A similar species, *E. rufescens*, occurs on the same type of algae. To about 1 1/2 in. Tropical seas worldwide. Photo: Black Point, O`ahu. 2 ft.

RINGED SAP-SUCKING SLUG

Plakobranchus ocellatus
Hasselt, 1824
Family Elysiidae
• This slug is common on silty sand from the shallows down to at least 50 ft. The dull tan body is covered with small ringlike marks of various colors; its two long, widely spaced rhinophores have been compared to ox horns. Like many sacoglossans, it feeds on algae by piercing individual cells and sucking out the contents. Gently spread apart the parapodia to see the vivid green of the retained chloroplasts. Solar-powered slugs such as these can survive for several months in a well-lit aquarium. To about 2 1/2 in. Indo-Pacific. Photo: Black Point, O`ahu. 2 ft.

SIDE-GILLED SLUGS. ORDER NOTASPIDEA

Side-gilled slugs have a set of long feathery gills between the foot and the mantle on the right side of the body. Also beneath the mantle are special acid-secreting glands that render the slugs distasteful to predators. Most of these animals are thought to feed on sponges or tunicates. Taxonomically the group is in disarray.

ORANGE GUMDROP

Berthellina citrina (Rüppell & Leuckart, 1831)
Family Pleurobranchidae
• This bright orange or occasionally yellow slug is seen at night crawling about along ledges and cliffs. By day it remains under rocks. Like all members of its family, it secretes an acid when disturbed. It probably preys on colonial tunicates. A common species in Hawai`i, it attains about 1 1/2 in. Indo-Pacific. Photo: Waimea Bay, O`ahu. 20 ft.

POLYGON PLEUROBRANCH • **naka-ōni`oni`o**

Pleurobranchus sp. 1
Family Pleurobranchidae
• Low, closely set, polygon-shape tubercles cover the back of this slug. Active by night, it is common on some shallow reef flats and also occurs along rocky cliffs and ledges in deeper water. This and similar pleurobranchs are often called *Pleurobranchus peroni*, but the identity of the true *P. peroni* remains unclear. The Hawaiian word **naka** means any sea creature, **ōni`oni`o** means "mottled." To about 3 in., but most are smaller. Photo: Pūpūkea, O`ahu. 20 ft.

TILED PLEUROBRANCH • **naka-ōni`oni`o**

Pleurobranchus sp. 2
Family Pleurobranchidae
• This striking slug is covered with red tubercles often surrounded by white, producing a tesselated, or "tiled," effect. It is nocturnal and occurs in silty harbors and lagoons. Eggs are laid in beautifully ruffled white strands. The specimen pictured was almost 8 in. long, fully extended. Photo: Magic Island boat channel, O`ahu. 15 ft.

FORSSKAL'S PLEUROBRANCH

Pleurobranchus forskalii (Rüppell & Leuckart, 1828)
Family Pleurobranchidae
• This large nocturnal slug is deep maroon. When it expands, as when disturbed or feeding at night, a highly contrasting white fish-scale pattern is revealed. Elsewhere in the Indo-Pacific the color may vary from almost black to cream with black markings. The name honors pioneering Danish naturalist Peter Forsskål, who collected marine organisms in the Red Sea in 1762 and 1763 and subsequently perished in the mountains of Yemen. To almost 8 in. Indo-Pacific. Photo: Hale`iwa Boat Harbor, O`ahu, 2 ft.

UMBRELLA SLUG • **kuapo`i**

Umbraculum umbraculum
Lightfoot, 1786
Family Umbraculidae

• One of the oddest of all opisthobranch molluscs, this large, sponge-eating, side-gilled slug carries a calcareous disk too small to cover itself completely. Even its naked gills are barely protected. The body (black, brown, yellow or orange) bears numerous white warts and pustules. Two eyes and a noselike penis form the rough semblance of a face. The slug occurs in a variety of habitats, from caves and crevices of exposed reefs to protected reef flats. The white or red egg coils attain the same diameter as the slug. The sole species in its genus, it has a wide Indo-Pacific and tropical Atlantic distribution but is uncommon in Hawai`i. To about 5 in. long with a shell 3 in. wide. Photo: Hālona Blowhole, O`ahu. 15 ft.

NUDIBRANCHS. ORDER NUDIBRANCHIA

Nudibranchs form the largest and best known of the opisthobranch orders and are the only marine slugs that always lose their shells as adults. (At least some species in other major opisthobranch orders retain some sort of external or internal shell.) Nudibranchs are carnivores. Most feed on sessile (fixed) organisms such as sponges, hydroids and corals, but a few are hunters, usually of other nudibranchs. Some zoologists compare nudibranchs to hawks and eagles because they sit at the top of their food chain and have few natural enemies. Others liken them to orchids or butterflies because of their fascinating colors and patterns.

A rare Hawaiian nudibranch, *Hypselodoris* sp. YO-257 wreck, off Waikīkī, O`ahu. 75 ft. (Steve Buck)

Brightly colored animals are often foul tasting, toxic, or venomous—their color serving as a warning to potential predators. Nudibranchs are no exception, but they rarely produce poisons or venoms of their own. Many recycle toxins found in their prey. Sponges for example (a favorite food) produce a wide assortment of irritants and poisons that the slugs concentrate in their bodies for their own defense. Nudibranchs that feed on hydroids or other stinging animals may ingest their prey's powerful stinging capsules (nematocysts) without discharging them. In an amazing biological feat, they pass these nematocysts through their bodies and store them, armed and ready to fire, at the tips of tentacle-like protuberances called cerata.

Most other nudibranchs rely on camouflage and nocturnal habits for protection. The Cup Coral Nudibranch *(Phestilla melanobrachia)* is difficult to see because in both form and color it closely resembles the Orange Cup Coral upon which it feeds. The Pitted Nudibranch *(Aldisa pikokai)* looks amazingly like a small encrusting sponge.

The order Nudibranchia is subdivided into four suborders, most important of which are the dorids (Doridacea) and the eolids (Aeolidacea). The two remaining groups, the dendronotids and arminids, are small and poorly represented in local waters. Most nudibranchs in Hawai'i are less than 1 in. long and uncommon to rare. Nevertheless they are popular among divers, who often make a special point of searching out these living jewels of the reef along the underwater cliffs, ledges and caves where they live.

DORIDS. SUBORDER DORIDACEA

Dorids, the largest group of nudibranchs, usually have a circlet of feathery gills on the back surrounding the anus. These exposed gills have given the whole order its name, Nudibranchia, meaning "naked gills." There are typically a pair of sensory tentacles (rhinophores) at the front. Dorid nudibranchs prey on a variety of organisms including sponges, bryozoans, tunicates, bristle worms and other sea slugs.

PITTED NUDIBRANCH

Aldisa pikokai Bertsch & Johnson, 1982
Family Aldisidae
• This tiny orange-red nudibranch resembles a small encrusting sponge, the three pits on its back mimicking excurrent pores. The gills are white. Active only at night, it remains hidden under stones or in crevices during the day, most often between 6 and 30 ft. and sometimes as deep as 80 ft. The species name comes from two Hawaiian words: **piko** ("navel") and **kai** ("sea"). The three pits reminded the zoologists who described this animal of pits pounded into **pāhoehoe** lava in ancient times, into which children's umbilical cords were deposited. A rare and much larger nudibranch of the genus *Sclerodoris* closely resembles this species but has only one pit. To about 1/2 in. Known only from the Hawaiian Islands. Photo: Mākua, O'ahu. 20 ft.

CLUMPY NUDIBRANCH

Asteronotus cespitosus (Hasselt, 1824)
Family Asteronotidae
• This large, shallow-water nudibranch occurs both on rocky substrate and on sand. It occasionally makes its way across the top of the reef in broad daylight. Greenish, gray-brown or yellowish, the surface of its rounded body is studded with irregular warts or bumps. The red Imperial Shrimp *(Periclimenes imperator)* sometimes lives commensally upon it (although it is more common on the Spanish Dancer, *Hexabranchus sanguineus*). The species name means "turf-colored" or "clumpy." To about 5 in. Indo-Pacific. Photo: Hanauma Bay, O`ahu. 3 ft.

GOLD LACE NUDIBRANCH

Halgerda terramtuentis Bertsch & Johnson, 1982
Family Asteronotidae
• Covered with ridges and bumps and criss-crossed with an irregular network of gold lines, this nudibranch's translucent body seems to glow from within. Reminded of a geodesic dome, some call it the "Buckminster Fuller nudibranch." The rhinophores and gills are white, speckled with black. Divers encounter this animal regularly throughout the main Hawaiian Islands, from depths of 15 to at least 100 ft., especially in or around caves. The species name, meaning "looking at the earth with care," honors the Earthwatch program, whose volunteers helped scientists Bertsch and Johnson in their pioneering research. To about 2 in. Known only from the Hawaiian Islands. Photo: Pūpūkea, O`ahu. 25 ft.

PALI NUDIBRANCH

Sclerodoris paliensis Bertsch & Johnson, 1982
Family Asteronotidae
• Deeply scalloped with ridges and depressions, the surface of this large nocturnal nudibranch is reminiscent of the fluted cliffs and valleys of the Hawaiian Islands. Its color is dirty yellow to orange-yellow. Bertsch and Johnson encountered most of their specimens at night at depths between 6 and 20 ft. The species name comes from the Hawaiian word **pali**, meaning "cliff." To about 2 1/2 in. Known only from the Hawaiian Islands. Photo: Scott Johnson.

KANGAROO NUDIBRANCH

Ceratosoma tenue Abraham, 1876
Family Chromodorididae
• Members of the large family Chromodorididae are typically small, soft and brightly colored. They feed primarily on sponges, a single species of nudibranch often preying exclusively on a single species of sponge. Toxins produced by the sponge are absorbed into the nudibranch's body, producing a toxic nudibranch whose bright colors warn away potential predators. This chromodorid, named for its long, heavy tail, is larger than most other Hawaiian species and occasionally hosts the small commensal shrimp *Periclimenes imperator* (p. 224). Orange-gold or brownish orange mottled with white, it has purple spots distributed about its body and one at the tip of each rhinophore. A fleshy projection rises behind the gills; a tail equal in length to the rest of the body trails behind. *Ceratosoma cornigerum* is a synonym. To at least 3 in., probably larger. Indo-Pacific. Photo: Hālona Blowhole, O`ahu. 60 ft.

WHITE-SPOTTED NUDIBRANCH

Chromodoris albopunctata (Garrett, 1879)
Family Chromodorididae
• This slug has a red or orange-yellow back densely speckled with tiny white spots and rings. Around the margin are three bands of solid color, the innermost orange-yellow, the outer bright blue, and the center band a dark mixture of the two. The rhinophores are purple-brown speckled with white. Rarely seen in Hawai`i, it has been found in both silty and clear water at depths between 10 and 30 ft. off O`ahu (at Ala Moana, Magic Island, Ewa and Honolulu Harbor). To about 2 1/2 in., but usually half that size. Tropical Pacific. Photo: Scott Johnson. Enewetak, Marshall Islands.

PURPLE-EDGED NUDIBRANCH

Chromodoris albopustulosa (Pease, 1860)
Family Chromodorididae
• This slug is light yellow with white tubercles; its perimeter is marked with purple spots that may merge into a continuous line. The rhinophores are brown with fine white lines and the gills are whitish. It is found from the shoreline to depths of at least 30 ft. in both exposed and sheltered waters. To about 1 in. Indo-Pacific. Photo: Magic Island boat channel, O`ahu. 10 ft.

PURPLE-SPOTTED NUDIBRANCH

Chromodoris aspersa (Gould, 1852)
Family Chromodorididae

• These white slugs are evenly covered by small, diffuse purple spots. Like many molluscs in Hawai`i, their numbers fluctuate considerably from year to year. During the 1970s, for example, they were among the most common chromodorids on O`ahu, found under rocks at low tide. Today (the late 1990s) they are seldom seen. The species name means "sprinkled." *Chromodoris lilacina* is a synonym. To slightly over 1 in. Indo-Pacific Photo: Scott Johnson. Kewalo, O`ahu.

DECORATED NUDIBRANCH

Chromodoris decora (Pease, 1860)
Family Chromodorididae

• The creamy white body of this slug is rimmed with a bright orange band that contains a scattering of purple dots. A narrow white stripe down the center also contains purple dots. It inhabits quiet harbors and lagoons. Discovered in Hawai`i, the animal is now known also from Japan and Australia. The species name means "decorated," "pleasing," "beautiful." To about 3/4 in. Indo-Pacific. Photo: Magic Island boat channel, O`ahu. 10 ft.

RED-SPOTTED NUDIBRANCH

Chromodoris sp.
Family Chromodorididae
• These medium-size slugs are mottled red and white with scattered red dots, mostly in the center of the back. The dots are ringed with white and occasionally form two loose rows. The margin is white with a gold edge. These animals occur at depths of about 20 ft. or more. Although colorful in photographs, on the reef at their natural depth (where red appears gray brown) they blend in well with their surroundings. To about 1 1/2 in. This species has been identified erroneously as *C. petechialis* but is probably undescribed. Photo: Pūpūkea, O`ahu. 20 ft.

TREMBLING NUDIBRANCH

Chromodoris vibrata (Pease, 1860)
Family Chromodorididae
• This attractive nudibranch's gills continually tremble and vibrate, hence the common and scientific names. (Many other chromodorids behave similarly.) The body is yellow with numerous white spots, some raised. The margin is dark purple; the rhinophores and gills also contain purple. This species occurs throughout the Hawaiian chain from sea level to depths of at least 80 ft. It is usually considered endemic but may also occur in Okinawa. To about 1 1/4 in. Photo: YO-257 wreck off Waikīkī, O`ahu. 80 ft.

SNOW-GODDESS NUDIBRANCH

Glossodoris poliahu Bertsch & Gosliner, 1989
Family Chromodorididae
• This nudibranch's light brown back appears dusted with frost. The frilly margin is white with a yellow edge. It is an uncommon species usually found at depths of 30 ft. or more. Because of the frosted appearance it has been named for the Hawaiian snow goddess **Poli`ahu**, who dwells on the summit of Mauna Kea. In Hawaiian mythology, **Poli`ahu** is sister to the volcano goddess **Pele**. To almost 1 1/2 in. Known only from the Hawaiian Islands. Photo: Pūpūkea, O`ahu. 40 ft.

WHITE-MARGIN NUDIBRANCH

Glossodoris rufomarginata (Bergh, 1890)
Family Chromodorididae
• These are among the most frequently seen nudibranchs in Hawai`i. Clustering in small groups under ledges or on vertical faces, they feed on a common gray sponge *(Cacospongia* sp.). Their white egg coils often lie nearby. The tan dorsal surface is finely speckled with orange-brown. It is surrounded by a conspicuous white margin that in turn has a scarcely visible orange-brown edge. (In some Indo-Pacific localities this edge is reddish and prominent, giving the species its scientific name, "red margin.") The gills and rhinophores are brown and white. Previous publications on Hawaiian nudibranchs have identified this species as *G. youngbleuthi, Chromodoris youngbleuthi* or *Chromolaichma youngbleuthi*, names now regarded as synonyms. To about 1 in., but usually half that size. Indo-Pacific. Photo: Pūpūkea, O`ahu. 30 ft.

TOM SMITH'S NUDIBRANCH

Glossodoris tomsmithi Bertsch & Gosliner, 1989
Family Chromodorididae

• This pretty slug is creamy white with porcelain white dots on the back; it has a ruffled margin bordered in yellow. The rhinophores are blue-black and the feathery circlet of gills shows some black as well. This species has been confused in the past with *C. albonotata*, a similar nudibranch from the South Pacific. It was named for biologist Bertsch's dive buddy, Tom Smith, who had a good eye for nudibranchs and often saved the day for his companion by supplying forgotten dive equipment. To about 1 in. Known from Hawai`i, the Marshall Islands and Okinawa. Photo: Pūpūkea, O`ahu. 15 ft.

ANDERSON'S NUDIBRANCH

Hypselodoris andersoni Bertsch & Gosliner, 1989
Family Chromodorididae

• These tiny slugs occur most often on the yellow sponge *Luffariella metachromia* (p. 5), common under overhangs and in caves. In years of abundance they sometimes mass on their prey like locusts. Their bodies are creamy white rimmed with blue or purple and striped lengthwise with exceedingly fine white lines. Although typically occurring under overhangs along exposed rocky shores at depths of 15 to at least 60 ft., the slugs have also been found in protected harbors and boat channels as shallow as 6 ft. The species was named for Roland Anderson of the Seattle Aquarium, who bid highest for the honor at a fund-raising auction for the Western Society of Malacologists. To 1/2 in., but usually smaller. Known only from the Hawaiian Islands. Photo: Scott Johnson, Mākua, O`ahu, 15 ft.

PAINTED NUDIBRANCH

Hypselodoris infucata (Rüppell & Leuckart, 1828)
Family Chromodorididae
• This nudibranch occurs primarily in silty, protected waters such as those of Pearl Harbor, the boat channel at Magic Island, or Kāne`ohe Bay, O`ahu. It varies from almost white to dusky gray, peppered with small, smudgy dark spots and even smaller yellow ones. The rhinophores and gills are predominantly orange-red. The species is presently recorded from East Africa and the Red Sea to Hawai`i and has entered the Mediterranean via the Suez Canal. To about 1 in. Photo: Mike Severns. Mokoli`i Islet (Chinaman's Hat), O`ahu. 1 ft.

MAGENTA-STRIPED NUDIBRANCH

Hypselodoris maridadalus Rudman, 1977.
Family Chromodorididae
• Nudibranchs of the genus *Hypselodoris* are often marked with longitudinal lines; this slug is a good example. Parallel magenta stripes run the length of its delicate white body. The rhinophores and gills are orange. The species was described from East African specimens and given the Swahili name "maridadi," meaning "beautiful." In Hawai`i it is known primarily from the boat channel at Magic Island, O`ahu, where it is occasionally seen by day crawling in the open. It ranges from East Africa to Hawai`i. To about 1 in. Photo: Magic Island boat channel, O`ahu. 15 ft.

DOT-AND-DASH NUDIBRANCH

Hypselodoris sp. 1

Family Chromodorididae

• This undescribed species of *Hypselodoris* has been incorrectly identified as *H. lineata* in several previous publications. It is light in color, marked with dark purple longitudinal streaks and spots. Between these "dots and dashes" meander fine longitudinal lines of bright frosty white. The rhinophores are bisected by an orange band and the gills show spots of orange. To about 1 in., but usually smaller. Known only from the Hawaiian Islands. Photo: Magic Island boat channel, O`ahu. 10 ft.

WHITE DOT NUDIBRANCH

Hypselodoris sp. 2

Family Chromodorididae

• This small, uncommon nudibranch is undescribed. Like many of its genus, it is marked with parallel longitudinal lines. The rhinophores are orange-red banded with white, the gills orange-red. Bright white dots front and rear give it its informal name. It is usually seen at depths of about 15 to 75 ft. To about 1/2 in. Known from the Hawaiian Islands as far north as Kure Atoll. Photo: Scott Johnson. Magic Island, O`ahu. 20 ft.

RARE HYPSELODORIS

Hypselodoris sp. 3

Family Chromodorididae

• Only four specimens of this flamboyant nudibranch have been recorded. It is creamy white with red lines and spots. The mantle is rimmed with gold; a purple line encircles the foot. The rhinophores are red, the gills white edged with red. All known specimens have come from depths of 75 ft. or more. It and other undescribed Hawaiian species of *Hypselodoris* are under study by Dr. Terrence Gosliner of the California Academy of Sciences. To about 1 1/2 in. Known only from Hawai`i. Photo: Scott Johnson. Off the reef runway, Honolulu Airport. 90 ft. (See also p. 157.)

JOLLY GREEN GIANT

Miamira sinuata (Hasselt, 1824)
Family Chromodorididae
• This unusual slug has a scalloped mantle with green, yellow and blue markings. First pictured under this catchy common name by Bertsch and Johnson in 1981, it has become a kind of Holy Grail among nudibranch hunters in Hawai`i because of its rarity. Despite the name, it is not particularly large. To about 1 1/2 in. Indo-Pacific. Photo: Scott Johnson. Waialua, O`ahu.

IMPERIAL NUDIBRANCH

Risbecia imperialis (Pease, 1860)
Family Chromodorididae
• These nudibranchs were abundant in Hawai`i in the 1970s, all but disappeared in the 1980s, and began to come back in the mid-1990s. Their white bodies, edged in dark purple, are covered with gold spots of varying sizes. The purple margin expands into irregular spots along the sides, encompassing one or more of the gold spots. The rhinophores are black with white flecks, the gills white edged with black. These slugs commonly occur in pairs and are active during the day. The genus name honors French zoologist Jean Risbec (1895-1964), who pursued dual careers in entomology and malacology. To about 2 in. Western Pacific. Photo: Hālona Blowhole, O`ahu. 70 ft. (See also p. 95.)

DANIELLE'S THORUNNA

Thorunna daniellae (Kay & Young, 1969)
Family Chromodorididae
• This small nudibranch has a slender white body rimmed with bright purple. Its rhinophores and gills are orange-red, and the gills tremble continuously as it crawls along. It is found in the open from shallow water to at least 80 ft. (most common at the deeper end of its range). The species is named for American malacologist Danielle Fellows. An undescribed species of *Thorunna* (below) has almost the same color pattern. Mimicry might be involved, or the pattern itself could be advantageous for reasons not yet understood. To about 1/2 in. Central and Western Pacific. Photo: Scott Johnson. Magic Island boat channel, O`ahu. 15 ft.

PINK-TINGED THORUNNA

Thorunna sp.
Family Chromodorididae
• Externally, this undescribed nudibranch almost exactly resembles *Thorunna daniellae* (above). Its back has a pinkish tinge, however, whereas that of *T. daniellae* is snow white. Microscopic examination of their respective radular teeth reveals significant differences. The habitats of the two overlap and this pinkish species is most common at the shallow end of the range. To about 1/2 in. Photo: Magic Island boat channel, O`ahu. 15 ft.

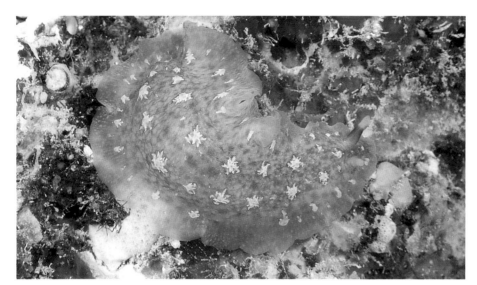

WHITE AND BROWN DENDRODORIS

Dendrodoris albobrunnea Allan, 1933
Family Dendrodorididae
• Nudibranchs of the family Dendrodorididae possess a long sucking tube with which they feed on sponges; there is no radula. Some are quite large. This species is of moderate size, usually brownish or grayish with a double row of almost star-like white blotches along its back and a ring of smaller white blotches at the edges. Outside Hawai`i it can be whitish with brown blotches. (These may be two distinct species.) It occurs from the shallows to at least 50 ft. To about 3 in. Indo-Pacific. Photo: Kea`au Beach Park, O`ahu. 40 ft.

BLACK DENDRODORIS

Dendrodoris nigra (Stimpson, 1856)
Family Dendrodorididae
• Shellers and tide-poolers who turn over rocks will occasionally find this slug, often with its bright yellow egg coils near-by. The soft, slimy body is black in adults, orange in juveniles. Some adults retain flecks of orange. To about 3 in., but usually smaller. This is a common sea slug throughout the Indo-Pacific. Photo: Kīpapa Island, O`ahu. 2 ft.

TUBERCULOUS NUDIBRANCH

Dendrodoris tuberculosa (Quoy & Gaimaird, 1832)
Family Dendrodorididae
• This amazing slug resembles a lady's flower hat of the last century. Rosette-like nodules in various subtle hues cover its rounded back. Sometimes, however, the color is almost uniform. This animal occurs on exposed rocky, sponge-covered shores such as those at Pūpūkea and Mākaha, Oʻahu, and is most active at night. It is reported to secrete a substance irritating to human eyes. To about 6 in. Indo-Pacific. Photo: Mākua, Oʻahu. 10 ft.

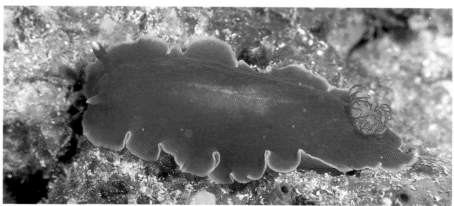

RED DENDRODORIS

Dendrodoris rubra (Kelaart, 1858)
Family Dendrodorididae
• This is a large red or brownish red dorid with a yellow margin, white-tipped rhinophores and whitish gills. It appears to be active both by day and night. In the Indian and Western Pacific oceans it is described as having black, gray or dark red blotches and no yellow margin. Specimens from Hawaiʻi and the Marshall Islands have a yellow margin but lack the blotches. Regional variation may account for this, or the species could be new. To about 2 in. Indo-Pacific. Photo: Waimea Bay, Oʻahu. 50 ft.

FELLOW'S NUDIBRANCH
Peltodoris fellowsi Kay & Young, 1969
Family Discodorididae
• Snow white with jet black gills and rhinophores, this slug is fairly common and easy to spot. Its firm body is slightly granular to the touch. Divers find it regularly in or at the entrances of caves at depths of 20 to 50 ft., usually on sponges. It was named for biologist David Fellows, who collected the first scientific specimen. A smaller, less common endemic, *Ardeadoris scottjohnsoni*, has similar coloration but is soft and smooth to the touch. To about 2 in. Endemic. Photo: Pūpūkea, O`ahu. 30 ft.

Spanish Dancer (a)

SPANISH DANCER
Hexabranchus sanguineus (Rüppell & Leuckart, 1831)
Family Hexabranchidae
• Spanish Dancers are the largest, most conspicuous nudibranchs in Hawai`i (and probably the world). At night they are often a blotchy pinkish red, like fatty ground beef; by day they are usually pink to deep crimson, frequently with yellow or whitish markings. The rhinophores and gills may have yellow or gold highlights. Usually furled out of sight, the mantle's margin is rimmed with white or yellow. If touched, the animal flares out in a colorful display, dramatically increasing its apparent size. These slugs can swim by flexing the body and undulating the outspread mantle (hence their common name). Beachgoers have seen them swimming near the surface in the middle of the day. Their conspicuous egg coils resemble pink or red roses attached to the substrate and contain the same sponge-derived poisons as the nudibranchs themselves. At least three similar species of *Hexabranchus* have been proposed; only *H. sanguineus* is now recognized. The small red Imperial Shrimp, *Periclimenes imperator* (p. 224), sometimes lives commensally on its body surface. The species name means "bloody." To at least 15 in., but usually smaller in Hawai`i. Indo-Pacific and tropical Atlantic. Photos: a) David R. Schrichte. Pūpūkea, O`ahu. (with eggs); b) Pūpūkea, O`ahu. 20 ft. (nighttime color) (See also p. 178.)

◄ Spanish Dancer (b)

VARICOSE PHYLLIDIA
Phyllidia varicosa Lamarck, 1801
Family Phyllidiidae
• The phyllidiids are a group of firm, tough-bodied dorids in which the usual circlet of gills around the anus are replaced by leaflike secondary gills under the mantle skirt. Their backs are covered with hard bumps and ridges, often brightly colored. Typically poisonous, they have no known predators and are probably the most frequently-seen tropical Indo-Pacific nudibranchs. This species, the most common and conspicuous member of the family in Hawai`i, has a black body covered with pale blue ridges upon which are rows of orange- or yellow-tipped bumps. It is sometimes called the "scrambled egg nudibranch." When disturbed it secretes a toxic and very pungent mucus. (Do not place this, or any other phyllidiid in your aquarium!) Specimens from O`ahu, and their prey sponges *Hymeniacidon* sp., contain a chemical poisonous to fish and crustaceans. Scientists named it 9-isocyanopupukeane after the dive site, Pūpūkea, on O`ahu's north shore. A very similar species of *Phyllidia* from Hawaiian waters remains unnamed. To at least 3 1/2 in. (but most are half that size). Indo-Pacific. Photo: David R. Schrichte. Pūpūkea, O`ahu.

POLKA-DOT PHYLLIDIA

Phyllidia polkadotsa
Brunckhorst, 1993
Family Phyllidiidae
• Conspicuous but rare, this nudibranch is bright yellow with about 10 round black dots. Ridges run along the back. It occurs on sponge-encrusted walls to depths of at least 75 ft. (A similar and equally rare Hawaiian species, *Phyllidia scottjohnsoni*, is white with black spots.) To about 1 in. At present known only from Hawai`i, the Banda Islands, Indonesia, and Okinawa. Photo: "Chromis Corridor" Midway Atoll, 50 ft.

PUSTULOSE PHYLLIDIA

Phyllidiella pustulosa (Cuvier, 1804)
Family Phyllidiidae
• The black body of this species is studded with pinkish nodules, which may appear green underwater. The second most common phyllidiid in Hawai`i, it sometimes occurs in areas of heavy coral growth and can usually be found along the coral-encrusted sides of the hot water outlet at Kahe Point, O`ahu. Typical of its family, it feeds on poisonous sponges, concentrating the toxin in its tissues as a defense against predation. The bright colors are believed to serve as a warning; as evidence of this, a dorid nudibranch and a flatworm that mimic this nudibranch occur in Papua New Guinea. To about 2 in. Indo-Pacific. Photo: Kahe Point, O`ahu. 20 ft.

ROSY PHYLLIDIA

Phyllidiella rosans (Bergh, 1873)
Family Phyllidiidae
• Long, rounded pink ridges on a black body identify this species. The ridges are smooth. The edge of the mantle is pink and the rhinophores are black. Because of its color this nudibranch somewhat resembles *P. pustulosa* (above), but it is smaller and less common. Like most members of its family, it is active both day and night. To almost 1 1/2 in. Indo-Pacific. Photo: Pūpūkea, O`ahu. 30 ft.

SPHINX PHYLLIDIA

Phyllidiopsis sphingis Brunckhorst, 1993
Family Phyllidiidae
• This striking nudibranch has a ground color of white, becoming vivid blue toward the margins. Black longitudinal stripes, a U-shape black band and black rays and spots complete the picture of this uncommon creature. It is presently known from Papua New Guinea, Guam and Hawai`i and will probably be found in other Indo-Pacific locations. Concerning the species name, Brunckhorst writes "Phyllidiids with their lumps, bumps, ridges and bright colours often give the image of tiny monsters; hence...the fun name derived from the Sphinx—the mythical monster of Thebes who posed riddles for people passing by and consumed them if they could not answer!" To about 1 in. Photo: Waimea Bay, O`ahu. 30 ft.

SPECKLED PLATYDORIS

Platydoris formosa (Alder & Hancock, 1866)
Family Platydorididae
• These large white dorids are unevenly speckled with fine brown and slightly larger red spots, producing a patchy appearance. Their stiff, flat bodies have a grainy texture. The rhinophores are bright red and surrounded at the base by a fleshy sheath. These uncommon slugs are seen at night on shallow reef flats and down to at least 30 ft. The species name means "beautiful" and at close range these animals are quite attractive. To about 3 in. Indo-Pacific. Photo: Waimea Bay, O`ahu. 20 ft.

Tambja morosa (Bergh, 1877)
Family Polyceridae
• These wrinkly skinned, dark olive-green nudibranchs appear almost black underwater. A bright blue margin, a blue head band, and about five round blue spots down the back (sometimes fused) add color and contrast. A large circle of branched gills sprouts at the center of the back. These slugs occur both in clear water and in silty harbors, probably feeding on bryozoans or tunicates. The species name, meaning "gloomy," most likely refers to the animal's dark color. To about 3 in. Indo-Pacific. Photo: Magic Island boat channel, O'ahu. 15 ft. (See also p. 93.)

DENDRONOTIDS. SUBORDER DENDRONOTACEA

Dendronotids are a small group of nudibranchs which usually have an oral hood and a series of paired protuberances (cerata) running down the back.

THROW-NET NUDIBRANCH

Melibe pilosa Pease, 1860
Family Tethyidae
• Like other members of its genus, this nudibranch is a carnivore, preying on small crustaceans captured in a large sac or "oral hood." In an action comparable to the casting of a throw net, the slug continuously expands and contracts its hood as it crawls along. While water is being forced out during contraction, fringing tentacles trap tiny crustaceans that are then swallowed whole. These partially transparent slugs, with their cryptic coloration and paddle-like cerata resembling tufts of algae, are almost impossible to see in the wild, but their feeding method is easily observed in the aquarium if brine shrimp are provided. Found on shallow reef flats among seaweeds, they can swim by flexing side to side. They readily shed their cerata if disturbed. A smaller, quite spectacular Hawaiian species, *Melibe megaceras*, occurs in the same habitat. To almost 5 in. Indo-Pacific. Photo: Ala Moana Beach Park, O'ahu. 2 ft.

EEL NUDIBRANCH

Bornella anguilla Johnson, 1985
Family Bornellidae
• When disturbed, this remarkable nudibranch swims with fluid side-to-side undulations like a tiny eel. It can sustain this motion for many minutes. The long, paddle-like extensions on the back shield the animal's gills and its pattern of pale-edged, yellow-brown spots somewhat resembles that of a giraffe. The species name means "eel." An uncommon Indo-Pacific slug, it is extremely rare in Hawai`i. This one was about 1.5 in. long. Photo: South Point, Hawai`i. 20 ft.

EOLIDS. SUBORDER AEOLIDACEA

Eolid nudibranchs lack gills. They are typically long, slender and "frilly," almost always bearing numerous pairs of tentacle-like protuberances (cerata) along the body. The head bears both rhinophores and cephalic tentacles. These slugs usually feed on hydroids, corals or other stinging animals, and many species incorporate the stinging capsules (nematocysts) of their prey for protection. Some also incorporate symbiotic algae (zooxanthellae) from their prey, thereby augmenting their food supply. A few eolids feed on barnacles or the eggs of other molluscs.

INDIAN NUDIBRANCH

Caloria indica (Bergh, 1896)
Family Facelinidae
• Like many eolids, this delicate-looking species is armed with stinging cells at the tips of its cerata, obtained second hand from the common hydroid *Pennaria disticha* (p. 20) upon which it feeds. The sting deters small predators but would probably not penetrate human skin. Although common in Hawai`i, this usually tiny nudibranch often escapes notice by divers and snorkelers. Its colors are somewhat variable but the cerata are always white at the tips followed by blue, then red from center to base. To 1 1/2 in., but most are far smaller. Indo-Pacific. Older books may list it as *Caloria militaris*, a non-Hawaiian species with which it has been confused. A similar eolid common in the Islands is *Facelinella semidecora*. Photo: Kahe Point, O`ahu. 20 ft.

177

Egg-eating Nudibranch (a)

Eggs of Egg-eating Nudibranch (b)

Favorinus japonicus Baba, 1949
Family Facelinidae
• These well-camouflaged eolids prey exclusively on the eggs of other sea slugs. They are most easily observed on the large red or pink coiled ribbons of the Spanish Dancer (p. 172). Gently spread the rosette of eggs to find them. Like worms in the heart of a rose, they devour the flower from the inside. Often their white egg coils can be seen as well— eggs on eggs. The animals are pink, whitish or greenish, depending on the color of the eggs they consume. Nudibranch eggs are often poisonous and these predators probably incorporate the toxins for their own defense. They have no sting. To about 1/2 in. Known from Japan, Guam and Hawai'i. Photo: a) Pūpūkea, O`ahu. 30 ft. b) Lehua Rock, Ni`ihau. 30 ft. (eggs)

BICOLOR NUDIBRANCH

Flabellina bicolor (Kelaart, 1858)
Family Flabellinidae
• This eolid nudibranch has 4 to 8 pairs of cerata, each dividing at the base into 2 or 3 long branches. The cerata are white, each branch bearing an orange ring. The slender body is white or purple. This animal is found in the open by day (sometimes by the dozens) from the intertidal zone to depths of at least 20 ft., apparently feeding on hydroids. To about 1/2 in., or slightly larger. Indo-Pacific. Photo: Scott Johnson. Kewalo, O`ahu.

MAN-OF-WAR NUDIBRANCH

Glaucus atlanticus Forster, 1777
Family Glaucidae

• This unusual predator drifts the oceans of the world in search of Portuguese Men-of-War *(Physalia physalis)*, By-the-Wind Sailors *(Velella velella)* and other stinging hydrozoans (p. 23-24). Bluish purple, much the same color as its prey, it hangs upside down on the surface and maneuvers into attack position by "rowing" with its lateral lobes. Like many other eolids it incorporates the powerful stinging capsules of its prey for its own defense. It occurs worldwide in tropical seas and is most often seen when blown ashore, along with its prey. To about 1 1/2 in., but most are half this size. Photo. Scott Johnson. Windward O`ahu.

BLUE DRAGON NUDIBRANCH

Pteraeolidia ianthina (Angas, 1864)
Family Pteraeolidiidae

• Like long frilly Chinese dragons, these nudibranchs (often several inches in length) wind their way over the substrate, fully in the open. Varying greatly in color (blue, purple, greenish or tan), they are sometimes conspicuous, sometimes hard to see. The pair of tentacles at the front are always encircled with two purple bands. Blue Dragon Nudibranchs feed on hydroids but are also sustained by microscopic algae (zooxanthellae) living in their tissues, in the manner of many corals and anemones. Differing kinds and numbers of zooxanthellae probably account for the wide variation in color. Because of the nutrients supplied by zooxanthellae, these attractive nudibranchs might survive for some time in a bright aquarium. This is an abundant species in Hawai`i. Its cerata carry stinging cells sometimes powerful enough to be felt by humans. The species name means "violet blue." The largest recorded was almost 6 in. long; most are half that size. Indo-Pacific. Photo: Lāna`i Lookout, O`ahu. 20 ft.

ANEMONE-EATING NUDIBRANCH

Berghia major (Eliot, 1903)
Family Aeolidiidae
• These slugs feed both on the Swimming Anemone, *Boloceroides mcmurrichi* (p 36), which they resemble, and on the common Glass Anemone, *Aiptasia pulchella* (p. 41). The latter, often a pest in home aquariums, can be controlled by these slugs (which are sometimes raised in captivity for that purpose). The slugs live on protected shallow reef flats, often among algae, where their dull colors make them hard to see. To almost 4 in. Indo-Pacific and Eastern Pacific. Photo: Scott Johnson. Kewalo, O`ahu. (with eggs)

CUP CORAL NUDIBRANCH

Phestilla melanobrachia Bergh, 1874
Family Tergipedidae
• This eolid feeds on cup corals of the family Dendrophylliidae. In Hawai`i this is usually the Colonial Cup Coral, *Tubastraea coccinea*, (p. 65) which occurs under ledges and in caves, especially in areas with some current. Active only at night, the slug matches its prey in color and form and is difficult to see. One way to find it is to look for the white skeletons of recently eaten polyps. The slug itself is often slightly darker and more intensely colored than the living polyps. In Australia these animals also feed on a species of dark green cup coral. Experimentally moved from orange to dark coral, they change color in about a week. The first specimens collected were probably dark, hence the species name meaning "dark arms." To about 1 1/2 in. Indo-Pacific. (The similar *P. lugubris* feeds on corals of the genus *Porites*. Although rarely noticed in the field, it is easy to culture and has been the subject of considerable research at the University of Hawai`i.) Photo: Pūpūkea, O`ahu. 20 ft.

BIVALVES

CLASS BIVALVIA (OR PELECYPODA)

A Ventricose Ark *(Arca ventricosa)* nestles in a rocky cliff in Waimea Bay, O`ahu. The small goby is *Priolepis aureoviridis.*

Bivalves are molluscs with two opposing shells. Clams, mussels and oysters typify this class. Hinged by strong ligaments at one end, the shells are connected on the inside by a pair of muscles. Examine an empty bivalve shell and you can often see marks where the muscles once attached. A thin layer of tissue called the periostracum may cover the outer sides of the shells. The cavity between the shells contains the mantle, gills, and other organs. There is no head.

Bivalves usually lie buried in sand or mud or attach to hard surfaces such as rocks or pilings, sometimes by filaments called byssal threads. Some bore into soft stone. The "shipworms" (bivalve molluscs, not worms) drill into wood. A bladelike foot enables the burrowers to move about; some can dig surprisingly quickly. A few of the attached (sessile) bivalves are capable of relaxing their hold and creeping about to a limited extent. Scallops and file shells can actually swim by clapping their two valves together.

Most bivalves feed by filtering water through their gills (used both for feeding and respiration). A layer of mucus traps suspended particles, which are passed to the mouth and consumed. Buried bivalves draw water into their gill chamber through a fleshy siphon. A few large tropical clams, exposed to bright sunlight on shallow Indo-Pacific reefs, gain partial nutrition from the photosynthetic activity of microscopic algae (zooxanthellae) living in their mantles and siphons.

Bivalves range in size from the 700 pound, 4-foot-wide giant clam *Tridacna gigas* (in Hawai`i found only at the Waikīkī Aquarium) to tiny animals 1/25 in. wide. Altogether, about 7,700 species are known, with 129 from local waters (66 endemic). Several edible oysters and clams from North America and Japan have been introduced successfully to Hawai`i and occur in limited numbers in shallow sheltered areas such as Pearl Harbor, Maunalua Bay and Kāne`ohe Bay, O`ahu. Gathering the Japanese Littleneck Clam *(Venerupis philippinarum)* was a popular annual event on mud flats in Kāne'ohe Bay until over-harvesting and heavy siltation nearly wiped out the beds after the 1969 season. The species had been introduced in 1920.

Although bivalve shells are abundant on many of the world's beaches, they are uncommon on Hawai'i's shores (some tiny species excepted). In old Hawai'i, bivalves were known in general as `ōlepe (although in some places this word specifically denoted clams). Only a few large species, such as pearl oysters and arks, were popular as food. Shells with a pearly inner lining were called **paua** if thick and heavy or **pā** if light.

ARK SHELLS. FAMILY ARCIDAE

These bivalves have solid ribbed shells that bear numerous fine teeth along the hinge line. Some burrow in soft bottoms, others attach to rocks with strong byssal threads. Hawai'i has at least 12 species.

VENTRICOSE ARK SHELL
• `ōlepe-pāpaua

Arca ventricosa Lamarck, 1819
Family Arcidae
• These odd bivalves lie "backwards" on rocky faces, their long straight hinges facing out and almost flush with the cliff. Facing inward and nestling in a cavity, the shells appear almost heart-shape to the viewer. These animals attach to the rock by strong byssal threads that pass through a large gap between the shells on the inward-facing side. When feeding, they tilt outward and open slightly, enabling them to draw in and expel water through the siphon; when approached, they clamp down firmly. The shells are usually covered with sponges and other growths, sometimes of contrasting colors. The first of these bivalves to be scientifically described was aptly named *Arca noae*—Noah's Ark. It is not found in Hawai'i. Of the 12 species that do occur here, this is the largest and most often seen. In old Hawai'i it was eaten. To about 3 in. Indo-Pacific. Photo: "Ewa Pinnacles," O'ahu. 80 ft. Sponges of different colors grow on each shell. (See also p. 181.)

MUSSELS. FAMILY MYTILIDAE

Typical mussels are dark colored intertidal bivalves that attach in great numbers to exposed rocky shores with strong byssal threads. Some bore into solid rock or coral, however, and others live commensally in the tests of tunicates. Hawai'i has about a dozen species.

HAWAIIAN MUSSEL • **nahawele li`i li`i, kio-nahawele**

Brachidontes crebristriatus (Conrad, 1837)
Family Mytilidae
• This mussel is known only from Hawai`i. Dense patches are common on some flat, wave-washed limestone shores at the low tide mark, usually half buried. The largest specimens grow along shores where fresh water mingles with salt. Individual mussels are usually from 1/4 to 1/2 in. long but can attain 1 in. or more in brackish environments. Endemic. Photo: Moku`auia (Goat Island), O`ahu.

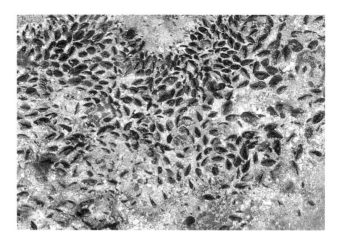

PEN SHELLS. FAMILY PINNIDAE

Pen shells are triangular, usually winglike or fanlike in shape. Most species lie partly buried in soft substrate, pointed end down, wide end just protruding. The shells themselves are thin, brittle and easily damaged. The animals often anchor themselves to buried rubble with byssal threads.

PRICKLY PEN SHELL

Pinna muricata Linnaeus, 1758
Family Pinnidae
• Vast beds of these bivalves used to exist off the leeward coasts of O`ahu at a depth of 80-100 ft., creating habitat for an assemblage of other small animals. **Kona** storms in 1980 damaged the beds, and Hurricane Iwa in 1982 finished them off; they have never recovered. Similar concentrations may survive elsewhere in the Hawaiian chain. A pair of commensal shrimps *(Conchodytes tridacnae)* sometimes inhabits living pen shells. To about 9 in. long. Indo-Pacific. Photo: Ed Robinson. Maui.

BAGGY PEN SHELL • **nahawele**

Streptopinna saccata (Linnaeus, 1758)
Family Pinnidae
• This bivalve's unusual shells are fused at the sides and open at the top; they cannot close. Common but seldom noticed, the animal occurs in crevices from shallow water to depths of at least 50 ft. The exposed ends of large specimens are often bent and distorted. To about 9 in. Indo-Pacific. Photo: Puakō, Hawai`i. 50 ft.

PEARL OYSTERS. FAMILY PTERIIDAE

Pearl oysters attach to the substrate with byssal threads and are thus not true oysters, which always have one shell cemented to a solid surface. Their shells have a pearly interior, but only a few species produce valuable pearls. Three species inhabit Hawaiian waters. Most important in ancient times was a small oyster called **pipi** (*Pinctada radiata*). Its pure white pearly shell was much prized for making fish hooks, the lustrous sheen of the shell itself serving as a lure. The pearls sometimes found in it were considered useless. Chants and stories related that **pipi** had to be gathered in total silence. If not, the wind might come up, spoiling the visibility, or the oysters might drop off their rocks into deep water. **Pipi** require some fresh water to survive and were once abundant near the mouth of the Pearl River in Pearl Harbor, O`ahu (known in old times as Pu`uloa). When King Kamehameha I learned that pearls were valuable to Europeans, he put a **kapu** on the oysters, intending to reserve this source of income for himself. His divers soon over-harvested the beds and they never recovered. Today the **pipi** is rarely seen.

BLACK-LIPPED PEARL OYSTER • **pā**

Pinctada margaritifera (Linnaeus, 1758)
Family Pteriidae
• This is the oyster cultured in the South Pacific to produce the expensive Tahitian black pearl. Pearls are extremely rare in wild oysters, however, and the animals are not abundant enough in Hawai`i to support a fishery. The lagoon at Pearl and Hermes Reef, Northwestern Hawaiian Islands, once held extensive beds of these oysters (some weighing up to 15 pounds when alive); they were destroyed by over-harvesting in the 1930s and have never recovered. The taking of this species anywhere in Hawai`i is now illegal. Commercial attempts to culture it offer new hope: cultured larvae have been released around the Islands in an attempt to bring the oysters back. Two commensals, a shrimp *(Conchodytes meleagris)* and a pearl fish *(Onuxodon fowleri)* may occupy the cavity between the shells. The squat, rounded shrimps can sometimes be seen through the shell's opening with an underwater light. To almost 1 foot in width. Indo-Pacific and Eastern Pacific. Photo: Kea`au Ledge, O`ahu. 50 ft.

Top picture on next page ➤

◄ Black-Lipped Pearl Oyster

WINGED PEARL OYSTER

Pteria brunnea (Pease, 1863)
Family Pteriidae
• Both family and genus names of this species mean "winged," referring to the pointed, winglike extension on the shells of many of their members. These cluster on black coral trees (*Antipathes* spp.) and occasionally on wire corals (*Cirrhipathes* and *Stichopathes* spp.), thereby increasing their exposure to plankton-bearing currents. They have occasionally been found on the wire mesh of old fish traps. Attachment is by byssal threads. The "wings" are usually oriented parallel to branches growing from the stem to which the oyster is attached. Bryozoans and other growths typically live on the shells. This species may be endemic to Hawai'i; similar ones occur elsewhere. Photo: "Black Coral Arch," east shore, Kaua'i. 80 ft. (on *Antipathes dichotoma*) (See also p. 70.)

PURSE SHELLS. FAMILY ISOGNOMONIDAE

Closely related to pearl oysters, purse shells are thin, flat and often irregular, with a flaky exterior and a pearly interior. They attach to rocks (usually in shallow water) by means of byssal threads, and typically occur in dense clusters. They are sometimes called "toothed pearl shells" because of the flattened teeth on the hinge line.

BLACK PURSE SHELLS
• **nahawele, pāpaua**

Isognomon californicum
(Conrad, 1837)
Family Isognomonidae
• These bivalves cluster in crevices around the high tide line, often in brackish environments and most abundantly on Maui and the Big Island. The dark massed shells are reminiscent of some temperate-water mussels. The species name *californicum* was applied in error; these animals do not occur in California and are likely endemic to Hawai`i. They attain a height of about 1 1/2 in. Photo: Keauhou Landing, Hawaii Volcanoes National Park, Hawai`i.

BROWN PURSE SHELLS • **nahawele, pāpaua**

Isognomon perna Linnaeus, 1767
Family Isognomonidae
• These flat scaly bivalves are common under rocks in shallow water. They are buff with radiating brown marks and have a pearly inner lining. The shells are of unequal size, the smaller usually covered by the larger. A white polyclad flatworm, *Pericelis hymanae* (p. 74), is often found under the same rocks as these bivalves and may be associated with them. The species name means "mussel." To about 2 in. Indo-Pacific. Photo: Pūpūkea, O`ahu. (tide pool)

SCALLOPS AND FILE SHELLS. FAMILIES PECTINIDAE AND LIMIDAE

Scallops (or pectens) are free-living bivalves that lie on the substrate always with the right shell down, often loosely attached by byssal threads. A row of tiny eyes fringes the mantle, each with lens, retina and optic nerve. Most scallops can swim by clapping their shells together. At least ten species occur in Hawai`i, most rare and small.

File shells are similar, living under rocks and boulders, loosely attached by byssal threads. (Some species actually construct nests of these threads.) They are characterized by thick fleshy bodies sandwiched between papery white shells that cannot close completely. Numerous tentacles protrude between the shells. Like scallops, file shells will swim when disturbed. Five species are known from Hawai`i.

JUDD'S SCALLOP

Haumea juddi Dall, Bartsch & Rehder, 1938
Family Pectinidae
• These animals are rarely seen; localized "runs" are said to occur about every 15 years, sometimes lasting several weeks. About the size of a penny, the scallops lie lightly covered with sand at depths of 30 ft. or more, their open shells facing the current. If disturbed they swim a short distance and settle again. Scallop fancier John Earle describes the animal as "tan and cream, with long tentacles extruded while feeding and more bright blue eyes than a classroom of Swedes." After storms, their beautifully ribbed shells sometimes pile up on beaches. To about 1 in. Endemic. Photo: John L. Earle. Kahe Point, O`ahu. 60 ft.

FRAGILE FILE SHELL

Lima fragilis Chemnitz, 1784
Family Limidae
• This is the most spectacular of Hawai`i's file shells because of its color and size. Body and tentacles are fire-engine red. The tentacles, extended to confuse predators, are sticky and detach readily (hence the common and scientific names). As a last resort, the animal is capable of swimming jerkily away, flapping its valves together much like a scallop. In the aquarium trade these and similar animals are known as "flame scallops." If you acquire one, touch the tentacles as little as possible—they will come off in your hand. Position the animal in a crevice near the front of the aquarium where with luck it will attach loosely with its byssal threads. File shells, like most bivalves, are filter feeders. To about 1 1/4 in. long, excluding the tentacles. Indo-Pacific. Photo: Koko Head, O`ahu. 45 ft.

THORNY OYSTERS. FAMILY SPONDYLIDAE

The outer shells of these bivalves are ribbed and in many species covered with long spines and projections. These may be white, yellow, orange, red or violet. Because they cement one shell to the substrate they are usually called "thorny oysters" or "spiny oysters." However, they are more closely allied to scallops and file shells. Like scallops, they often have colorful mantles bearing many small eyes at the margin, and as juveniles they are free-living. "Rock scallops" might be a better name. There are at least four species in Hawai'i.

SPINY OYSTER

Spondylus nicobaricus Schreibers, 1793
Family Spondylidae
• The long fragile spines of this and similar species may serve to keep gastropod predators, such as murexes, from drilling into the main shell. They also form good points of attachment for tunicates, bryozoans and sponges which camouflage it. This species (*S. hystrix* is a synonym) occurs from 3 to 400 ft. The similar but often more colorful *S. linguaefelis* is known from depths of about 100 ft. or more. Both prefer well protected holes and caves. Naturalist Mike Severns reports the most common colors of the latter to be orange, white and wine, with occasional color combinations in a single shell. Maximum length about 4 in. Photo: Lāna'i Lookout, O'ahu. 20 ft.

CLIFF OYSTER • `okupe

Spondylus violacescens Lamarck, 1819
Family Spondylidae
• Cliff faces and the undersides of overhangs are the habitat of this heavy bivalve. One shell is cemented firmly to the substrate; the second opens outward. Encrusted with annelid worm tubes, sponges and other growths, they resemble lumpy projections of the stone itself. Divers usually notice the shells only when they close. (The colorful mantle can sometimes be glimpsed before they snap shut.) If the animal dies, the outer shell falls off revealing the white interior and attractive purple or yellow-brown margin of the attached one. The thick shells with their encrustations can weigh well over 2 pounds and grow to a diameter of over 4 in. Ribs are visible only on small specimens. At high tide during a full moon these bivalves sometimes spawn, opening and closing like bellows to pump out a milky fluid containing eggs or sperm. Indo-Pacific. *S. tenebrosus* is a synonym. Photos: a) "Fantasy Reef," off Kāhala, O'ahu. 45 ft.; b) Hālona Blowhole, O'ahu. 30 ft.

Top picture on next page ➤

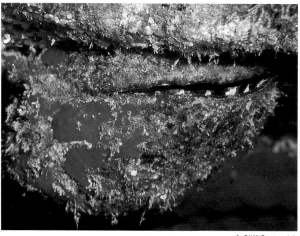 

◄ Cliff Oyster (a)   ◄ Cliff Oyster (b)

TRUE OYSTERS. FAMILY OSTREIDAE

True oysters always have one shell firmly cemented to the substrate. Both shells are typically thick and irregular, the attached shell usually larger. The second shell seats tightly to the first, sealing in the animal and protecting intertidal species from drying out at low tide. Six species occur in Hawai`i, several from deep water. Two of the shallow-water species, the American or Blue-Point Oyster *(Crassostrea virginica)* and the Japanese Oyster *(C. gigas)*, are introductions. A number of fossil oysters also occur, including a massive extinct species, *Pyncnodonta kamehameha*, found in ancient upraised reefs on O`ahu.

JAPANESE OYSTER

Crassostrea gigas (Thunberg, 1793)
Family Ostreidae
• Introduced from Japan to Hawai`i in 1926, this oyster has become especially abundant in Kāne`ohe Bay, O`ahu, on pilings, walls and other solid surfaces between the high and low tide marks. Its gray-white shell is long, narrow and somewhat irregular. Introduced also to the west coast of North America, the species now occurs from Canada to California. It is a major food source in both Japan and the U.S. The species name means "giant." To about 13 in. Photo: Moku o Lo`e (Coconut Island), O`ahu.

189

Ostrea sandvicensis Sowerby, 1871
Family Ostreidae
• Clusters of these oysters are common on reef tops in Kāne`ohe Bay, often growing among branches of Finger Coral. They also grow on rocks and pilings in other protected locations around the Islands, sometimes in brackish water. The edges of the shells are scalloped, making the species easy to recognize. One valve is considerably larger than the other. These oysters are preyed upon by a flatworm, *Stylochus* sp. Maximum length about 2 in., but usually smaller. Endemic. Photo: Pearl Harbor, O`ahu. 2 ft.

COCKLES AND TELLINS. FAMILIES CARDIIDAE AND TELLINIDAE

Cockles, or heart shells, are active burrowing bivalves with shells of equal size that usually bear radial ribs. The closed shells viewed endwise form a heart shape. The animals burrow in sand using their long foot. Three species occur in Hawai`i. Tellins are similar to cockles but usually have concentric ridges or lines on their shells. They take food particles from the surface with their long inhalant siphons. Nine species inhabit the Islands.

ROUNDED COCKLE • **pūpū kupa**

Trachycardium orbita (Sowerby, 1833)
Family Cardiidae
• These bivalves occur on rubble bottoms either partly buried or just under the surface. The shells are attractively ribbed and their edges scalloped. The living animals can hop about to some extent on their long foot. Of the three *Trachycardium* species in Hawai`i, this is the largest. To about 3 in.; most specimens are smaller. Indo-Pacific. Photo: Molokini Islet, Maui. 100 ft.

RASP TELLIN

Tellina scobinata Linnaeus, 1758
Family Tellinidae
• Tellins burrow in sand or rubble, using their long siphons to feed on organic material lying on the surface. Some are quite colorful. The white shells of this species are completely covered with sharp, rounded scales, rendering them rough and rasplike. Length about 2 in. Hawaiian Islands and Kiribati. *Tellina elizabethae* is a synonym. Photo: Lori Kane. Lāna`i Lookout, O`ahu. 30 ft.

OCTOPUSES and SQUIDS

CLASS CEPHALOPODA, ORDERS OCTOPODA and TEUTHOIDEA

Ornate Octopus *(Octopus ornatus)* Mākua, O`ahu. 15 ft.

No science-fiction creatures are stranger than the earth's own cephalopods—octopuses, squids, and their relatives. Consider the octopus, a baglike animal equipped with eight sucker-lined arms and a beak like a parrot. It can ooze through cracks a fraction of an inch wide, and if attacked ejects a blob of ink as a "smokescreen," changes color, and jets invisibly away. To top it off, the octopus is intelligent; its memory and problem solving abilities approach those of some birds and mammals.

Now picture the squids: living jet engines equipped with forward stabilizing fins and trailing, sucker-lined arms—voracious hunters capable of out-swimming mackerel and tuna, killing

Day Octopus *(Octopus cyanea)* peering from its lair. Hanauma Bay, O`ahu, (D.R. Schrichte).

them with a venomous bite, and devouring everything but the head. Most squids, admittedly, are too small to fit this description, but some exceed it. The 12-foot-long Humboldt Current squids *(Dosidicus gigas)*, whose suckers are ringed with sharp, clawlike hooks, feed on sharks and have attacked divers. The famed giant squids (*Architeuthis* sp.), 60 ft. long and weighing up to a ton, grapple with whales in the dark abyss. That such fearsome predators are related to cockles and clams only reflects the incredible diversity of the phylum Mollusca.

All cephalopods are marine. The name, meaning "head-footed," comes from the set of ten-tacular arms (often called tentacles) surrounding the mouth and probably derived from the ante-rior end of the molluscan foot. Other characteristics of cephalopods are a tubular funnel, (which directs a jet of water expelled from the mantle cavity and can be pointed forward or backward), a chitinous beak, and image-forming eyes. In addition to octopuses and squids (orders Octopoda and Teuthoidea), the group includes cuttlefishes (Sepioidea), chambered nautiluses (Nautiloidea) and the strange deep-sea vampyromorphs (Vampyromorpha). All except nautiluses have 8-10 arms and either an internal shell or no shell at all. Nautiluses have an external shell and numer-ous short tentacles. All can swim backward by jet propulsion—taking water into their mantle cav-ity and ejecting it through the funnel near the mouth. (Octopuses, however, scuttle along the bot-tom more often than they swim.) Cephalopods are found from shallow tide pools to abyssal depths. Squids, pelagic octopuses and vampyromorphs inhabit open water; all others are close-ly tied to solid substrates.

Cephalopods are predators. Octopuses feed mainly on crabs, other molluscs, and occa-sionally fishes. A pile of discarded crab and snail shells often marks the entrance to their lairs. Squids prey on fishes. Both use venoms or poisons to subdue their prey. The bite of the tiny but beautiful Blue-Ringed Octopus *(Hapalochlaena maculosa)* can kill a human. Its saliva contains tetrodotoxin, the same compound that makes pufferfishes (fugu) deadly. This octopus is not recorded from Hawai`i.

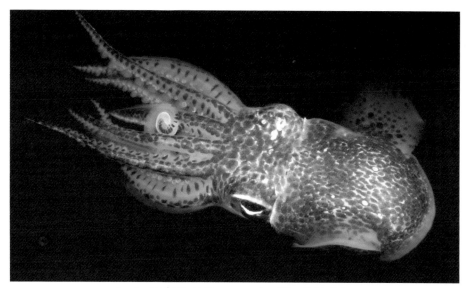

Hawaiian Bobtail Squid *(Euprymna scolopes)* Hawai`i Kai, O`ahu.

For defense, cephalopods rely heavily on camouflage and the quick escape. Octopuses and cuttlefishes, which live on or near the bottom, can change their color patterns and skin texture to match the background almost instantaneously. If surprised or attacked they eject a blob of thick black fluid that expands in a cloud to resemble their body, hopefully confusing the predator while they flee. (In old Hawai`i, when a clever rascal made mischief and escaped, people would say **"kūkae uli"**—"octopus ink." During whaling days the expression became synonymous with "prostitute" because of these women's ability to escape precarious situations.)

Squids also can change colors and are especially known for their rippling, iridescent mating displays. Some that live in mid-ocean hide their dark silhouettes from predators below with light-emitting organs that match the surface light.

Cephalopods have separate sexes and complex mating behaviors. Males transfer sperm by inserting a specially modified arm (hectocotylus arm) into the mantle cavity of the female. Females usually lay eggs in clusters on the bottom and, in the case of some octopuses, carefully tend and guard them until they hatch. The young are typically larval and pelagic, but some species emerge as miniature adults, ready to jet away into the blue or start life on the bottom, as the case may be. Most species mate once and then die, living only a year or two.

The largest cephalopods are the giant squids, with a body length of about 35 ft., and a total length of over 60 ft. Nothing of this size has been reported in Hawai`i, but in 1981 two O`ahu fishermen accidentally snagged a 20-foot squid off Kāne`ohe Bay. While being brought to the boat the squid was attacked by a pod of false killer whales that tore away large chunks of its tentacles and body. According to the fishermen, the squid became "molten red" while struggling with its attackers. Its remains weighed almost 200 pounds; the beak, the fishermen said, was the size of a "small football."

The largest octopus reportedly measures 30 ft. across its outstretched arms. No Hawaiian shoreline octopus approaches this size, but in 1996 the remains of large pelagic octopus were found floating off Kailua-Kona, on the island of Hawai`i. It was estimated to have an arm span of about 10 ft. and was identified as *Haliphron mollis*. The smallest cephalopods are less than 1 in. long.

There are 73 cephalopod species recorded from Hawaiian waters; most live offshore or at great depths. In shallow shoreline waters one squid and at least nine octopus species are known (some undescribed). A second small squid species may also exist. Cuttlefishes are not found in Hawai`i, although a small relative about 2 in. long occurs in protected sandy habitats. No true nautiluses inhabit Hawaiian waters. The paper nautiluses or argonauts, whose shells sometimes wash up on windward beaches, are more closely related to the octopuses.

Along Hawaiian shores octopuses are far more common than squids, although local people use the word "squid" for both. In the Hawaiian language octopuses and squids are **he`e** and **mūhe`e**, respectively. **He`e** (octopuses) are most abundant during the summer months and were an important food in old Hawai`i, eaten raw, dried or cooked. In the shallows they were speared and in deeper water caught with cowry shell lures. "Only shells with small red spots breaking through a reddish-brown ground have an attraction for the hee, and it will not rise to any other kind," wrote J.N. Cobb around the turn of the century. "Shells which have suitable spots but unsuitable background are given the desired hue by steaming them over a fire of sugar-cane husks." In order to see clearly into the depths, fishermen using such lures would chew up **kukui** nuts and spit the oily meat on the water, thus calming the surface.

The Hawaiian observer S. Kamakau wrote around 1869 that "the squid [i.e. octopuses] today number about fifty percent ... from old times." In those days, he writes, "they were count-less. ... When the tide was low, the whole floor was furrowed as if rooted up by hogs, with the burrows scattered in every direction, and the squid would be lying spread but flat, like lumps of dark earth, moving the head about slightly. If they saw a man they would squirt and he had to run to escape." Perhaps the decline Kamakau chronicles came from the breakdown of the tradi-tional Hawaiian fishing **kapu**. In the old days the taking of **he`e** was seasonally regulated by the chiefs, probably on pain of death.

Today, as in ancient times, most people think of octopuses mainly as food. Aquarists and a few divers, however, know them as resourceful and intelligent creatures that deserve our closer attention. Consider this story by local diver Dorothy Wendt, whose dive companion caught a large octopus and placed it in a zippered bag hanging from his belt. Some minutes later her buddy was banging on his scuba tank in distress.

"Octopus arms were all over him" she writes. "One had pulled his face mask loose so that it was flooded. Another was doing something to his ear, while a third seemed to be pulling his hair. I sort of half expected to see an octopus arm turning off the scuba air valve.... While three arms were harassing its captor, a fourth was delicately sliding the zipper catch down the track."

The late Jacques Yves Cousteau, in his book **Octopus and Squid: the Soft Intelligence**, offers this view: "One must have lived in the water with octopuses for months, swum in the same waters, brushed past the same rocks and the same algae, in order to be able truly to appreciate the beauty of the octopus. In the water, the octopus looks like a silken scarf floating, swirling, and settling gently as a leaf on a rock, the color of which it immediately assumes. Then it disap-pears into a crack which appears to be hardly large enough to accommodate one of its arms, let alone its entire body. The whole process is reminiscent of a ballet. It is somehow ethereal and, at the same time, elaborate, elegant, and slightly mischievous."

DAY OCTOPUS ▪ **he'e mauli**   Day Octopus (a)

Octopus cyanea Gray, 1849
Order Octopoda. Family Octopodidae

• This is the most frequently-seen Hawaiian octopus. Active by day, especially during the early morning and late afternoon, it occurs from the shallows to depths of 150 ft. or more, and is most often observed peering from its lair or scuttling over the reef in search of food. Like most octopuses it can assume a variety of color patterns and skin textures both as camouflage and to communicate its "emotional" state. Grayish brown is normal, but a startled octopus may flush deep reddish brown, and an aggressive or threatened one bleaches to ashen white with a dark circle (ocellus) on each side. A hiding octopus can adopt a range of mottled patterns matching the color of almost any background; its skin may pucker into complex bumps, warts and ridges, merging with the texture of the reef. If detected, it jets away, often changing color several times to confuse the predator. Even while resting, subtle colors play almost continually along the body of this amazing animal.

The Day Octopus is solitary and spends much of its time half in and half out of its lair, often surrounded by scattered fragments of crustacean shells. If closely approached it withdraws but almost always emerges in a minute or two to look at a diver who waits quietly by. If harassed in its hole it squirts a powerful current of water at the intruder through its funnel. If that fails, it wraps itself in its arms, suckers out, or uses its suckers to pull a wall of stones between it and its attacker. At night it barricades itself in its hole with stones in a similar manner. Only as a last resort, or when suddenly surprised, does it eject its ink.

This species eats mostly crabs; in deeper water it may possibly take molluscs and occasionally fish. Like most octopuses, it captures a prey animal by pouncing on it with the web between its arms spread wide, thus enclosing it. The octopus may inject a dose of poisonous saliva to weaken or kill the animal before it is consumed. Tough shells are no barrier; the octopus can drill straight through with its radula. Octopuses are strongly attracted to certain cowries, which were used in old times as lures. (In those days, when a man looked lecherously at young girls, it was said "the octopus notices the little cowries.")

Octopus cyanea is widespread in the Indo-Pacific. It attains about 3 ft. with arms spread, with a weight of 4-5 pounds. Like many cephalopods, its life span is only about a year. It was named for Cyane, a nymph in Greek mythology who was turned into a fountain. Photos: a) D.R. Schrichte. Hanauma Bay, O`ahu. 30 ft.; b) & c) Lehua Rock, Ni`ihau. 15 ft. (Scuttling across the reef in the open, this octopus froze in place when it saw the photographer, adopting a camouflage pattern and texture. The two pictures were taken only seconds apart.)

◄ Day Octopus (b)

◄ Day Octopus (c)

ORNATE OCTOPUS * he`e pūloa, he`e mākoko

Octopus ornatus Gould, 1852
Order Octopoda. Family Octopodidae
• This species is known locally as the "night octopus." Reddish to orange brown with buff or white stripes and spots, it is usually identifiable by color alone. The long, slender arms and elongate, pointed body (mantle) are also distinctive, although the mantle may take a bulbous shape, like that of the Day Octopus. Hiding by day and emerging at dusk, this animal is usually found flattened against the bottom, snaking its arms into crevices and holes in search of prey. It is rarely seen swimming. Like all octopuses, this species is capable of biting divers who toy with it. Persons sensitive to the venom may suffer swelling, aching in the joints, and general discomfort lasting sometimes for weeks. In old Hawai`i this species was sometimes used medicinally. **Pūloa** means "long head" and **mākoko** means "reddish." To about 2 ft., with arms spread. Indo-Pacific. (*O. luteus*, a nocturnal species superficially similar in size and color but with smaller white spots that can be elevated into papillae, also occurs in Hawai`i.) Photo: Mākua, O`ahu. 15 ft. (See also p. 191.)

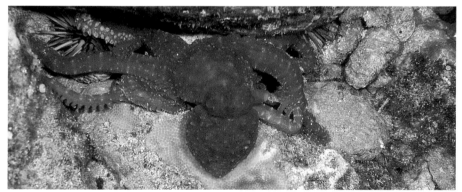

HAWAIIAN OCTOPUS

Octopus hawiiensis (Souleyet, 1852)
Order Octopoda. Family Octopodidae
• This small, dark brown octopus with short arms and relatively smooth skin is seen most often at night on rocky substrate at depths of 15-30 ft. Like all octopuses, it can change color, often developing whitish patches. Little is known of its habits. The species was first collected off the Big Island in 1837 by the French naturalist F.L.A Souleyet, whose published description bore the apparently misspelled species name *hawiiensis*. The rules of zoological nomenclature, however, state that the originally published name must not be "corrected." Over a hundred years passed after Souleyet's discovery before this octopus was seen again by scientists, presumably because it is nocturnal and small. It is not, however, particularly rare. Known only from the Hawaiian Islands. Photo: Mākua, O`ahu, 15 ft.

197

Octopus sp. 1
Order Octopoda. Family
Octopodidae
• This animal is typically reddish
brown to cream with many small
white spots that create a mottled,
almost reticulated pattern. Other
patterns include darkening the
eyes and lightening the body and
arms, or darkening the entire
body except for the arms. It lives
in sand adjacent to the reef.
Known only from Hawai`i, this
undescribed species is under
study by Dr. Eric Hochberg of the
Santa Barbara Museum of Natural
History. Photo: Mike Severns.
Mākena, Maui. 60 ft.

LONG-ARM SAND OCTOPUS

Octopus sp. 2
Order Octopoda. Family Octopodidae
    These small octopuses occupy holes in silty sand, sometimes old mantis shrimp burrows. They can swim over the bottom by holding their arms together and undulating, much like a flatfish. Little is known about this species and it is probably undescribed. Photo: Bruce Carlson. Kahe Point, O`ahu. 50 ft.

ROUND-SPOT OCTOPUS

Octopus sp. 3
Order Octopoda. Family Octopodidae
• Small, shallow-water octopuses of two almost identical species, known informally as "crescent-spot" and "round-spot" octopuses, have been known to scientists in Hawai`i since the 1970s but remain undescribed. The "round-spot" is pictured here. Although fairly common in rubble on shallow semi-protected reef flats, and sometimes in tide pools, these octopuses hide until sunset and are infrequently seen. Their normal color is brown or brown and cream. The common names are derived from two small round or crescent-shape marks often present about half way between the eyes and the tip of the mantle. These animals grow to about the size of a golfball with a maximum weight of about 3 ounces. A Honolulu aquarist reports that at least one of these species will lure crabs or small fish by darkening the body and wiggling a white arm tip. Known only from the Hawaiian Islands. Photo: Black Point, O`ahu. 3 ft.

HAWAIIAN BOBTAIL SQUID

Euprymna scolopes Berry, 1913
Order Sepioidea. Family Sepiolidae
• This tiny, cuttlefish-like animal has a short, thick, rounded body with a semicircular fin on each side. There are eight arms and two slightly longer tentacles. The overall color is brown, with a few iridescent spots of green and blue that can expand and contract. These animals have been collected from the surface and trawled from the bottom 100 miles offshore, where they are apparently abundant. The shallow protected mud and sand flats of Kāne`ohe Bay, O`ahu, harbor numbers of them. They bury themselves by day and emerge at night to feed on shrimps (in Kāne`ohe Bay largely *Palaemon debilis*, p. 222) and small worms. On moonlight nights organs containing luminescent bacteria *(Vibrio fischeri)* help the swimming squids hide their dark silhouettes from predators below. Reflective tissues direct the light downward and lenslike tissues diffuse the glow over the entire underside. A diaphragm controls the amount of light emitted. The symbiotic relationship between squid and bacterium is easily studied because both organisms can be cultured in the laboratory. Although generally less than 2 in. end to end, specimens more than twice that length have been found. The species is known only from the Hawaiian Islands and surrounding waters. Photo: Ron Holcom. Aquarium. Specimen collected near surface at Hawai`i Kai, O`ahu. (See also p. 193.)

199

OVAL SQUID • **mūheʻe**

Sepioteuthis lessoniana Lesson, 1830
Order Teuthoidea: Family Loliginidae
• Squids are schooling, open-water animals not generally found close to coral reefs. This is the species most likely to be encountered by divers and snorkelers in Hawaiʻi; it is seen infrequently. Like all squids, it has eight arms and two longer tentacles, used to capture prey. This particular genus is characterized by two expanded fins forming an oval and extending almost the length of the mantle. By rippling these fins the squid can swim forward or backward; for speed it jets backward. In old Hawaiʻi it was said that the squid "moves two ways, like a crab." This species is of commercial importance; in Japan it has been found to mature in less than 100 days, females growing to about 10 in., males to about 14 in. It has a life span of 9-10 months and is one of the few squids that have been successfully maintained in captivity. Eastern Indian Ocean to Hawaiʻi. Photo: Honolua Bay, Maui. 15 ft.

CRUSTACEANS
PHYLUM ARTHROPODA
SUBPHYLUM CRUSTACEA

Regal Slipper Lobster *(Arctides regalis)*. Pūpūkea, Oʻahu. 30 ft.

The crustaceans are a subgroup within the vast assemblage of animals with jointed appendages called arthropods—the largest single group of animals on earth. Probably over a million species of arthropods exist, most of them insects. In addition to crustaceans and insects, the phylum Arthropoda includes spiders, mites, ticks, scorpions, centipedes, millipedes and the extinct trilobites. Some specialists regard crustaceans as constituting a phylum in their own right.

Like insects on land, crustaceans are ubiquitous in the aquatic realm, having penetrated all ocean, river, lake and stream habitats. Most are small or microscopic organisms that constitute a vital part of the food chain for larger fishes and marine mammals. One third belong to the order Decapoda, which includes almost all of the larger crustaceans including the familiar lobsters, shrimps and crabs.

Crustaceans (and arthropods in general) boast an amazing variety of jointed appendages used not only to move about and capture food, but also to eat, breathe, sense the environment and reproduce. In addition to the obvious walking legs and claw-bearing limbs (often called chelipeds), all crustaceans have two pairs of antennae and numerous small appendages around the mouth (mandibles, maxillae and maxillipeds) for holding, biting, grinding and chewing. Appendages under the abdomen called pleopods or swimmerets are used for swimming, holding masses of eggs (females), or grasping the female during mating (males). Even the gills are connected to, or arise from, the walking legs and other limbs.

Crustaceans are also characterized by an exterior skeleton, or shell, composed of plates and rings of a flexible, horny substance called chitin. In the larger crustaceans this shell is often calcified, making it thick and hard. Both a blessing and a curse, the shell provides protection and a rigid framework to which muscles can attach, but must be shed periodically to allow for the animal's growth. Molting occurs every few weeks or months (depending on species and size), during which time the animal is especially vulnerable. In the larger species this soft shell stage lasts 24 hours or more.

Hiatt's Hinge Beak Shrimp *(Cinetorhynchus hiatti)*, Kailua-Kona Harbor, Hawai`i. 15 ft.

Molting in crustaceans is governed by complex hormone interactions. After a period of growth, the animal releases substances that cause its shell to split. Crawling out, it leaves a perfect but empty replica of itself. Later, in hiding, it releases other hormones that cause its body to swell with water and secrete a new shell. When the shell hardens, yet another round of hormones cause the body to shrink back to normal size. Its shell is now a bit too large, leaving room for future growth.

Although arthropod bodies are generally divided into three regions (head, thorax, and abdomen or tail), in crustaceans the head and thorax are usually united, forming a cephalothorax. The cephalothorax is covered by a single plate, the carapace. Crabs have a flattened carapace, typically broader than long, with a short, inconspicuous tail folded forward under the body. Lobsters and shrimps have a cylindrical carapace with a long muscular tail. The tail in all crustaceans is segmented, each segment protected by a ringlike plate.

Crustaceans have well-developed organ systems, including compound eyes, a primitive brain and a heart that pumps fluids around the body spaces. Instead of hemoglobin, crustacean blood contains the light blue oxygen-carrying compound, hemocyanin. The sexes are usually separate. Eggs are fertilized by copulation, the male transferring sealed packets of sperm to the female. In most cases the female broods her eggs, holding them in a mass under her tail until they hatch. Newly hatched larvae drift with the plankton for weeks or months, going through a number of complicated growth stages before finally settling to the bottom (in most cases) to mature. Crustacean larvae of various kinds constitute a large proportion of the plankton.

Of the many classes of crustaceans (6 to 11, depending on whom you consult), only 2 are covered here: the small class Cirripedia (barnacles), and the large, diverse class Malacostraca, which includes shrimps, lobsters, hermit crabs, true crabs, mantis shrimps, mysid shrimps, and a host of less conspicuous groups such as isopods and amphipods. Of the approximately 38,000 crustacean species known worldwide, about 970 are recorded from Hawai`i.

BARNACLES

SUBPHYLUM CRUSTACEA. CLASS CIRRIPEDIA

Hembel's Barnacle *(Euraphia hembeli)* Makapu`u, O`ahu. 3 ft.

Barnacles are strangely modified crustaceans that spend their adult lives attached permanently to rocks, pilings, floating objects or other animals. An oft-quoted phrase, usually attributed to the great naturalist Louis Agassiz (1807-1873), describes a barnacle as "a little shrimplike animal standing on its head in a limestone house and kicking food into its mouth." A similar comparison occurs in the writings of Sir Thomas Huxley (1825-1895). Whoever penned it first, the idea is not inaccurate.

Like most other crustaceans, barnacles begin life as free-swimming larvae. When ready to mature they cement themselves head down to a suitable surface by means of cement-secreting glands in their specialized first pair of antennae. (Barnacle cement is so strong and durable it was once proposed for dental work.) Fixed in place, young barnacles typically form a number of calcareous plates to enclose and protect the limbs and body organs. Because their shells resemble those of limpets and bivalves, early scientists classified barnacles as molluscs. Only after 1830 when their larvae were discovered and examined under a microscope were these animals fully accepted as crustaceans.

Adult barnacles feed by rhythmically sweeping the water with featherlike jointed appendages (cirri) to collect drifting food particles. When active, barnacles extend their cirri through an opening in the shell. When resting, exposed at low tide, or disturbed they draw in their cirri and seal their "houses," usually with two pairs of hinged plates.

The barnacles most likely to be noticed by divers, snorkelers and beachgoers fall into two groups: rock barnacles and goose barnacles. Rock (sessile) barnacles form broad volcano- or barrel-shape shells that usually attach directly to the substrate, looking somewhat like limpets. Goose (pedunculate) barnacles usually form shells resembling those of bivalve molluscs and attach to the substrate by means of a tough, fleshy stalk (peduncle). Although goose barnacle stalks may resemble the necks of geese, the name originates in the medieval English idea that baby geese and ducks "hatch" from the somewhat egglike shells of stalked barnacles. Conveniently, this belief allowed the consumption of geese and ducks on fasting days.

Rock barnacle shells, sharp and often densely packed, typically grow on rocky shores, piers, pilings and the bottoms of boats (from which they must be periodically scraped). Some species attach to turtles, crabs, lobsters, whales or other marine animals. Humpback whales arriving in Hawaiian waters in October and November are each estimated to bear up to a ton of barnacles (mostly *Coronula diadema*) accumulated in their rich arctic feeding grounds. The whales' great breaches and subsequent thunderous crashings into the sea may partly serve to dislodge these unwanted hitchhikers.

Some goose barnacles attach to driftwood and other floating objects, even floating feathers. Others are symbiotic with a variety of other animals, including crustaceans. The shells and jointed appendages of the latter may be much reduced; in some cases their hard shells are lost entirely. In temperate waters, goose barnacles attached to drifting wood attain lengths of over 30 in. None in Hawaiian waters approach this size.

Crustaceans must mate to reproduce. Because barnacles are fixed in place, coming together presents a problem. Those species that settle in close proximity solve it by means of a long, extensible penis. Such barnacles are generally hermaphrodites, ensuring that any neighbor is a potential mate. Widely spaced or solitary species (such as the turtle barnacle *Chelonibia*) require a different solution. Adults, which are either hermaphrodites or females, are accompanied by tiny degenerate males that attach outside or inside their shell, almost as parasites.

Rock barnacles and goose barnacles belong to the order Thoracica. Two more orders deserve mention. The burrowing barnacles, order Acrothoracica, are mostly minute naked creatures that dig into the shells or skeletons of corals, molluscs and even other barnacles. The highly specialized parasitic barnacles, order Rhizocephala, invade the soft tissues of crabs, shrimps, and other crustaceans. Rhizocephalans, as adults, lose all traces of exoskeleton, appendages and segmentation that might mark them as arthropods. In some cases these nightmarish parasites castrate or sterilize their hosts and force them, male or female, to care for and ventilate their egg masses, which grow under the unfortunate host's tail.

Roughly 50 barnacle species occur in Hawai`i, many brought here on ships from other parts of the world. Most native Hawaiian barnacles are small and inconspicuous, parasitic, or from deep water. The largest *(Euraphia hembeli)* grows to about 3 in. across.

The Hawaiian word for barnacles is **pī`oe`oe**. In old Hawai`i, persons constantly pursued by the opposite sex were described as "clung to by barnacles." Europeans have similar ideas. In 1904, when Irish writer James Joyce first dated his future wife, Nora Barnacle, his father predicted "She'll never leave him."

AMPHITRITE'S ROCK BARNACLE

Balanus amphitrite Darwin, 1854
Family Balanidae
• The shell of this barnacle is smooth or slightly ribbed vertically, whitish, and marked with vertical violet stripes. Large portions of the shell are often eroded. The animal is found just below the low tide line along protected shores (especially harbors), attached to rocks and pilings. Part of a widespread species complex common in tropical and warm temperate seas, it is named after Amphitrite, wife of the ancient Greek sea god Poseidon. This barnacle was probably introduced to Hawai`i; the subspecies *hawaiiensis* is no longer considered valid. The similarly marked *B. reticulatus* and the entirely white *B. eburneus* (both likely introduced) are common in the same habitat, although the former is usually restricted to harbors. Typical specimens are about 1/2 in. across the base. Photo: Moku o Lo`e (Coconut Island), O`ahu. Intertidal.

TRIGONATE BARNACLE

Balanus trigonus Darwin, 1854
Family Balanidae
• This red-and-white striped barnacle has an unusual triangular oriface, hence the name. Strictly a marine species, it occurs from shallow water to depths of several hundred feet. In Hawai`i, Kāne`ohe Bay is a good place to see it. It often occurs with sponges, corals, and even on mollusc shells. Although many of Hawai`i's barnacles are introduced, fossils indicate that this species was present here 100,000 years before the coming of humans. Native to the Indo-Pacific, it was spread to the Atlantic Ocean by shipping and is now patchily distributed in warm temperate and tropical seas around the world. To almost 1 in. across. Photo: William A. Newman. Santa Cruz Island, California.

TURTLE BARNACLE

Chelonibia testudinaria (Linnaeus, 1758)
Family Balanidae
• Barnacles of the genus *Chelonibia* usually attach to other animals. This species specializes in sea turtles, in Hawai`i principally the Green Turtle (*Chelonia mydas*). The barnacle's low profile minimizes drag as its host swims about, while its thick plates resist damage when the turtle scrapes against overhangs or other objects on the reef. Sometimes several barnacles occur on one turtle, but many turtles carry none. Although hermaphroditic like most barnacles, the sometimes widely scattered individuals of this genus must carry small "complementary males" to ensure adequate cross-fertilization. The species name derives from *testudo*, Latin for "tortoise." It occurs in all warm seas in which turtles still survive, attaining as much as 2 in. diameter. (The related *C. patula* attaches to the shells of spiny lobsters and large swimming crabs and has recently been found on floating plastic.) Photo: Nahuna Point ("Five Graves"), Kīhei, Maui. 40 ft.

PROTEUS' ROCK BARNACLE

Chthamalus proteus Dando & Southward, 1980
Family Chthamalidae
• Unreported in Hawai`i until 1995, this intertidal species was evidently brought by ships from the tropical Western Atlantic sometime since 1973. Prior to its introduction, few barnacles colonized the zone between low and high tide in the Islands. In the placid waters of Kāne`ohe Bay, O`ahu, and in lagoons and harbors throughout the islands as far north as Midway, these barnacles are now conspicuous at low tide on pilings and rocks, in some places almost completely obscuring the substrate. Occupying the high intertidal zone, they may not be submerged at every high tide. The genus *Chthamalus* (can be pronounced without the initial "ch") is rarely found on oceanic islands, thus its introduction to Hawai`i is notable. This variable species is named for the Greek sea god Proteus, who was capable of changing his form. Typical specimens are about 1/3 to 1/4 in. across the base. Caribbean, tropical Western Atlantic, Hawai`i, Guam and Japan. Photo: Kāne`ohe Bay, O`ahu.

PURPLE ROCK BARNACLES
• **pī`oe`oe**

Nesochthamalus intertextus
(Darwin, 1854)
Family Chthamalidae

• These small barnacles cluster high on shoreline rocks on semi-exposed shores where they are submerged only at high tide or by periodic surge. They vary from white to purple and can be recognized by their jagged bases, contorted to fit irregularities of the substrate. The sutures (lines where the shell plates join) are also highly irregular. The species was first described by the great naturalist Charles Darwin (1809-1882). Although better known for his theory of evolution, Darwin was also an ardent student of barnacles. The species name means "interwoven." Typically about 1/4 in. across the base. Indo-Pacific. Photo: Lāna`i Lookout, O`ahu.

HEMBEL'S ROCK BARNACLE • **pī`oe`oe**

Euraphia hembeli (Conrad, 1834)
Family Chthamalidae

• Hawai`i's largest barnacles (and the largest in their family), these are most common on exposed offshore sites such as the rocks on the west side of Waimea Bay, O`ahu, or the back wall of Molokini Islet, Maui (where dense colonies occur from the low tide line to depths of about 40 feet). A few individuals may settle above the low tide mark. Smaller barnacles of several species sometimes grow on their shells. Beautiful, uneroded specimens (frequently covered with tiny red tunicates) occur in wave-scoured cave entrances along O`ahu's north shore at depths of 15-20 feet, and snorkelers at Hanauma Bay can find a few on the outer side of the boulder reef. Like most barnacles they are hermaphrodites, and individuals must be closely spaced for cross-fertilization to occur. While large specimens are almost 3 in. broad at the base and 2 in. high, most are half that size or less. Indo-Pacific. Photo: Pūpūkea, O`ahu. 15 ft. (in cave) (See also p. 203.)

GOOSE BARNACLE

Lepas anserifera Linnaeus, 1767
Family Lepadidae
• Two species of stalked barnacles frequently wash up on windward Hawaiian beaches attached to driftwood and other floating objects. *Lepas anserifera*, the original "goose barnacle" (*anser* is Latin for "goose"), can have a stalk up to 3-4 in. long. *Lepas anatifera* (from *anas*, the Latin word for "duck") usually has a shorter stalk. These barnacles drift the seas worldwide on flotsam and jetsam, growing considerably larger in cool, temperate regions. Photo: Waimānalo Beach, Oʻahu.

PARASITIC LOBSTER BARNACLE

Paralepas sp.
Family Heteralepadidae
• These parasitic barnacles with the striking red median stripe superficially resemble *Paralepas palinura urae* Newman, 1960 but may represent a new species. Like others of their kind they lack shells. Attaching around the joints and mouthparts of spiny lobsters (pictured here on *Panulirus penicillatus*), they grasp and rasp their host's food with specially modified cirri. They may also settle on the abdomens of female lobsters to feed on eggs. Other species of semi-parasitic barnacles may live on the lobster's gills where they filter-feed in the water circulated by their host. Slipper lobsters and crabs often bear similar parasites. Photo: Mōkapu Rock, Molokaʻi.

MANTIS SHRIMPS
SUBPHYLUM CRUSTACEA. ORDER STOMATOPODA

Shortnose Mantis Shrimp (*Odontodactylus brevirostris*) peering from its burrow. Maui. (David A. Fleetham)

Mantis shrimps (stomatopods) are evolutionarily advanced crustaceans only distantly related to true shrimps (order Decapoda). Their intelligence, learning ability and sensory faculties (especially vision) place them well above other arthropods, analogous perhaps to octopuses in the phylum Mollusca. All are voracious predators that either spear or smash prey with a pair of raptorial arms resembling those of their namesake, the terrestrial praying mantis.

These animals are bottom-dwellers, their elongate bodies widest at the posterior end. They usually run along the bottom but are capable of swimming. Their eyes, set on stalks, are independently movable. Of eight pairs of legs, only the last three are used for walking. The third, fourth and fifth pairs bear small oval "claws" used for feeding; the first pair, slender and hairy, are cleaning appendages. The greatly enlarged second pair of legs form the long raptorial arms.

Like the praying mantises, mantis shrimps ordinarily keep their arms folded like a jackknife, the slender last segment of each arm fitting into a groove on the previous segment. In species that normally spear their prey (the majority), the last segment is equipped with multiple barbed spines. By snapping out an arm, spearers impale soft-bodied worms, shrimp or fish. Smashing species, which usually lack such spines, break apart crab or mollusc shells with powerful underarm blows from the thickened joint (the "elbow") at the base of the last segment. Smashers usually stalk their victims; most spearers ambush their prey from a concealed position.

The spearing action of mantis shrimps is completed in less than 5 milliseconds, and the impaled victim is quickly conveyed to the mouth to be ripped apart by the mandibles and maxillipeds. A smashing stomatopod, however, may take its time, wedging a snail or hermit crab against an "anvil" rock and systematically destroying its shell (taking several hours, if necessary). Crabs are quickly disabled by knocking off their legs and claws; they are then dragged into the mantis shrimp's den to be further battered and consumed. A large smasher, whose blow approaches in power a small caliber bullet, can crack a heavy clam shell in a single blow. Some smashers can also spear. These animals are highly territorial and do not hesitate to attack their own kind. However, ritualized fights and warning displays using brightly colored spots on the inner surface of the raptorial arms (meral spots) help keep mortality to a minimum.

Mantis shrimps occupy burrows in sand or mud or inhabit coral cavities. Unlike most other crustaceans, females do not hold their eggs under the tail but deposit them at the bottom of their burrows where they must be constantly aerated. These crustaceans are common in all warm seas and often active by day, but divers and snorkelers seldom notice them.

Needless to say, mantis shrimps are usually unwelcome in aquariums, where they are sometimes introduced by accident in "live rock." Spearers may slice up your fishes and even small smashers can crack the glass walls of your tank. Keeping one as a special pet, however, could be rewarding; their fluid movements are fascinating, and the alert, independently movable eyes give the appearance of unusual intelligence. On the reef, some species are curious and will interact with humans. Snorkelers, divers and aquarists, however, should avoid handling stomatopods, or even rocks containing them. If you pick up a piece of coral and accidentally cover a mantis shrimp's cavity you could get painfully cut. Some fishermen call these crustaceans "thumb-splitters," and large ones can amputate a human finger.

Hawai`i has 17 known species of stomatopods, of which perhaps 3 are endemic. They range in length from about 1 to 12 in. The Hawaiian name for all mantis shrimps is **aloalo**.

CILIATED MANTIS SHRIMP

Ciliated Mantis Shrimp (a)

Pseudosquilla ciliata (Fabricius, 1787)
Family Pseudosquillidae

• These are the mantis shrimps most likely to be encountered by snorkelers in Hawai`i. Common on shallow reef flats and down to at least 60 ft., they inhabit U-shape burrows, often under rocks. Juveniles might occur in dead coral cavities, but the aggressive Philippine Mantis Shrimp (below), introduced accidentally in the 1950s, has largely taken over this habitat. In color they match their surroundings with greens, tans, and browns, sometimes mottled or striped. Unusual yellow-orange specimens, noted 50 years ago off Waikīkī by C.H. Edmondson of the Bishop Museum, can still be seen there today. Prey is captured by spearing. To about 3 1/2 in. All warm seas except the Eastern Pacific. Photos: a) Roy L. Caldwell; b) Wai`alae Beach Park, O`ahu. 3 ft.

◄ Ciliated Mantis Shrimp (b)

PHILIPPINE MANTIS SHRIMP

Gonodactylaceus mutatus (Lancaster, 1903)
Family Gonodactylidae

• This is a smasher that preys on other crustaceans and molluscs, breaking open their shells with its thickened "elbows." It was first noticed locally in the 1950s; some researchers considered it a newly discovered endemic, which they named *G. aloha*. It is now thought to have been introduced to Hawai`i from the Philippines after World War II. Common in Kāne`ohe Bay, the animal is rapidly spreading about O`ahu and the other islands, to some extent displacing the native Ciliated Mantis Shrimp (above) on shallow reef flats. It lives in coral rubble and within cavities in dead coral heads, usually in shallow water. Males are dark green, females often tinged with orange. (The similar but smaller *Gonodactylellus hendersoni* may also be present in the same habitat. Its meral spots are white, those of *G. mutatus* yellow to orange.) To almost 2 1/2 in. Indo-Pacific. Photo: Roy L. Caldwell.

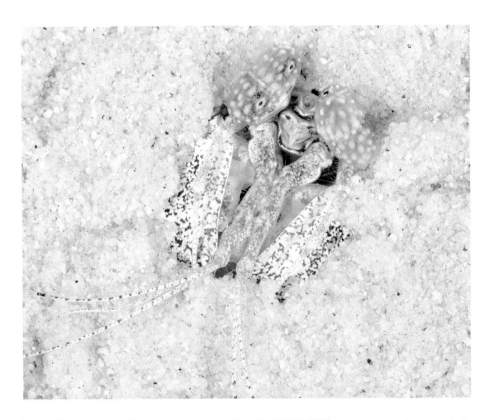

GIANT MANTIS SHRIMP • **aloalo**

Lysiosquillina maculata
(Fabricius, 1793)
Family Lysiosquillidae
• Monogamous pairs of these strikingly banded spearers inhabit burrows in protected sandy or muddy bottoms, often only a few inches deep, where they may live together 15 years or more. They are most commonly seen peering from their burrow entrances, which may be 4 in. across. In old Hawai`i it was said that if a female were caught, the male would soon follow her out, but if the male were to be caught first, the female would never emerge. These large stomatopods can seriously injure the hands of anyone attempting to grab them. The heavy, ivory-like claws have been used for making carved ornaments. Up to 14 in. long, these may be the largest of all stomatopods. Indo-Pacific. Photos: Roy L. Caldwell. a) Typical prey capture position; b) Subadult (about 3 in., aquarium photo)

WHITE MANTIS SHRIMP • **aloalo**

Oratosquilla calumnia (Townsley, 1953)
Family Squillidae
• This species lives in muddy, sometimes brackish areas in U-shape burrows and was once common in Waikīkī's Ala Wai Canal. Because of its habitat and nocturnal nature, few people ever see it. The body is light translucent gray, tan or reddish brown, marked with sharp longitudinal ridges and iridescent streaks. There are dark marks on the flattened appendages (uropods) on either side of the tail. It captures its prey by spearing and is similar to *O. oratoria*, a crustacean now cultured in Japan for making shako sushi. To about 6-8 in. Indo-Pacific. Photo: Mike Miller. Philippines.

SHORTNOSE MANTIS SHRIMP • **aloalo**

Odontodactylus brevirostris (Miers, 1884)
Family Odontodactylidae
• This smasher lives in burrows in rocky habitats from the intertidal zone to depths of 100 ft. Active by day and brightly col-ored orange and white, it is sometimes seen by divers peering from its burrow or darting back from a foraging trip. The eyes contain a horizontal band with a dark spot above and below. It is curious by nature and not difficult to approach. To about 2 1/2 in. Indo-Pacific. Photo: David A. Fleetham. Maui. (See also p. 209.)

URCHIN-TAIL MANTIS SHRIMP

Echinosquilla guerini (White, 1861)
Family Protosquillidae
• Only night divers are likely to see these secretive and unusual mantis shrimps, whose tail fans (telsons) resemble a sea urchin of the genus *Echinometra* (p. 316). They inhabit cavities and old worm tubes in coral, rock and rubble and often sit partly emerged. If disturbed they turn around and close the entrance with their urchin-like tail. These animals are not uncommon at depths of 50-120 ft. and have been dredged from depths of cver 600 ft. They are reddish overall except for the spiny white tail. Reflective gold eyes make them fairly easy to spot at night by divers with lights. Prey is captured by smashing. To almost 3 in. Indo-Pacific. Photo: Mike Severns. Molokini Islet, Maui.

DECAPODS

SUBPHYLUM CRUSTACEA. CLASS MALACOSTRACA, ORDER DECAPODA

Squat Anemone Shrimp *(Thor amboinensis)* in Sand Anemone *(Heteractis malu)*. Aquarium photo.

The decapods—shrimps, lobsters, crayfishes, anomuran crabs, and true crabs—form the largest crustacean order with about 10,000 known species. Almost all large crustaceans and virtually all commercially important ones are decapods. The word decapod means "ten footed" and these animals, whatever their shape or size, have five pairs of walking legs. One of the first three pairs often bears enlarged pincers, or claws. (Enlarged pincer-bearing limbs are called chelipeds; the pincers themselves, chelae.) Also characteristic are three pairs of small appendages flanking the mouth, the maxillipeds or "foot-jaws". These lie one on top of the other, the third pair outermost. As in other crustaceans, the head and thorax are fused into a cephalothorax covered by a single plate, the carapace. In decapods the carapace extends down the sides of the cephalothorax to cover the gills, located at the bases of the walking legs.

The most primitive of the decapods are the penaeid shrimps (suborder Dendrobranchiata) which have branched gills and release their eggs into the sea. Many penaeids are pelagic and some are of commercial importance. All other decapods, including other shrimps, have unbranched gills and brood their eggs under their tails. Members of this second group (suborder Pleocyemata) are typically bottom-dwellers.

The great majority of decapods are marine. Only a few freshwater crabs complete their life cycles out of the sea. Most freshwater decapods and all terrestrial ones return to the sea to reproduce and have a marine larval stage.

Interestingly, the direction of decapod evolution seems to be from long, well-developed tails (shrimps and lobsters) through progressively shorter tails (hermit crabs, porcelain crabs, squat lobsters) to very short, almost vestigial tails (true crabs).

SHRIMPS. SUBPHYLUM CRUSTACEA. CLASS MALACOSTRACA
ORDER DECAPODA. INFRAORDERS STENOPODIDEA AND CARIDEA

Shrimps are decapods with cylindrical carapaces and long muscular tails ending in a tail fan, or telson. Their shells are thin, flexible and never calcified like those of some crabs and lobsters. Most have a pair of whiplike antennae and a rostrum, a prolongation of the carapace between the eyes that is often sharply pointed. Unlike most decapods, many shrimps can swim or hover using their legs or small appendages under the tail called pleopods or swimmerets. When threatened, most can shoot backwards by rapidly flexing their powerful tails. Although some shrimps are pelagic, those of the coral reef are more likely to walk or cling to the substrate than to swim.

Shrimps bear pincers on their first two or three pairs of legs, the size and location of which are useful for their classification. Pincers of about equal size on all first three legs indicate the primitive suborder Dendrobranchiata. These include the commercially important penaeid shrimps, many of which are pelagic. Shrimps of this group are only distantly related to other decapods.

All other shrimps belong to the suborder Pleocyemata, which includes all other decapods. Most of these shrimps are bottom-dwellers. Those with enlarged pincers on either the first or second legs are known as caridean shrimps (infraorder Caridea). Those with enlarged pincers on the third pair of legs belong to a small but conspicuous group, the stenopodidean shrimps (infraorder Stenopodidea).

Large shrimps (often called prawns) attain a length of about 8 in. Small species are less than 1/4 in. long. The average size is from 1 to 3 in. The general name for shrimps in Hawaiian is `ōpae. As well as eating them, ancient Hawaiians used shrimps in sorcery to cast out evil spirits. The term **po`o `ōpae** ("shrimp head") was an insult.

Because of their graceful, upward-curving antennae and pale color, Flameback Coral Shrimps *(Stenopus pyrsonotus)* are often called Fountain Shrimps or Ghost Shrimps. South Point, Hawai`i. 20 ft. (See p. 219.)

Of approximately 2,000 species of shrimps, about 190 are known from Hawai`i; 38 are described below. For convenience, the tiny non-decapod mysid "shrimps" are also included in this chapter.

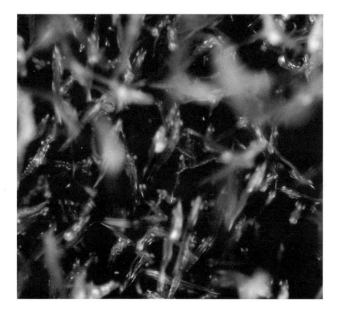

MYSIDS
Order Mysidacea
• These tiny "shrimps" are not decapods but belong to an order of their own. About 1/8 to 1/4 in. long, they swarm by the thousands around divers' lights at night. Attracted to certain hotel spotlights along the Kona coast of the Big Island, they are preyed upon by manta rays and thus create the basis of the flourishing "manta ray night dive" industry. Mysids are sometimes called "opossum shrimps" because females carry developing embryos in a brood pouch. Many mysids remain buried in the sand by day. About 450 living species are known. No identification from this photo is possible. Photo: David B. Fleetham. Hawai`i.

FAMILY PENAEIDAE

Shrimps of this family are only distantly related to other shrimps. They have branched gills and do not hold their eggs under the tail. Some species are pelagic; many others dwell in and on soft bottoms. Most commercially important shrimps belong to this family.

MARGINED SHRIMP • `ōpae lōlō

Melicertus marginatus (Randall, 1840)
Family Penaeidae
• Hawai`i's largest shallow-water shrimp, this pinkish species is seen most often at night swimming at the surface of protected bays and harbors. It may be more abundant in deep water (down to at least 1,000 ft. in some parts of the world). During the day it probably buries itself in the sand. Attaining 6 in., occasionally almost 8, it was at one time harvested for food in Hawai`i. The original Hawaiian name might have been **`ōpae loloa**, meaning "long shrimp." Although it has a wide Indo-Pacific distribution, Hawai`i may be the only locality where it is common. Photo: Hawai`i Kai, O`ahu. (surface)

MINSTREL SHRIMP

Metapenaeopsis hilarula (De Man, 1911)
Family Penaeidae
• One of several sand-dwelling penaeid shrimps occasionally encountered by divers or snorkelers in Hawai`i, *M. hilarula* is seen only at night, usually at the surface of sand patches and often half buried. When approached it digs itself in and disappears. Relatively slow-moving, specimens are easy to pick up or scoop out of the sand. The species name comes from *hilaros*, a Greek word meaning "cheerful" or "merry." To almost 3 1/2 in. Indo-Pacific. Photo: "Hale`iwa Trench," O`ahu. 70 ft.

BICOLOR SAND SHRIMP

Metapenaeopsis sp.
Family Penaeidae
• In its habits this sand shrimp is much like the species above. Although it can change its pattern, it usually displays a red carapace and abruptly lighter tail, giving it a distinctive two-color appearance. Even the legs show the two colors. It lives in rubbly sand and is active at night, Known only from Hawai`i, it may be undescribed. The specimen pictured was about 1 1/2 in. long. Photo: Hālona Blowhole, O`ahu. 50 ft.

RICHTER'S SAND SHRIMP

Trachypenaeopsis richtersii (Miers, 1884)
Family Penaeidae
• If sufficiently disturbed, this sand shrimp swims high into the water column to foil sand-grubbing predators. It can be abundant in some rubbly sand bottoms adjacent to the reef. To about 2 in. Indo-Pacific. Photo: Mākaha, O`ahu. 20 ft.

CORAL SHRIMPS. FAMILY STENOPODIDAE

Any coral reef shrimp with an enlarged pair of pincers on the third pair of legs belongs to the family Stenopodidae (infraorder Stenopodidea). This small group contains some of the most conspicuous and colorful of the reef-dwelling shrimps. When the opportunity arises most will act as cleaners, grooming larger fishes to remove parasites and dead skin. The gracefully curving antennae, usually white and very long, signal the shrimp's presence to its "customers." Stenopodidean shrimps are sometimes called "boxing shrimps" because they habitually hold their large pincers extended.

BANDED CORAL SHRIMP

Stenopus hispidus (Olivier, 1812)
Family Stenopodidae
• Most snorkelers and divers in the tropics are familiar with these striking red and white shrimps, also known as Barber Pole Shrimps and Bandana Prawns. Clinging upside down to the undersides of ledges and coral heads, almost always in pairs, they can usually be spotted by their prominent white antennae, which signal their presence to other reef-dwellers. Although these shrimps will groom eels or other fishes, in Hawai`i they are rarely seen doing so, perhaps because they are most active at night. Females are slightly larger than males and frequently carry a light greenish blue egg mass under their abdomens, held in place by their swimmerets. Fascinating aquarium animals, Banded Coral Shrimps will feed each other, guard each other during molting, perform intricate courtship dances, and breed readily in captivity. They should be kept only as mated pairs or single individuals; unmated pairs will fight. Like many shrimps, they shed their claws if stressed, and even in the wild are often seen with only one claw. (The claws grow back in the next molt, although several molts may be necessary to regain full size.) In captivity they require a recess or cave where they can shelter from the bright aquarium lights. The species name means "bristly" or "spiny" because the upper surfaces of the body and legs are covered with small hooked spines, forward-pointing at the front of the body, backward-pointing at the rear. They occur in all warm seas, from tide pools to depths of 100 ft. or more, attaining about 2 in. body length. Photo: Lāna`i Lookout, O`ahu. 25 ft.

FLAMEBACK CORAL SHRIMP

Stenopus pyrsonotus Goy & Devaney, 1980
Family Stenopodidae
• These are larger than Banded Coral Shrimps, less common, and usually found at depths of 30 ft. or more, typically upside down on the ceilings of caves. Like others of their family, they are cleaners (often of eels), although this behavior is not frequently observed in Hawai`i. A red stripe on the back of the tail gives them their species name, meaning "fire back." They are also called "Ghost Shrimp" and "Fountain Shrimp" because of their pale white color and prominent, upward-curving white antennae. To about 2 in. body length. Indo-Pacific. Photo: Mākua, O`ahu. 50 ft. (See also p. 216.)

EARLE'S CORAL SHRIMP

Stenopus earlei Goy & Randall, 1984
Family Stenopodidae
• Tiny compared to other Hawaiian *Stenopus*, these reclusive shrimps are known at present only from Hawai`i and the Comoro Islands, Western Indian Ocean. They are considered extremely rare. The body is white. A pair of red stripes originating at the sides of the carapace converge in a "V" at the tail. Like most others in their genus, these shrimps live in pairs under coral slabs and in crevices and caves. Judging from the prominent long white antennae they may be cleaners, although such behavior is not yet documented. Hawaiian specimens have been found off Mākua and Pūpūkea, O`ahu at 130 and 30 ft. respectively, and off Lāwa`i, Kaua`i, at 50 ft. The largest was slightly less than 1 in. total body length. In the mated pair pictured here the male (right) is larger than the female, unusual in *Stenopus*. Bluish green eggs held under the abdomen of the female impart a blue-green tinge to her body. The name honors Honolulu naturalist John L. Earle, who captured the first scientific specimens. Photo: Aquarium specimens collected off Wai`anae, O`ahu.

FAMILY ATYIDAE

This family of shrimps occurs in brackish or fresh water. It and all the remaining families in this section belong to the group known as caridean shrimps (Infraorder Caridea), which have enlarged pincers on either the first or second pair of legs. Most species of shrimps are carideans.

RED POND SHRIMP • `**ōpae ula**

Halocaridina rubra Holthuis, 1963
Family Atyidae
• These tiny shrimps, along with several similar species, live in anchialine pools on the Big Island, Maui, Kaho`olawe, Moloka`i and O`ahu. Anchialine pools are brackish ponds with indirect underground connections to the sea, usually found near the coast in lava flows or limestone formations. (Although most are natural, a site on Kaho`olawe lies in a bomb crater and one at Kahuku, O`ahu, is in an aquaculture well.) The shrimps feed principally on algae and bacteria and are at times numerous enough to give the pool a reddish hue. A Maui legend relates that water in the cave at Wai`ānapanapa reddens periodically to commemorate the death of the chiefess Pōpō`alaea, who hid there from her jealous husband but was found by him and slain. Specimens from O`ahu have the least color; those from the Kona coast of the Big Island are most intense. Unfortunately, this endemic species is rapidly disappearing due to coastal development (loss of habitat) and the introduction of exotic fishes, mainly *Tilapia*. The Hawaiian name means "red shrimp." They were also called `**ōpae hiki**, or "shrimp that appears." To about 1/2 in. Endemic. Photo: `Āhihi-Kīna`u, Maui.

Top picture on next page ➤

◀ Red Pond Shrimps

FAMILY PALAEMONIDAE

These caridean shrimps have enlarged pincers on the second pair of legs. Two subfamilies exist: Palaemoninae and Pontoniinae. Members of the former usually live on rocks in shallow water, sometimes in sizable numbers. They have a long, pointed rostrum that is serrated on the upper and lower sides. Shrimps of the subfamily Pontoniinae have short flattened rostrums and are usually commensal with larger invertebrates. A commensal relationship is one in which one organism (in this case the shrimp) is benefited while the other (the host) is neither benefited nor harmed.

TWO-CLAW SHRIMP

Brachycarpus biunguiculatus
(Lucas, 1849)
Subfamily Palaemoninae
• With two large pincers, this shrimp resembles a tiny clawed lobster. It is brownish overall with bands on the pincer-bearing limbs. It can be seen in tide pools at night with a flashlight and also occurs down to at least 30 ft. along rocky walls. *Palaemonella orientalis* is a synonym. To about 1 in. All warm seas. Photo: Yokohama Beach, O`ahu. 20 ft.

FEEBLE SHRIMP • `ōpae huna

Palaemon debilis (Dana, 1852)
Subfamily Palaemoninae
• These shrimps live near the water line along protected rocky shores, sometimes in considerable numbers. Almost transparent, with rows of white spots and various black lines and blotches, they have a long, upward-curving rostrum and bulbous eyes. They can tolerate and may even prefer brackish and anchialine environments and are common in the ponds at Ala Moana Beach Park, O`ahu. In Kāne`ohe Bay, O`ahu, they are one of the principle foods of the Hawaiian Bobtail Squid (p. 200). The species name means "weak." To about 1 1/4 in. Indo-Pacific. Photo: Keauhou Landing, Hawaii Volcanoes National Park, Hawai`i.

TIGER SHRIMP

Palaemon pacificus Stimpson 1860
Subfamily Palaemoninae
• Similar to the species above, these shrimps prefer slightly deeper water near shore and grow larger. The translucent body and limbs are marked with narrow dark bands. To about 2 in. Indo-Pacific. Photo: Bruce Mundy. Waimea Bay, O`ahu.

Tiger Shrimp

CLEAR CLEANER SHRIMP

Urocaridella antonbruunii (Bruce, 1967)
Subfamily Palaemoninae
• These delicate shrimps have almost perfectly transparent bodies marked with a few bright red spots. The legs and pincers are banded white and red; the antennae are light orange. The tail is white at the tips and outlined in red. They are nocturnal cleaners usually seen on eels or sleeping fish and are capable of hovering in midwater or swimming short distances from the bottom up to the fishes they clean. In Hawai`i they appear to be most common in harbors and on well-protected leeward shores. The species is named for the oceanographic research vessel *Anton Bruun*. It appears in many books under the synonym *Leandrites cyrtorhynchus*. To about 1 in. Indo-Pacific. Photo: Aquarium (cleaning a Convict Tang, *Acanthurus triostegus*). Specimen collected at Hawai`i Kai, O`ahu.

Top picture on next page ➤

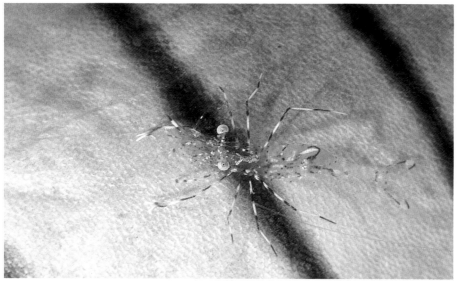

◄ Clear Cleaner Shrimp

FLATTENED CORAL SHRIMP

Harpiliopsis depressa (Stimpson, 1861)
Subfamily Pontoniinae

• Living deep within branching corals, these small shrimps are seldom seen except at night, when they venture onto the outer branches. Their flattened (depressed) bodies, greenish or yellowish and almost transparent, sometimes have thin bluish stripes along the back. They appear to feed on coral mucus and zooxanthellae and are thus dependent upon the living coral for both food and shelter. In Hawai`i they colonize Cauliflower Coral (*Pocillopora meandrina*). The number of shrimps present depends on the size of the coral head. Other species of *Harpiliopsis* are probably present in Hawai`i as well, in the same habitat. To about 1 in., but usually smaller. Indo-Pacific and Eastern Pacific. Photo: "Hale O Honu," south shore, Kaua`i. 20 ft.

223

IMPERIAL SHRIMP

Periclimenes imperator Bruce, 1967
Subfamily Pontoniinae
• In Hawai`i these red and white commensal shrimps usually live on the surface of the similarly colored Spanish Dancer nudibranch, *Hexabranchus sanguineus* (p.172), often in pairs. They also occur on other large nudibranchs such as *Dendrodoris tuberculosa* and *Asteronotus cespitosus* and on sea cucumbers of the genera *Stichopus*, *Bohadschia* and *Opheodesoma*. Like many commensal shrimps, their color varies to match the animal on which they live. Occasionally, however, they retain their imperial red regardless of the color of the host. They are uncommon and you may have to inspect many large nudibranchs and sea cucumbers before you find one. Rarity, in this case, may promote survival, the search being too costly to a predator to be worthwhile. To about 3/4 in. Indo-Pacific. Photo: Honokōhau, Hawai`i. 40 ft.

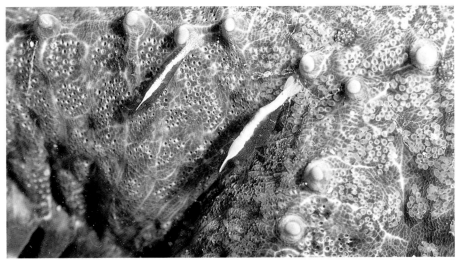

Sea Star Shrimp (a)

SEA STAR SHRIMP

Periclimenes soror Nobili, 1904
Subfamily Pontoniinae
• This tiny shrimp is most easily found by examining the underside of the Cushion Star, *Culcita novaeguineae* (p. 301). Turn the star over and the shrimp will scoot away, usually to the opposite side. Colored to match its host—red, yellow, orange, maroon, brown, or even blue—it may be difficult to spot until it moves. The body is often finely spotted with white, with a broad white stripe down the back. In Hawai`i these commensal shrimps also occur on the sea stars *Acanthaster planci*, *Linckia multifora*, *Asteropsis carinifera* and *Mithrodia* spp. Although as many as 24 shrimps have been found on a single star, 1 or 2 is more usual. To about 1/2 in. total length; most are smaller. The species name means "sister." Indo-Pacific. Photos: a)Hekili Point, Maui. 3 ft. (on *Asteropsis carinifera*); b) Palea Point, O`ahu. 40 ft. (on *Culcita novaeguineae*)

◄ Sea Star Shrimp (b)

BARRED WIRE CORAL SHRIMP

Pontonides sp. 1
Subfamily Pontoniinae
• These small shrimps have not yet been scientifically described. They live commensally on the wire coral *Cirrhipathes anguina*, possibly subsisting on their host's mucus or planktonic organisms caught in its tentacles. Their pattern of yellowish green bars separated by transparent areas blends remarkably well with the coral's polyps. Occurring in pairs (females twice the size of males), they probably spend their entire adult lives on a single coral colony, sometimes sharing it with a pair of small gobies (usually *Bryaninops yongei*). Some books confuse this species with the closely related *P. unciger*, known only from the Red Sea. Look-alike shrimps of the genera *Dasycaris* and *Miropandalus* also live on *Cirrhipathes* but are not known from Hawai`i. To about 1/2 in. total length. Indo-Pacific. Photo: Lāna`i Lookout, O`ahu. 40 ft.

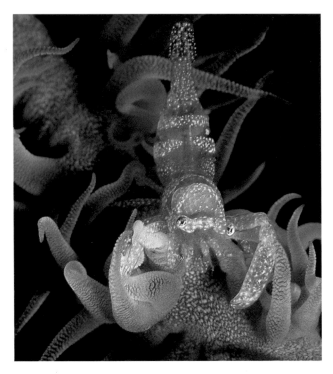

FLECKED WIRE CORAL SHRIMP

Pontonides(?) sp. 2
Subfamily Pontoniinae
• This rare shrimp is known only from a few photographs taken in Hawai`i and the Philippines. All photos show it on a species of wire coral (in Hawai`i, *Cirrhipathes anguina*). The shrimp's clear body contains yellow-orange flecks that match the similarly colored coral polyps. Photo: Ken Yates, Kona, Hawai`i. 30 ft.

WHITE-STRIPE URCHIN SHRIMP

Stegopontonia commensalis Nobili, 1906
Subfamily Pontoniinae
• These unusually slender shrimps spend their adult lives on long-spined sea urchins, attaching head inward to the long primary spines. Their bodies are purplish black with a narrow white stripe along each side that breaks up the shrimp's outline. Occurring alone or in pairs (females larger), they are commensal with Hawaiian urchins of the family Diadematidae (but uncommon on Hawai`i's dominant diadematid, *Echinothrix calamaris*, possibly because it almost always harbors parasitic crabs). They are most easily found on light-colored hosts such as *Astropyga* and *Leptodiadema*, although these urchins are uncommon. In captivity the shrimps require no special care as long as their host urchin is happy. They appear not to leave the urchin, and it is unclear on what they feed. To almost 1 1/2 in. Indo-Pacific. Photo: Mike Severns, Molokini Islet, Maui. (on *Astropyga radiata*)

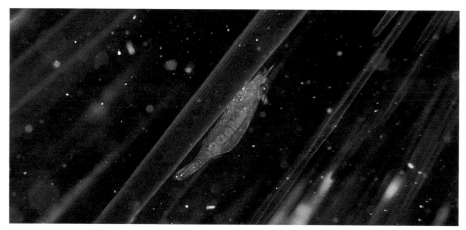

HOLTHUIS' URCHIN SHRIMP

Tuleariocaris holthuisi Hipeau-Jacqotte, 1965
Subfamily Pontoniinae
• On dark urchins these commensal shrimps are dusky purple, almost black by day, changing to red at night. On light-colored urchins they are reddish pink. They cling head inward to the long primary spines of the host, in Hawai`i usually *Echinothrix diadema* and *Astropyga radiata*. Agile swimmers, they leave the urchin if disturbed but soon return. Smaller than the White-Stripe Urchin Shrimp (above), they attain at most about 1/2 in. The species name honors Dutch crustacean specialist L.B. Holthuis. Indo-Pacific. Photo: Magic Island boat channel, O`ahu. 15 ft. (on *Echinothrix diadema*)

FAMILY GNATHOPHYLLIDAE

Members of this small family of caridean shrimps usually associate with echinoderms (sea urchins, sea stars, etc.), either as predators or commensals. Their bodies are typically short and stout with a flattened rostrum. Both the first and second pairs of legs bear small pincers, the second pair slightly enlarged. There is little difference between this family and the pontoniine shrimps (above) except that the third maxillipeds (outermost feeding appendages on either side of the mouth) are expanded into leaflike plates (the family name means "leaf-jaws").

MINER'S URCHIN SHRIMP

Gnathophylloides mineri Schmitt, 1933
Family Gnathophyllidae
• This tiny shrimp is commensal with short-spined sea urchins. Hosts in Hawai`i are *Pseudoboletia indiana* and *Tripneustes gratilla* (most commonly the former, as pictured here). The shrimp's back is pink marked with thin, closely spaced dark lines similar to those on the spines of its host. The sides bear a wide white stripe. The under-surface varies, matching the color of the urchin. The shrimp clings to the spines on the oral (under) side of its host, its head usually facing outward. It feeds on detritus caught on the urchin's spines, on the skin covering of the spine itself, and on plankton. To about 1/3 in. It occurs in the Caribbean as well as the tropical Pacific and Indian oceans. (*Periclimenes insolitus*, a smaller, more slender shrimp discovered off Waikīkī, is also commensal with the urchin *P. indiana*.) Photo: Mike Severns. Maui.

BUMBLEBEE SHRIMP

Gnathophyllum americanum
Guérin-Méneville, 1856
Family Gnathophyllidae
• These tubby little shrimps are not uncommon under stones in some areas. They occur from wading depth to at least 50 ft. The pattern of alternating light and dark lines on the body is unmistakable. Agile and quick to hide, they are rarely seen, even when specifically looked for. Captive specimens at the Waikīkī Aquarium have fed on the tube feet of sea urchins. Other observers report them feeding on the papillae of sea stars. In captivity they will also accept frozen foods and can be kept singly or in pairs. The pincer-bearing limbs of males are longer than those of females. The species occurs in tropical seas around the world. Body length of Hawaiian specimens probably does not exceed 1/2 in. *Gnathophyllum fasciolatum* is a synonym. Photo: Hālona Blowhole, O`ahu.

HAWAIIAN CAVE SHRIMP

Gnathophyllum precipuum
Titgen, 1989
Family Gnathophyllidae
• This secretive shrimp is rarely seen, generally only in caves at night (most sightings at about 30 ft.). Light yellow and glossy with numerous irregular reddish brown spots, it cannot be confused with any other Hawaiian species. The tail is clear, accentuating its tubby appearance. Although an individual has been photographed crawling about on the sea cucumber *Stichopus horrens*, no definite association is known; it may prey on tube feet like others of its genus. The species name, meaning "extraordinary," refers to an unusual pattern of teeth on the rostrum. This shrimp probably grows to no more than 1 in., total length. Known only from Hawai`i. Photo: Scott Johnson. Puakō, Hawai`i. 30 ft.

HARLEQUIN SHRIMP

Hymenocera picta Dana, 1852
Family Gnathophyllidae
• Looking more like orchids than shrimps, these fantastic animals have white and cream bodies covered with large, wine-red spots (brown with bluish edges in the Indian Ocean). The pincers expand into broad, brightly patterned leaflike plates held face out in front of the head; one pair of antennae bear flaglike extensions. The legs are banded. Thin plates extending down the sides give the effect of a gaily caparisoned medieval horse. These animals usually occur in pairs, feeding on sea stars, which they turn over and devour alive over a period of days. (Sometimes, however, a star will escape, leaving behind one of its arms.) They attack most species of sea stars occurring in their habitat, including on occasion the notorious Crown-of-Thorns. Single shrimps or mated pairs can be successfully kept in aquariums if provided with live food. A 4-in. star will keep a pair of large shrimps occupied for about a week. Females are larger than males. Wild pairs appear to be territorial, sometimes remaining in the same area for months or years. Flamboyant, slow-moving animals such as these are often toxic or bad tasting; it is possible that Harlequin Shrimps incorporate toxins from their prey. To about 2 in. Indo-Pacific and Eastern Pacific. Photo: Portlock Point, O`ahu. 30 ft.

RED PENCIL URCHIN SHRIMP

Levicaris mammillata
(Edmondson, 1931)
Family Gnathophyllidae
• These uncommon little shrimps exactly match the color of their host, down to the thin dark lines on the spines. They usually occur in pairs (the females larger), orienting themselves head outward. The pincers are relatively long and heavy. The species name comes from the name of its host urchin, *Heterocentrotus mammillatus* (p. 317). Long considered endemic to the Hawaiian chain, these shrimps are now known also from the Ryukyu and Ogasawara (Bonin) Islands of Japan. To about 2/3 in. Photo: David Fleetham. Maui.

SNAPPING SHRIMPS. FAMILY ALPHEIDAE

Members of this large caridean family are called "snapping shrimps" or "pistol shrimps." All have a greatly enlarged pincer modified to produce a loud pop or crack when snapped shut. Although the resulting shock wave can reportedly stun small prey animals, the noise is probably used most often to define or maintain the animal's territory. The continuous crackling sounds commonly heard underwater around reefs are produced by these animals. During the early days of submarines these clicks and pops frequently confused sonar operators. Although abundant, snapping shrimps are rarely seen by snorkelers and divers; typically they live in burrows, inside sponges, or deep between the branches of coral.

ORANGE-BANDED SNAPPING SHRIMP

Alpheus paracrinitus Miers, 1881
Family Alpheidae
• This snapping shrimp lives under coral slabs and stones, occurring from the shoreline to depths of at least 40 ft. The body is grayish white, marked with orange bands. Photo: Pearl Harbor boat channel. 40 ft. (identity not confirmed)

LOTTIN'S SNAPPING SHRIMP

Alpheus lottini Guérin-Méneville, 1830
Family Alpheidae
• The largest and most colorful of Hawai`i's snapping shrimps, this crustacean inhabits the spaces between branches of Cauliflower Coral, *Pocillopora meandrina*, (p. 50) and possibly other members of the genus *Pocillopora*. Its body is bright orange-red or whitish, with dark spots or stripes on the back. The pincers are finely spotted with orange. Feeding on tiny invertebrates, algae and coral mucus, the shrimp is dependent on the living coral for its survival. To about 1 1/2 in. Indo-Pacific and Eastern Pacific. (Another snapping shrimp, the orange-red *Synalpheus charon*, often lives in crevices at the bases of the same coral heads and attains about 3/4 in.) Photo: Kahe Point, O`ahu. 45 ft.

230

PETROGLYPH SHRIMP

Alpheus deuteropus Hilgendorf, 1878
Family Alpheidae
• Although these little snapping shrimps are never seen, their presence is evident on virtually any Hawaiian reef. Pairs create and inhabit the conspicuous dark fissures or channels common on the surface of massive or encrusting corals, usually *Porites lobata*. They also live in *P. evermanni*, *P. rus*, *Montipora capitata*, *M. patula* and *M. flabellata*. Their channels, (up to about 10 in. long and often branched), are sometimes reminiscent of ancient Hawaiian petroglyphs. The outer edges are lined with tiny hydroids of several species. The shrimp, which "farm" filamentous algae in the channels, live in burrows deep inside. There are usually two shrimp, a male occupying one burrow and a female the other. The shrimps are laterally compressed and mostly colorless with scattered red spots on the carapace. They may excavate the coral by chemical means. Indo-Pacific. To about 1 1/4 in. Photo: Kahe Point, O`ahu. 20 ft. (in *Porites lobata*) (See also p. 57.)

GOBY SHRIMP

Alpheus sp.
Family Alpheidae
• Two similar species of snapping shrimps (*Alpheus rapax* Fabricius, 1798, and *A. rapacida* De Man, 1911) are symbiotic with the little Hawaiian Shrimp Goby *(Psilogobius mainlandi)*. The shrimp digs a burrow that it shares with its partner. The goby, with its better eyesight, keeps watch at the entrance while the almost blind shrimp labors underground clearing and extending passages. Arriving at the surface, pushing a load of rubble like a little bulldozer, the shrimp keeps one antenna in contact with the goby's tail fin. At a hint of danger the goby twitches its tail in warning and both disappear instantly into the hole. Oddly, both species' names derive from the Latin *rapax*, meaning "grasping" or "greedy." Shrimp-goby pairs are abundant on shallow reef flats in Kāne`ohe Bay, O`ahu, and occur down to at least 70 ft. in silty sand in other calm protected areas. To about 1 1/2 in. total length. Although the goby is endemic, both shrimps have Indo-Pacific distributions. Photo: John E. Randall. Moku o Lo`e (Coconut Island), Kāne`ohe Bay, O`ahu. 3 ft.

HUMP-BACKED SHRIMPS. FAMILY HIPPOLYTIDAE

Because many shrimps of the caridean family Hippolytidae have a hump on their tail they are occasionally called "hump-back" or "broken-back" shrimps. (Many other shrimps share this characteristic.) Some hippolytid shrimps are cleaners, a few live commensally with anemones or sponges, and the rest are not directly associated with other animals.

Marbled Shrimp (a)

MARBLED SHRIMP

Marbled Shrimp (b)

Saron marmoratus (Olivier, 1811)

Family Hippolytidae

• Abundant in Hawai`i, these bizarre shrimps occur from tide pools to scuba depths but are usually active only at night. They are especially common in areas where lush coral growth provides ample refuge. Males have absurdly long claw-bearing limbs (exceeding body length) used for grappling and sparring with each other. On older males these limbs become stout and thickened. Females bear strange tufts of bristles on their bodies and first pair of legs, earning them the common name "Fuller Brush Shrimp." Because of a pronounced hump on their tails they are also called "Camel Shrimp." Individuals of both sexes are greenish by day, reddish at night. Although hardy in captivity, they hide by day and may require night feedings. Two animals of the same sex should not be kept together. Large males attain a body length of slightly over 3 in. Indo-Pacific. Photos: a) Honokōhau, Hawai`i (female); b) Hanauma Bay, O`ahu. 45 ft. (males)

EYESPOT SHRIMP

Saron neglectus De Man, 1902
Family Hippolytidae
• Uncommon in Hawai`i, this species has banded green legs, a pair of dark spots on the hump of its tail, and a smaller pair on its tail fan. The false eyespots possibly serve to confuse or discourage predators. Tufts of bristles on its back and underside, characteristic of the genus *Saron*, are much smaller than those of the Marbled Shrimp (above). Males have elongated claw-bearing limbs. These shrimps have been found inhabiting heads of branching coral. Maximum size is probably about 2 in. Indo-Pacific. The specimen illustrated was collected from under a boulder at Kewalo Park, O`ahu. 3 ft.

SQUAT ANEMONE SHRIMP

Thor amboinensis (De Man, 1888)
Family Hippolytidae
• Marked with large pale spots rimmed with blue or violet, these little shrimps are easily recognized. When disturbed they rapidly whip their tails up and down or assume a comical "tails up" stance. Shrimps of this species usually associate with anemones, corals and hydroids, but in Hawai`i such associations are not always evident. They sometimes occur with the anemones *Heteractis malu* and *Telmatactis decora* (see p. 39) but other specimens live under dead coral with no obvious host. Females, larger than males, attain about 1/2 in. Although rare in Hawai`i, this is a common Indo-Pacific and Caribbean species. Named for the Indonesian island of Ambon. Photo: Kahe Point, O`ahu. 45 ft. (See also p. 214.)

CANDY CANE SHRIMP

Parhippolyte mistica (Clark, 1989)

Family Hippolytidae

• Five or six red bands on a whitish translucent body give these shrimps their common name. The bright yellow ovaries of females are often visible through the top of the carapace. The legs are spindly and the antennae long. Inhabiting the deepest recesses of caves, usually in small groups, they venture toward the entrance only at night. The species name, meaning "mixed," refers to a combination of characters shared with related genera. Until recently this shrimp and the almost identical *P. uveae* were much confused. The latter inhabits cracks and crevices in anchialine ponds and is active by day. To at least 2 1/2 in. body length. Indo-Pacific. Photo: Kea`au Beach Park, O`ahu. 50 ft.

SCARLET CLEANER SHRIMP

Lysmata amboinensis (De Man, 1888)

Family Hippolytidae

• Although other cleaner shrimps occur in Hawai`i, none engage so actively in their trade as these flame-backed beauties. Divers extending an ungloved hand toward them will almost always get their fingernails cleaned. If given the chance, the shrimps will even clean a diver's teeth, picking and probing with their delicately clawed front legs. Most common on the leeward sides of the Islands, they inhabit crevices and holes at depths of 40 ft. or more, often in the company of large eels. They are relatively hardy aquarium animals and can be kept in a small group, accepting ordinary invertebrate foods. Named for the Indonesian island of Ambon. To about 2 1/2 in. body length. Indo-Pacific. (*Lysmata grabhami* from the tropical Atlantic is almost identical). Photo: Mākua, O`ahu. 40 ft.

STARRY LYSMATA

Lysmata sp.
Family Hippolytidae
• Species of the genus *Lysmata* are cleaners. Three are recorded from Hawai`i. This, however, corresponds with none of them and probably represents a new species. Shown here cleaning a Mustache Conger Eel *(Conger cinerascens)*, it is reddish brown, liberally speckled with tiny white spots, and marked with several transverse rows of larger brilliant blue-white spots. A pair has also been observed at Midway Atoll with the eel *Gymnothorax albimarginatus.* Photo: Jerry Kane. Pūpūkea, O`ahu. 50 ft.

HINGE-BEAK SHRIMPS. FAMILY RHYNCHOCINETIDAE

In this family of largely nocturnal caridean shrimps the spikelike rostrum between the eyes is hinged; in most other shrimps it is unmovable. Other characteristics include large bulging eyes and a humped back. Hinge-beak shrimps are often brightly colored, usually in shades of red with transverse bands. Like many crustaceans they can somewhat alter their color patterns. Although abundant in Hawai`i, they spend the day hidden deep within the reef; most are seen only by night divers. Their nocturnal habits and a tendency to attack living coral make them poor candidates for living-reef aquariums.

UNIFORM HINGE-BEAK SHRIMP

Cinetorhynchus concolor (Okuno, 1994)
Family Rhynchocinetidae
• This striking reddish to orange-red shrimp is marked with four diagonal white bands, two on the tail and two on the carapace, the lower extending to the eye. Viewed from above, the upper carapace bands form a forward-pointing "V" (sometimes indistinct). The long jagged rostrum is red with a white tip. [*C. hiatti* and *C. fasciatus* (described below), both of which it resembles, have 5 or 6 transverse white bands on the tail, two extending diagonally part way onto the rear of the carapace.] The species name means "colored uniformly." Attaining about 3 in., it is one of the largest members of its family in Hawai`i. Western and Central Pacific. Photo: Hōnaunau, Hawai`i. 30 ft.

BANDED HINGE-BEAK SHRIMP

Cinetorhynchus fasciatus Okuno & Tachikawa, 1997
Family Rhynchocinetidae
• This shrimp is red overall with five bands on the abdomen that vary from white to almost colorless and indistinct. Two bands extend diagonally onto the rear of the carapace and two cross the tail fan. The similar *C. hiatti* (p. 237) has a solid red tail fan. Known at present only from Hawai`i and the Ogasawara (Bonin) Islands, this species is very close to the Indo-Pacific rhynchocinetid shrimp *C. striatus*. To about 2 in. Photo: Hālona Blowhole, O`ahu. 15 ft.

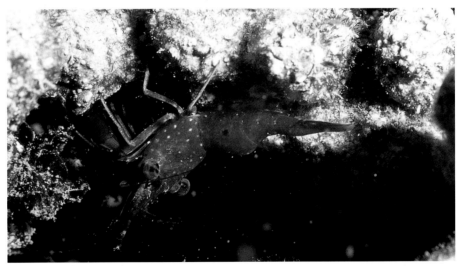

HAWAIIAN HINGE-BEAK SHRIMP

Cinetorhynchus hawaiiensis Okuno & Hoover, 1998
Family Rhynchocinetidae
• Dive at night around Finger Coral *(Porites compressa)* and you will often see dozens of eyes glowing in the beam of your flashlight. Many will belong to these small shrimps, which typically inhabit spaces between dead branches at the base of the colonies. Although abundant on many Hawaiian reefs, they have only recently been described and named. They are pale to deep red marked with scattered blue-white spots of various sizes, sometimes faint, sometimes conspicuous. There is a prominent dark spot below and just forward of the abdomen's hump. The legs may be white or reddish, sometimes faintly banded. The rostrum is reddish or clear near the base, becoming banded toward the tip. The claw-bearing limbs of old males become elongated and thickened. Known only from Hawai`i. To about 1 1/4 in. Photo: Mākaha, O`ahu. 15 ft.

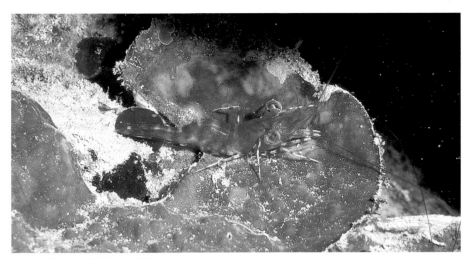

HENDERSON'S HINGE-BEAK SHRIMP

Cinetorhynchus hendersoni (Kemp, 1925)
Family Rhynchocinetidae
• The rostrum and legs of this shrimp are banded white and red. The body is whitish or almost colorless with reddish mottling on the carapace and red saddles and bands on the tail; in other parts of the world the entire body may be mottled with red or brown. The claw-bearing limbs of old males become greatly elongated and thickened. These shrimps are abundant on some shallow sheltered reefs, especially off the Kona coast of Hawai`i (often in the coral *Porites lobata*). Named for Scottish crustacean specialist John Robert Henderson (1863-1925), the species is known from Japan to the Pacific coast of Colombia. To about 1 in. Photo: Hōnaunau, Hawai`i. 30 ft.

HIATT'S HINGE-BEAK SHRIMP

Cinetorhynchus hiatti (Holthuis & Hayashi, 1967)
Family Rhynchocinetidae
• These shrimps are common in cracks and crevices of protected reefs, in harbors and along the calm sides of breakwaters and sea walls. On dark nights they sometimes crawl within a few inches of the surface and are easily seen from shore with a flashlight. The carapace and tail are red or orange-red; five or six transverse bands on the tail vary from bright white bordered by intense red to almost colorless and indistinct. The first two of these bands extend diagonally onto the rear of the carapace. The band at the base of the tail often consists of a ring of spots. The tail fans are almost uniform red, distinguishing it from the similar *C. fasciatus* (p. 236). The name honors Robert W. Hiatt (1914-1997), professor of zoology, dean and acting president at the University of Hawai`i. Hiatt was instrumental in developing the Hawai`i Marine Laboratory and the U.H. Department of Zoology. To about 1 1/2 in. Indo-Pacific. Photo: Kailua-Kona Harbor, Hawai`i. 10 ft. (see also p. 202.)

RETICULATED HINGE-BEAK SHRIMP

Cinetorhynchus reticulatus Okuno, 1997
Family Rhynchocinetidae
• · With its blotchy red-brown body and banded legs, this shrimp resembles the smaller *C. hendersoni*. Curiously, its external anatomical characters are almost identical to *C. hawaiiensis*. Less common in Hawai`i than others of its family, it is seen most often on vertical walls near caves and deep undercuts, usually in small aggregations. To about 2 1/2 in. Indo-Pacific. Photo: "Hale`iwa Trench," O`ahu. 20 ft.

RATHBUN'S HINGE-BEAK SHRIMP

Rhynchocinetes rathbunae Okuno, 1996
Family Rhynchocinetidae
• This newly described Hawaiian endemic was reported as *R. rugulosus* in earlier publications. It is easily recognized by its pattern of double red lines interspersed with white dots and lines. The ground color varies from gray-green to reddish (usually the latter). The legs are primarily white. It inhabits rocky substrate from 10 to 50 ft., and is occasionally found in the coral environment. The name honors American zoologist Mary J. Rathbun (1860-1943), who described crabs and other crustaceans from Hawai`i and the Pacific around the turn of the century. Rathbun examined the first scientific specimen of this shrimp and suggested it might be new. It sometimes enters the aquarium trade under the name "Mandarin Shrimp." Photo: "Hale`iwa Trench," O`ahu. 10 ft.

LOBSTERS.SUBPHYLUM CRUSTACEA. CLASS MALACOSTRACA
ORDER DECAPODA. INFRAORDERS ASTACIDEA, PALINURIDEA AND
THALASSINIDEA

Lobsters are decapod crustaceans with cylindrical or flattened carapaces and long muscular abdomens ending in a broad tail fan, or telson. Lobsters grow larger than shrimps and are always bottom-dwelling (benthic). Their carapaces do not project between the eyes into a distinct rostrum, and their shells are typically thick and hard. Lobsters divide easily into two groups: those with enlarged pincers on the first pair of legs, and those without (infraorders Astacidea and Palinuridea). True lobsters and their relatives form the first group; spiny, slipper, and deep-sea lobsters make up the second.

Best known of the true lobsters (family Nephropidae) is the American Lobster, *Homarus americanus*, frequently seen in restaurant tanks. The largest on record weighed 48 pounds. Lobsters of this family do not occur in Hawai`i. However, the related reef lobsters (family Enoplometopidae) are common. Even more abundant are the spiny lobsters (family Palinuridae) and slipper lobsters (family Scyllaridae). Hawai`i has at least 16 lobster species of which 12 are illustrated here. Also discussed are two species of lobster-like burrowing shrimps of the infraorder Thalassinidea.

All lobsters are known in Hawaiian as **ula**. In olden times, as in the present, they were prized as food and were eaten raw, cooked, or partly decomposed and running out of the shell. Lobsters were sometimes substituted for pigs in sacrifices to the gods.

Banded Spiny Lobsters *(Panulirus marginatus)*. Waimea Bay, O`ahu. 20 ft.

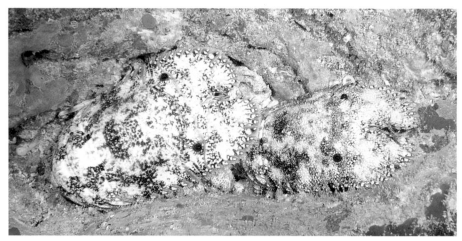

Sculptured Slipper Lobsters *(Parribacus antarcticus)*. Lāna`i Lookout, O`ahu. 20 ft.

REEF LOBSTERS. FAMILY ENOPLOMETOPIDAE

Two enlarged pincers on the first pair of legs place these small lobsters in the infraorder Astacidea along with the true lobsters of culinary fame. These animals, however, are neither large nor common enough to be of commercial importance. There are seven known species, all from the tropics.

DEBELIUS' REEF LOBSTER

Enoplometopus debelius Holthuis, 1983
Family Enoplometopidae
• This splendid reef lobster is covered with red or purple spots; the two pincers are solid purple, banded with orange at the ends. The first scientific specimen was collected by Alex Kerstitch in 1981 off Makapu`u, O`ahu, in coral rubble at a depth of 70 ft. at night; it has not been reported from the Islands since. It is either extremely rare here or its primary habitat has yet to be found. More common in the Western Pacific, it occasionally enters the aquarium trade from Indonesia and the Philippines. The name honors German author, underwater photographer and aquarist Helmut Debelius, who drew scientific attention to this unusual species after seeing it in an aquarium dealer's tank. To about 4 in. total length. Indo-Pacific. Photo: Helmut Debelius. Aquarium.

RED REEF LOBSTER

Enoplometopus occidentalis (Randall, 1840)
Family Enoplometopidae
• Rarely seen by day, this lobster is common under ledges and in caves from the shoreline to at least 50 ft. It is bright orange to purple red with white spots and markings. Long, stiff sensory hairs cover much of its body and claws. The name "Hawaiian Lobster," used in the aquarium trade, is misleading—the species occurs throughout the tropical Indo-Pacific. In captivity these animals make a good addition to a large, fish-only aquarium, but would be highly disruptive in a carefully tended mini-reef setup. They need a place to hide, and usually do so during daylight hours. They molt about every six weeks. The species name, meaning "western," resulted from a labeling error: a specimen collected in Hawai`i was marked "California." To about 5 in. Indo-Pacific. Photo: Waimea Bay, O`ahu. 30 ft.

BULLSEYE REEF LOBSTER

Hoplometopus holthuisi (Gordon, 1968)
Family Enoplometopidae
• This species is considerably slimmer than the Red Reef Lobster (above). It is orange or red with white spots and bands, especially on the legs. Each side of the carapace is marked with a white circle sometimes surrounding a central white spot—the bullseye. In Hawai`i it is most plentiful along the Kona coast of the Big Island; on O`ahu and Kaua`i it is rare. Like other members of its family this lobster inhabits crevices and caves, seldom straying far from shelter and almost never emerging by day. The name honors Dutch crustacean specialist L.B. Holthuis. To about 5 in. Indo-Pacific. Photo: Hōnaunau, Hawai`i. 50 ft.

SPINY LOBSTERS. FAMILY PALINURIDAE

The many species of spiny lobsters lack massive pincers on the first pair of legs. Most are equipped with stout, forward pointing spines on the carapace and antennae, and one pair of antennae is often greatly enlarged. These animals inhabit crevices and caves, where they can be abundant. In pristine areas dozens may be seen on a single ledge, scrambling and sliding all over each other. Spiny lobsters feed at night, often foraging widely over the sandy bottom adjacent to the reef. Easy to catch in traps and tangle nets, these large crustaceans are especially vulnerable to over-fishing. The commercial catch in Hawai`i declined precipitously from almost 40,000 pounds in 1948 to 4,800 pounds in 1968. To allow them to reproduce, the State of Hawai`i prohibits taking spiny lobsters during June, July and August. In addition, the taking of undersized lobsters and "berried" females (females with eggs) is prohibited at any time. Lobster size is measured by carapace length (from the base of the eyestalks to the rear margin of the carapace); the minimum legal size is 3 1/4 in. Spearing lobsters is illegal because it is difficult to measure the animal or determine if it is berried before taking it.

About 40 percent of adult female lobsters in Hawai`i are berried at any given time. They spawn up to four times a year, large females producing up to half a million eggs each time. Females hold their bright orange eggs in a mass under the tail (abdomen) using special appendages called swimmerets. The swimmerets also circulate oxygen-bearing water around the eggs. When about three weeks old the eggs turn brown; in another week they hatch. Development thereafter is complicated, with numerous larval stages that take almost a year to complete before a recognizable lobster settles on the reef. Juveniles a few inches long are especially colorful and can easily be kept in aquariums.

Of the three species of spiny lobsters in Hawai`i, only two are of commercial value. All three are discussed below. Also included is the uncommon Mole Lobster, a member of the family Synaxidae which scientists place between the spiny and slipper lobsters.

BANDED SPINY LOBSTER • **ula**

Panulirus marginatus (Quoy & Gaimard, 1825)
Family Palinuridae
• This endemic spiny lobster is the darker of the two commercial species occurring here. The carapace (brown, reddish or purplish) is bejeweled with dark and light nodules and spines. The spiny bases of the antennae are pink, reddish, or purple. The reddish or purplish tail segments each have a narrow white band on the rear margin, giving the species its common and scientific names. The legs vary from almost solid black to dark purplish blue, often with orange joints and faint white lines running lengthwise. The outspread tail fan is light blue edged with black and white. This lobster occurs from depths of a few feet to at least 600 ft. The old Hawaiian names **ula poni** and **ula hiwa** probably referred to it; **poni** means "purple," while **hiwa** signifies "black." Older publications identify this species as *P. japonicus*, a similar lobster not found in Hawai`i. To about 16 in. Endemic. Photo: Pūpūkea, O`ahu. 30 ft. (See also p. 239.) This species is protected by law. See Appendix B (p. 346).

TUFTED SPINY LOBSTER • **ula**

Panulirus penicillatus (Olivier, 1791)
Family Palinuridae
• Underwater this spiny lobster is immediately recognizable by the conspicuous yellow-white stripes running lengthwise along its green or brown legs. The carapace is greenish tan to bluish gray with some light blue markings, especially about the bases of the antennae. These bases themselves are brown, the spines mostly white. The segments of the tail vary from greenish to bluish gray and are densely covered with tiny light spots; there are no white bands. The outspread tail fan is almost solid green edged with yellow or orange. There are four spines below the eyes and between the antennae (the Banded Spiny Lobster has only two). Some studies indicate this species is most abundant in Hawai`i at 20 ft. or less, but it also occurs in deeper water. Throughout its range it shares habitat with the Banded Spiny Lobster, the two frequently occurring together. The species name means "brushlike" or "tufted," a reference to tufts of sensory bristles on the tips of the walking legs. To about 16 in. Indo-Pacific and Eastern Pacific. Photo: Pūpūkea, O`ahu. 30 ft. This species is protected by law. See Appendix B (p. 346).

LONG-HANDED SPINY LOBSTER

Justitia longimanus (H. Milne Edwards, 1837)
Family Palinuridae
• Spiny lobsters lack an enlarged pincer-bearing first pair of legs. At first glance, however, this species seems an exception: its first two legs are elongated and flattened, and each ends in a curved "false claw." The long "hands" are banded red and white on most individuals, solid red on others. The carapace is red with yellow markings. An uncommon lobster, it is most likely to be encountered at depths of 60 ft. or more in crevices and holes, often in the roofs of caves. Adults appear to be solitary, but smaller individuals are sometimes found together. *Justitia mauritiana* is a synonym. To about 8 in. Indo-Pacific, Caribbean Sea and tropical Western Atlantic. Photo: Lehua Rock, Ni`ihau. 50 ft.

FAMILY SYNAXIDAE

This small family of lobsters falls somewhere between the spiny lobsters (Palinuridae) and the slipper lobsters (Scyllaridae)

MOLE LOBSTER

Palinurellus wieneckii (De Man, 1881)
Family Synaxidae
• This unusual crustacean has a flattened orange body covered with short brown bristles that give it a furry appearance at close range. Living deep within submarine caves and lava tubes, it ventures near the entrance only after dark. It enters the aquarium trade on occasion and makes a hardy aquarium animal that is best kept with fishes rather than other invertebrates. The Copper Lobster (*P. grundlachi*) is a closely related Caribbean species. To about 4 in. Indo-Pacific. Photo: Kea`au Beach Park, O`ahu. 50 ft.

SLIPPER LOBSTERS. FAMILY SCYLLARIDAE

Sometimes called shovel-nosed lobsters, members of this family have flattened bodies and antennae reduced to thin plates, appearing as a pair of rounded, sometimes colorful "shovels" in front. They lack pincers and long spines, relying on camouflage and armor for protection. Like almost all decapods, slipper lobsters hide in crevices and holes during the daylight hours and forage at night. In Hawai`i they are usually found on hard substrates where they blend in well and are hard to spot. Smaller specimens do well in captivity. Five species occur within sport diving depths in Hawai`i. Their Hawaiian name, **ula-pāpapa**, means "flat lobster." Slipper lobsters are protected by law. See Appendix B (p. 346).

REGAL SLIPPER LOBSTER • **ula-pāpapa**

Arctides regalis Holthuis, 1963
Family Scyllaridae
• Before the advent of scuba this colorful and rather common lobster was considered rare. The antennae consist of flattened gray-blue plates, bright red at the periphery and equipped with a few forward-pointing, yellow-tipped spines. The rough gray carapace is covered with tubercles, some tipped with red or yellow. The beautifully sculptured tail segments are primarily orange-red. The legs are banded yellow and orange. Although appearance alone might justify the name "regal," the species is named for Mary Eleanore King, a passionate shell collector active for many years in Honolulu. "Mariel" King sponsored a number of scientific expeditions in her research vessel *Pele*, making substantial contributions to Pacific marine zoology. The first scientific specimen of this lobster was collected near Moku o Lo`e (Coconut Island), Kāne`ohe Bay, O`ahu. Known also from Easter Island, New Caledonia, and Réunion, it appears to prefer cooler waters at the edges of the tropics (a pattern of distribution common to many Hawaiian marine animals and known as "antitropical"). To about 7 in. Indo-Pacific. Photo: Palea Point, O`ahu. 40 ft. (See also p. 201.)

Top picture on next page ➤

◄ Regal Slipper Lobster

SCULPTURED SLIPPER LOBSTER
• **ula-pāpapa**

Parribacus antarcticus (Lund, 1793)
Family Scyllaridae

• These lobsters occur in warm seas throughout the world, not in the Antarctic as the species name suggests. The animal is flat and wide (almost oval when viewed from above) and easily recognized by shape alone. The rough carapace and antennae are notched and bordered by spines and bristles. It is grayish to yellowish brown, with darker mottling, often with violet marks about the eyes and spines. This is Hawai`i's most common slipper lobster. Occurring in the open on the tops of reefs (sometimes in small groups), it is less associated with caves and overhangs than others of its kind, blends in well with rocky substrate, and is easy to overlook. To about 7 in. total length, but most are smaller. Indo-Pacific, Caribbean and tropical Atlantic. Photo: Kahe Point, O`ahu. 30 ft. (See also p. 240.)

RIDGEBACK SLIPPER LOBSTER • **ula-pāpapa**

Scyllarides haanii (De Haan, 1841)
Family Scyllaridae
• This heavily armored, slow-moving crustacean is the largest Hawaiian slipper lobster. Carapace and tail are covered with low tubercles. A series of humps down the center of the back distinguish this animal from the similar Scaly Slipper Lobster (below). The flattened antennae bear no notches or spines at the edges. It is patchy gray to orange-brown without prominent markings, although the edges of the platelike antennae tend to pink or purple. Like others of its kind, it seldom ventures into the open by day. To about 20 in. Indo-Pacific. Photo: "Treasure Islands," Kona coast, Hawai`i. 45 ft. This species is protected by law. See Appendix B (p. 346).

SCALY SLIPPER LOBSTER • **ula-pāpapa**

Scyllarides squammosus (H. Milne Edwards, 1837)
Family Scyllaridae
• Smaller specimens of this lobster are often encountered on the roofs of caves, especially where orange cup coral is abundant. Adults may congregate in hollows and crevices on top of the reef. The heavily armored carapace and tail are covered with low tubercles and mottled brown and red, sometimes with whitish patches. The edges of their flattened antennae are pinkish and smooth, without notches or spines. This species feeds primarily on bivalve molluscs, which it cracks open by wedging. The second largest of Hawai`i's slipper lobsters, it attains a total length of about 16 in. Indo-Pacific. Photo: Pūpūkea, O`ahu. 30 ft. This species is protected by law. See Appendix B (p. 346).

HAWAIIAN LOCUST LOBSTER

Scyllarus sp.
Family Scyllaridae
• The genus *Scyllarus* contains over 40 small species, often called "locust lobsters." This one is undescribed and occurs along rocky Hawaiian shores, especially near crevices and caves and sometimes in living coral. It is usually solitary, but large aggregations have been observed. The species is under study by Robert B. Moffitt of the National Marine Fisheries Service Honolulu Laboratory. To about 1 1/2 in. Known to date only from Hawai`i. Photo: "Hale`iwa Trench," O`ahu. 20 ft. (At least four other locust lobsters occur in Hawai`i: *S. modestus* and *S. demani*, occur at 30-40 fathoms, while *S. aurora* and *S. aureus* are found from 80-200 fathoms. *S. timidus*, reported from Hawai`i in several publications, is a Philippine species similar to *S. aurora*.)

GHOST SHRIMPS AND AXIID SHRIMPS.
FAMILIES CALLIANASSIDAE AND AXIIDAE.

The lobster-like shrimps of the infraorder Thalassinidea are burrowing decapods with soft vulnerable abdomens and large pincers of unequal size. Ghost shrimps are common on shallow reef flats. They live in long burrows, typically with side chambers and two entrances, which they seldom if ever leave. Temperate-water species are usually translucent whitish, giving rise to the common name. Those from Hawai`i, however, are more likely to be red. Most are filter feeders. Seen less often, axiid shrimps build branching burrows in rubble bottoms. About eight species of thalassinids occur in the Islands.

BORRADAILE'S GHOST SHRIMP

Corallianassa borradailei (De Man, 1928)

Family Callianassidae

• Plentiful on certain shallow reef flats, these crustaceans live in perfectly round, smoothly lined burrows. Massive red claws, spread wide at the entrance, are the only body parts normally seen. They appear to feed exclusively on bits of drifting algae, which they drag into their burrows and consume. Most active at night, they can sometimes be tempted part way out with a dangled piece of seaweed. *Callianassa oahuensis* is a synonym. The name honors British zoologist L.A. Borradaile (1872-1945), who studied and classified crustaceans of this family. To about 4 in. total length. Indo-Pacific. Photo: Kualoa, Oʻahu. 3 ft.

SERRATED AXIID SHRIMP

Axiopsis serratifrons (A. Milne Edwards, 1873)

Family Axiidae

• These lobster-like shrimps live in burrows in coarse sand mixed with rubble. Unlike ghost shrimps, they leave their burrows at night but do not venture far. During the day they can sometimes be exposed by turning over stones. In Hawaiʻi they occur at scuba-diving depths. Elsewhere in their wide range they may occur on shallow reef flats. The top of the carapace bears lines of fine, forward-pointing serrations, hence the name. To almost 2 1/2 in. Indo-Pacific and tropical Atlantic. Photo: Palea Point, Oʻahu. 50 ft.

HERMIT CRABS AND ALLIES. SUBPHYLUM CRUSTACEA
CLASS MALACOSTRACA. ORDER DECAPODA. INFRAORDER ANOMURA

Hermit crabs and their allies—porcelain crabs, squat lobsters and mole crabs—are a diverse group of decapods collectively known as anomurans. Their tails are shorter than those of lobsters but longer than those of true crabs (brachyurans), and their last pair of legs is always much reduced and sometimes turned upward. The word "anomura" means "irregular tail." Hermit crabs—the largest anomuran group—have asymmetrical tails that they usually coil into empty snail shells. The tails of all other anomurans are symmetrical, folding forward under the body. Most anomurans have chelipeds—an enlarged first pair of legs bearing strong pincers, often of unequal size. Approximately 1,575 anomuran species are known, almost all marine. A few inhabit fresh water, and some live on dry land as adults. Forty-four species are known from Hawai`i.

Jeweled Anemone Crab *(Dardanus gemmatus)* in Partridge Tun shell. Koloa Landing, Kauai 20 ft.

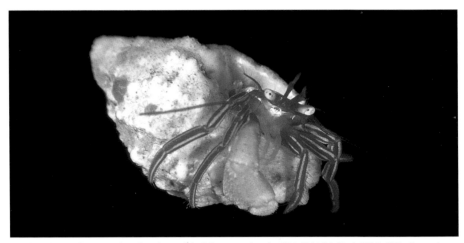

A rarely-seen deep-water hermit crab, possibly *Pylopaguropsis zebra.* Molokini Islet, Maui. 190 ft. (Mike Severns)

HERMIT CRABS. FAMILIES DIOGENIDAE AND PAGURIDAE

Anyone who has explored Hawai`i's tide pools is familiar with the hermits, the little crabs that live in snail shells. Although their front parts are covered by an external skeleton like other crabs, their long soft tails (abdomens) are typically unprotected. For this reason, most hermit crabs occupy an empty snail shell, coiling their tails deep inside and holding on from within with the help of their tiny, specialized last pair of legs. Not only is the tail protected by this arrangement, the entire animal can withdraw into the shell when threatened. It may seal the shell's opening from the inside with an oversize claw, or retreat so far that no body part is visible. Hermit crabs leave their shells only when changing to another. (They can pull out their tail and back into a new shell in the blink of an eye.) Otherwise, they are almost impossible to remove; in most cases the body will tear before they let go.

As hermit crabs grow they need ever-larger quarters. Seldom satisfied with their current home, they will try on any suitable empty shell they find. If it happens to be occupied by another crab (as is usually the case), a ritualistic fight may ensue. The attacker raps rapidly with its own shell on the shell it wants. The occupying crab, if smaller or of a less aggressive species, usually vacates, and the two exchange shells. If snail shells are in short supply, other hollow objects, such as the spent shell of a bullet, are sometimes substituted. Species of hermit crabs have evolved to fit shells of all sizes and shapes (including bivalves), from the giant Triton's Trumpet to tiny shells 1/8 in. long. An entire genus of hermits uses hollow pieces of bamboo; another group has traded mobility for a fixed existence in worm tubes or other small cavities in the coral. Members of the family Lithodidae, which includes the famous Alaskan King Crab *(Paralithodes camtschatica),* are technically hermits but carry no shell at all.

Although hermit crabs have five pairs of legs like other decapods, only the second and third pairs (periopods) are typically used for walking. The first pair (chelipeds) bear claws. The reduced fourth pair grip the edge of the shell when the crab extends forward, and the tiny fifth pair help wedge the coiled tail in the inner whorls of the shell. To extend beyond the shell's upper edge, the eyes are set on stalks.

In Hawai`i, hermit crabs occur from the shoreline to depths of 1,000 ft. or more. Some species are gregarious, occurring by the dozens or even hundreds in pools or under stones. In other parts of the tropics hermits have invaded dry land. The Coconut, or Robber, Crab of the Indo-Pacific *(Birgus latro)* is the world's largest land crab. Only the young inhabit shells; adults develop tough, leathery abdomens that need no protection. Weighing several pounds each, these crabs often climb coconut palms and are said to cut down the nuts, although this behavior has not been confirmed. In any case, their extremely powerful claws can easily break plastic pens and could undoubtedly nip off a human finger. By contrast, most Hawaiian hermit crabs are small, timid creatures that can be safely picked up even by children.

Hermit crabs typically withdraw into their shells when disturbed, but they usually emerge in a few minutes and begin crawling about. Small ones make ideal pets for a beginner's aquarium. Given several extra shells, three or four hermit crabs can provide hours of entertainment. Most species will eat standard fish or invertebrate food, scavenging for scraps left by other tank inhabitants. Many will also eat algae growing in the aquarium. Hermit crabs are most active after dark; in the still of the night they can make considerable noise in a small tank or fishbowl, clacking and clattering against the glass. Although small hermits are easy to keep and make useful scavengers, the larger species can wreak havoc in an aquarium not designed for them, attacking other invertebrates and moving rocks.

Of the six hermit crab families, only three are commonly seen. The Diogenidae, to which most Hawaiian species belong, are marine hermits whose left claws are usually larger than their right. (The family name recalls Diogenes, an ancient Greek philosopher who lived in the streets and was said to shelter in a large tub.) The Paguridae consist of marine species whose right claws are usually larger than their left. The Coenobitidae are the land hermits, not present in Hawai`i. At least 23 species of marine hermit crabs occur in local shoreline waters; 17 are described here. The general Hawaiian name for hermit crabs is **unauna**.

HOPPER'S HERMIT CRAB

Aniculus hopperae McLaughlin & Hoover, 1996
Family Diogenidae
• This recently described hermit lives in caves and under ledges along exposed rocky shores at depths of 3 to at least 70 ft. The walking legs and claw-bearing limbs are encircled with ringlike scales. The bright red claws have black tips; the two segments above the claws (carpus and merus) are dark brown, roughly crosshatched with white. The legs are red, the last segment (dactyl) speckled with white. The upper segments of the legs have transverse white bands and a longitudinal purple stripe. All limbs bear tufts of white-tipped bristles. The eyes are black, the eyestalks and the antennae orange. The few crabs of this species collected to date have all occupied round-mouthed drupe or triton shells. Two given to the Waikīkī Aquarium refused to eat in captivity, behavior unusual for a hermit crab. Their feces contained sponge spicules, a possible indication of diet. The name honors Dr. Carol N. Hopper of the Waikīkī Aquarium for her work with young people. To at least 1/2 in. carapace length. Known to date only from the Hawaiian Islands. Photo: Pūpūkea, O`ahu. 20 ft.

HAIRY YELLOW HERMIT CRAB

Aniculus maximus Edmondson, 1952
Family Diogenidae
• Many hermit crabs are hairy, but none more magnificently endowed than this colorful giant. Its walking legs and claw-bearing limbs are intense golden or orange-yellow with crimson trim. Ringed by white-edged scales, they appear highly segmented. From under the scales sprout dense tufts of long, off-white bristles. This crab usually occurs at depths of 50 ft. or more, typically under ledges or in caves and often occupying a large Triton's Trumpet shell *(Charonia tritonis)*. It grows to the size of a man's hand and can wreak havoc in an aquarium. Discovered in the Hawaiian Islands, it was scientifically described by the great student of crabs Charles H. Edmondson (1876-1970) of Honolulu's Bishop Museum; he called it *maximus*, meaning "the greatest." It has an Indo-Pacific distribution. Photo: "Black Coral Arch," east shore Kaua`i. 70 ft.

CONE SHELL HERMIT CRAB

Ciliopagurus strigatus
(Herbst, 1804)
Family Diogenidae
• The walking legs and claw-bearing limbs of this unusual hermit crab are ringed with vivid orange and red scales. The eyes, eyestalks and antennae are orange. In total contrast, the carapace (usually hidden within the shell) is chalky white. The crab's flattened body and limbs enable it to inhabit cone shells, into which most other hermits do not easily fit. Rarely, it will occupy other snail shells. This species occurs most frequently at depths of 20 ft. or more. Until recently known as *Trizopagurus strigatus*, it still appears under that name in many books. The maximum carapace length is probably about 1 in. Indo-Pacific. Photo: Kahe Point, O`ahu. 20 ft. (in a Textile Cone, *Conus textile*)

ARGUS HERMIT CRAB

Calcinus argus Wooster, 1982
Family Diogenidae
• *Calcinus* is a large genus of small, often colorful hermits. Nine species are known from Hawai`i; eight are illustrated here. The Argus Hermit Crab, although uncommon, is perhaps easiest to identify at a glance: its walking legs and claw-bearing limbs are maroon with white spots. The eyes are black with a few white spots; antennae and eyestalks are brownish. The left claw is larger than the right. The carapace is white. This hermit occurs on reef flats and to depths of at least 40 ft. and is named after the hundred staring eyes of the mythical monster Argus. To about 1/3 in. carapace length. Known from Guam, the Mariana Islands and Hawai`i. Photo: Mākua, O`ahu. 40 ft.

ELEGANT HERMIT CRAB

Calcinus elegans (H. Milne Edwards, 1836)
Family Diogenidae
• This colorful species is found from tide pools to depths of at least 30 ft. It is most common below the intertidal zone, with larger individuals occurring at the deeper end of the range. The walking legs are dark brown with bright orange bands and the last segments are bright orange with black spots. (Orange coloration occurs only in Hawaiian specimens. Elsewhere in the Indo-Pacific it is replaced by bright turquoise blue.) The two almost equal-size claws are brownish speckled with white and have white tips. The eyes and eyestalks are bright blue, the antennae orange, the back white. Attractive and moderately large (but not too big to cause damage), this is a good hermit for the aquarium. The maximum carapace length is about 3/4 in. Indo-Pacific. Photo: Hekili Point, Maui. 3 ft. (in triton shell)

GUAM HERMIT CRAB
Calcinus guamensis Wooster, 1982
Family Diogenidae
• In Guam where it was discovered, this hermit lives intertidally and on shallow reef flats where there is clear, moving water. In Hawai`i it is subtidal, occurring as deep as 100 ft. or more. The predominant color is light olive drab on large individuals, light yellowish on small ones. The two claws are of almost equal size. The black segment above each claw (carpus) is speckled with white, much like those on *C. latens*. The legs are uniform in color except for a narrow black band on the final segment. There is a black band around the middle of each eyestalk. The eyes are black with tiny white spots. Although the maximum carapace length is about 1/4 in., these hermits are usually far smaller. Known from Guam, the Mariana Islands, Hawai`i and the Marquesas. Photo: Pūpūkea, O`ahu. 20 ft.

HAIG'S HERMIT CRAB
Calcinus haigae Wooster, 1982
Family Diogenidae
• This uncommon hermit has bright purple legs with a narrow white band at the tips. The terminal leg segments (dactyls) are spotted with orange or crimson, as are parts of the segments immediately above. The two claw-bearing limbs are dark purple, fading to lighter purple and finally white toward the tips of the claws. The eyestalks are brownish purple with a narrow white band just below the eyes, which are black with white spots. It occurs at depths of 20 ft. or more in heads of branching coral (genus *Pocillopora*) or on hard or rubbly substrate. This species resembles two other Hawaiian hermits: the more common *C. hazletti* (below) has reddish brown legs with completely white dactyls, and *Pylopaguropsis keijii* (p. 261) has purple legs and a large, light tan right claw. The species name honors Janet Haig (1925-1996), longtime student of hermit crabs and porcelain crabs. Maximum carapace length is probably about 1/4 in. Indo-Pacific. Photo: Mākaha, O`ahu, 30 ft.

HAZLETT'S HERMIT CRAB

Calcinus hazletti Haig & McLaughlin, 1983
Family Diogenidae

• Living well below the intertidal zone, this is one of several Hawaiian hermit crabs that remained unnoticed by scientists until the 1970s. It often occurs in heads of branching coral (genus *Pocillopora*). At the slightest threat it releases its hold, dropping deep between the branches where it is safe from predators (a tactic common to many hermit crabs). Its legs are reddish brown except for the terminal segments (dactyls), which are almost all white. The claws are white, the segments above the claws dark reddish brown. The eyes are black with white spots, the eyestalks dark orange-red with a broad white band next to the eye, and the carapace magenta flecked with white. This is a very common crab off the Wai`anae coast of O`ahu at depths of 20 to 50 ft. or more, frequently inhabiting the Prickly Horn Shell *(Cerithium echinatum)*. At night dozens of these little crabs are sometimes found together, perhaps congregating both to feed and exchange shells. The name honors Brian Hazlett, who discovered this species (and *C. laurentae*, below) while studying the shell-fighting and mating behaviors of Hawaiian hermit crabs. Carapace length to about 1/5 in. Known only from the Hawaiian Islands. Photo: Kahe Point, O`ahu. 20 ft.

LEFT-HANDED HERMIT CRAB

Calcinus laevimanus
(Randall, 1839)
Family Diogenidae

• This is one of the most common shoreline and tide-pool hermit crabs in Hawai`i. It does not occur in water more than a few inches deep and prefers areas of low to moderate wave action. The left claw, mostly white with some dark brown, is enormous compared with the right; it is used to block the shell's opening when the crab retreats inside (a behavior typical of many hermit crabs). The legs are brown except for the tips, which are white with a black band. The eyes and lower half of the eyestalks are bright blue, the remainder of the eyestalk is orange. In Hawai`i it most often occupies turban, top or nerite shells. The pair illustrated, however, are inhabiting Swollen Bubble Shells *(Hydatina amplustre)*. Earlier books may use the synonym *C. herbstii* for this species. It attains about 1/4 in. carapace length. Indo-Pacific. Photo: Kīpapa Island, O`ahu.

HIDDEN HERMIT CRAB

Calcinus latens (Randall, 1839)
Family Diogenidae

• The legs and claws of this crab are purplish green becoming lighter at the tips. The segments above the claws (carpus and merus) are black speckled with white (as in the less common *C. guamensis*). The eyes are black with white or light blue spots, the eyestalks greenish to whitish, the antennae yellowish. The best identifying feature, although hard to see on such a small animal, is the bright purple band (actually composed of short blue and violet vertical stripes) near the tip of each walking leg. This crab occurs from the shore down to at least 30 ft. The species name means "hidden" or "secret." To about 1/2 in. carapace length. Indo-Pacific. Photo: Kewalo Park, O`ahu. 2 ft.

LAURENT'S HERMIT CRAB

Calcinus laurentae Haig & McLaughlin, 1983
Family Diogenidae

• This hermit crab is common at depths of 20 ft. or more in heads of branching corals of the genus *Pocillopora* or on rocky substrate. The legs are red, the last segments pink with little black tips. The claw-bearing limbs are brown, darkest close to the body and becoming white at the tips. The eyes are black with white spots, the antennae yellow-orange. The species is named for Michèle de Saint Laurent of the Muséum National d'Histoire Naturelle, Paris, who assisted authors Haig and McLaughlin in their original scientific description. To about 1/5 in. carapace length. Known only from Hawai`i. Photo: Kahe Point, O`ahu. 25 ft.

SEURAT'S HERMIT CRAB

Calcinus seurati Forest, 1951
Family Diogenidae
• These gregarious hermit crabs live in rocky pools of the splash zone just above sea level among the periwinkles and ner-ites, whose empty shells they usually occupy. They can tolerate warm stagnant water and will sometimes crawl about above the waterline on wet, algae covered rock. Bold black and white bands on the legs make them obvious as they move. The claws are gray. The left claw, much larger than the right, is used to block the entrance to the shell when the crab withdraws. Like many crabs of the genus *Calcinus*, the eyes are bright blue, the eyestalks orange. These hermits seem to be most com-mon on the windward sides of the Islands, where the sea is rough and the waves splash high. They are named for Gaston Seurat, who first collected them in French Polynesia between 1901 and 1905. To about 1/4 in. carapace length. Indo-Pacific. Photo: Makapu`u, O`ahu. Tide pool.

ZEBRA HERMIT CRAB

Clibanarius zebra (Dana, 1851)
Family Diogenidae
• Hermit crabs of the genus *Clibanarius* have claws that open horizontally instead of vertically; they are typically less color-ful than species of *Calcinus*. Many, this species included, tolerate slightly brackish conditions. The overall color of the Zebra Hermit is blackish gray; the carapace displays several longitudinal white stripes, accounting for its common and scientific names. The legs have faint white longitudinal stripes and speckles, except for the bluish terminal segments. The antennae are orange. This is a small intertidal species, occurring in sheltered or semi-sheltered locations such as the lagoon at Honolulu's Magic Island and Kāne`ohe Bay, O`ahu, where it is by far the most common intertidal hermit crab. Sometimes a dozen or more of these tiny crabs can be found under one rock; other rocks nearby may be devoid of them. This is the only known *Clibanarius* species in Hawai`i. To about 1/4 in. carapace length, but often much smaller. Indo-Pacific. Photo: Kīpapa Island, O`ahu.

JEWELED ANEMONE CRAB

Dardanus gemmatus (H. Milne Edwards, 1848)
Family Diogenidae
• These hermits usually carry anemones of two species on their shells. Up to half a dozen or more large *Calliactis polypus* typically attach to the outside, sometimes covering the shell completely. When disturbed they eject bright pink stinging threads (acontia). Several tiny white anemones (*Anthothoe* sp.), usually perch on the upper lip of the shell above the crab. In Hawai`i both these anemones almost always occur with a crab, gaining mobility and possibly scraps of food. A flatworm *(Stylochoplana inquilina)*, an amphipod *(Elasmopus calliactis)* and, in deeper water, a stalked barnacle *(Koleolepas tinkeri)* often live commensally on the crab/anemone pair. The crab itself is predominantly brown to purplish orange. The eyestalks are red, banded with white like a barber's pole, and the eyes are greenish or silver. The outer surface of the large left claw is completely covered with small rounded bumps (tubercles). This animal occurs from the intertidal zone to 100 ft. or more, hiding by day and foraging by night. The species name comes from *gemma* ("gem" or "jewel") and probably refers to the tubercles on the large left claw. (The similarly colored but uncommon *D. pedunculatus*, also recorded from Hawai`i, bears anemones and has a large left claw only half covered with tubercles.) To about 2 in. carapace length. Indo-Pacific. Photo: Hōnaunau, Hawai`i. 15 ft. (See also p. 249.)

PALE ANEMONE CRAB

Dardanus deformis (H. Milne Edwards, 1836)
Family Diogenidae
• Less common than *D. gemmatus* (above), this anemone-bearing hermit is predominantly light gray-brown and cream. The legs are banded. The outer surface of the large, cream-colored left claw lacks tubercles except for a single row along one side. The almost luminous green eyes are set on short white stalks that bear a dark brown transverse band. The species name means "misshapen" or "ugly." These crabs usually carry the same anemone species as *D. gemmatus* and occur on shallow reef flats. They attain at least 3/4 in. carapace length. Indo-Pacific. Photo: Ala Moana Beach Park, O`ahu. 2 ft.

WHITE-SPOTTED HERMIT CRAB

Dardanus megistos (Herbst, 1812)
Family Diogenidae
• This is probably Hawai`i's largest hermit crab and one of the largest in the world. In Hawai`i it seldom occurs above depths of 150 ft.; elsewhere in the Indo-Pacific it often inhabits shallow water. The color is unmistakable: legs and claws are reddish to bright orange-red and covered with white spots ringed in black. The carapace, when visible, is yellowish with similar spots. Tufts of dark red bristles sprout from most of the spots. The left claw is larger than the right. It usually inhabits a Triton's Trumpet shell *(Charonia tritonis).* The species name means "largest" or "greatest." The entire body of this crab may be 12 in. long. (*D. brachyops,* another large, deep-water hermit is buff with orange bristles and bears anemones.) Indo-Pacific. Photo: Mike Severns, Molokini Islet, Maui. 180 ft.

Bloody Hermit Crab (a)

Bloody Hermit Crab (b)

BLOODY HERMIT CRAB

Dardanus sanguinocarpus Degener, 1925
Family Diogenidae

• Bright red and purple splotches on the segments above the claws (carpus and merus) give this hermit its common and scientific names. Each walking leg has a large black spot on the second segment (red in deeper water). The above markings may be faint or absent on specimens living on sand or rubble bottoms below about 50 ft. The predominant color is brownish gray; the limbs are covered with black bristles tipped with greenish yellow. The left claw is slightly larger than the right. The eyes are black, the eyestalks yellow. This crab is active by day, inhabiting a variety of shells including cones and cowries. The species name, "bloody carpus," refers to the segment immediately above the pincers, analogous to a wrist. Carapace length to about 1 in. Described from Hawai`i, it is known also from the Marquesas Islands and possibly the Marshall Islands. Photos: a) Honokōhau, Hawai`i. 20 ft. (shallow-water coloration, in *Latirus nodatus*); b) "Hale`iwa Trench," O`ahu. 70 ft. (deep-water coloration, in *Mitra papalis*)

NOMURA'S HERMIT CRAB

Pagurixus nomurai Komai & Asakura, 1995
Family Paguridae
• These small hermits occur under stones along protected shores, often in groups. Their right claw is larger than their left (characteristic of the family Paguridae). The claws and walking legs are light reddish brown with white bands on the upper segments; the antennae are light reddish brown. The species is named for Mr. Keiichi Nomora of the Kushimoto Marine Park, Japan, collector of the first scientific specimens. In Japan this species occurs to depths of about 40 ft. The first known Hawaiian specimens were collected in 2 ft. of water at Ala Moana Beach Park by Mr. Darrell Takaoka in 1998. Maximum carapace length is probably about 1/5 in. Known to date only from Hawai`i and Japan's Ryukyu and Izu Islands. Photo: Ala Moana Beach Park, O`ahu. 2 ft.

KEIJI'S HERMIT CRAB

Pylopaguropsis keijii McLaughlin & Haig, 1989
Family Paguridae
• Only dedicated hermit hunters are likely to notice this small colorful crab, which lives in coral rubble from depths of about the shoreline to over 200 ft. Its walking legs and small left claw are bright purple. Its light tan right claw is enormous compared to the left claw. The name honors the eminent Japanese crab scecialist Dr. Keiji Baba. Maximum carapace length is probably about 1/5 in. Indo-Pacific. Photo: Magic Island, O`ahu. 20 ft. (In addition to this species and the one above, Hawai`i has at least three other shallow-water hermits with a large right claw; all are tiny and rarely seen.)

Squat Lobster (unidentified)

SQUAT LOBSTERS. FAMILY GALATHEIDAE

These small anomurans have oval carapaces drawn out in front into a sharp point (rostrum). The tail is bent under the carapace, giving the animal a short, squat appearance. Two lobster-like claws usually extend forward. A small fifth pair of legs are carried under the carapace. Hawaiian species are small and inconspicuous. Elsewhere in the Indo-Pacific, colorful squat lobsters live commensally with crinoids and other invertebrates. This Hawaiian species (identity unknown) was photographed at night at "Mākaha Caverns," off Mākaha Beach, O'ahu. 30 ft.

PORCELAIN CRABS. FAMILY PORCELLANIDAE

Porcelain crabs are anomurans closely related to squat lobsters. Species of the genus *Neopetrolisthes* (commensal with large sea anemones and not found in Hawai'i) are marked with colorful spots on a porcelain-white body. This, and the tendency of porcelain crabs to "break easily" (i.e., to shed their claws when stressed), may account for the unusual common name. Although they resemble true crabs, porcelain crabs have whiplike antennae, a muscular forward-folded tail (used for swimming backward, shrimp-fashion), and a reduced fifth pair of legs (used for cleaning and grooming). Filter feeders, they trap food particles or plankton in the water with their specialized third maxillipeds, which are held upright near the mouth.

Red Porcelain Crab (a)

RED PORCELAIN CRAB
• **kūmimi māka`o**

Petrolisthes coccineus Owen, 1839
Family Porcellanidae
• These active little crabs occur from shallow reef flats to depths of at least 50 ft., generally in crevices and often in groups. They often work the bottom with their large flattened pincers, stirring up detritus that they filter from the water with their bristly third maxillipeds. The species name means "red like a berry." To about 1/2 in. carapace width. Indo-Pacific. Photos: a) Kailua-Kona Harbor, Hawai`i. 15 ft. (feeding); b) Waikīkī, O`ahu. 1 ft.

Red Porcelain Crab (b)

MOLE CRABS. FAMILY HIPPIDAE

These small sand-dwellers have an almost cylindrical carapace with a symmetrical tail flexed beneath. There are no large pincers. The fifth legs are reduced and usually folded beneath the carapace. Some mole crabs are filter feeders (using their modified second antennae), others are predator/scavengers.

PACIFIC MOLE CRAB • **pe`eone**

Hippa pacifica (Dana, 1852)
Family Hippidae
• Living in the wave wash zone of sandy beaches (coarse sand preferred), these scavenger/predators remain buried, often with their antennae and eyestalks slightly protruding. When something edible washes by, such as a beached Portuguese Man-of-War, they grab what they can (usually a tentacle) and drag it under the sand to be eaten. If a piece of carrion is too large they will sometimes eat it at the surface. So completely adapted to digging are these animals that they cannot walk. They closely match the local sand color from white (at Lanikai, O`ahu) to reddish brown (at Kīhei, Maui) to black (at Punalu`u, Hawai`i). If experimentally moved from one beach to another they change color incrementally as they molt. Mole crabs are favorites of children, who can easily dig them up and examine them. They are often called "sand turtles" because of their oval, convex bodies. The Hawaiian name means "sand hiding." To about 1 1/2 in. Indo-Pacific. Photos: Susan Scott, Kailua Beach, O`ahu. a) feeding on beached filefish; b) on hand.

TRUE CRABS. SUBPHYLUM CRUSTACEA. CLASS MALACOSTRACA. ORDER DECAPODA. INFRAORDER BRACHYURA

True, or brachyuran, crabs—as opposed to hermit crabs and their allies (anomorans)—are the typical crabs of the seashore. Their broad flattened carapaces and sideways scuttle are familiar to most beachgoers. These crustaceans have a short, forward-folded tail visible only if the animal is turned on its back. (The word "brachyura" means "short tail.") The tails of males are narrow and pointed; those of females are broader, to hold eggs. Both tuck snugly into a hollow under the body. True crabs have a pair of chelipeds—enlarged pincer-bearing limbs—and four pairs of walking legs. In the large family Portunidae, however, the last pair of legs are flattened into paddles for swimming. True crabs are the most evolutionarily advanced decapods, accounting for about one third of all decapod species. Their abbreviated tails give many of them an agility that long-tailed shrimps and lobsters cannot match, and their flattened compact bodies fit easily into small crevices and under stones.

True crabs vary greatly in size. The world's heaviest, *Pseudocarcinus gigas* from southern Australia, weighs up to 35 pounds; the famous Japanese Spider Crab *(Macrocheira kaempferi)*, its body 18 in. wide, measures an astounding 12 ft. between its outstretched claws! The smallest known crabs, commensals on sand dollars, are about 1/16 in. across. Crabs may be predators, scavengers or algae eaters. A number live in association with other animals. Most brachyuran crabs are marine; some inhabit fresh water. A few spend their adult lives on dry land, returning to the water only to spawn. Of some 4,500 species, about 190 are known from Hawai`i; 48 are shown here.

In Hawai`i, small crabs are common along rocky shores and under stones in shallow water. Larger ones of commercial value usually inhabit muddy, brackish, or sandy environments. One of the largest Hawaiian species, the deep-dwelling Japanese Homolid Crab *(Paromola japonica)*, has a carapace about 5 in. wide and a leg span of about 3 ft. Specimens are occasionally brought up from depths of 30 to 500 fathoms, entangled in fishing lines.

Unidentified swimming crab, perhaps of the genus *Charybdis*, with Reticulated Cowry *(Cypraea maculifera)*. Kea`au, O`ahu. 50 ft.

The general name for crabs in Hawaiian is **pāpaʻi**, but many groups had special names such as **kūmimi** (inedible crabs), **kukuma** (grapsid crabs), **ʻelekuma** (small xanthid crabs) and **pokipoki** (box crabs). Prominent or unusual species had unique names. In old times as now, crabs were prized for their flavor. Hawaiians living inland, deprived of daily seafood, were said to get "crab-hungry." Depending on the variety, crabs were eaten raw, salted, or cooked. Some inedible species were used in sorcery and medicine.

SPONGE CRABS. FAMILY DROMIIDAE

Sponge crabs are considered the most evolutionarily primitive of the true crabs. Most are slow-moving and nocturnal. Their two hindmost legs, permanently bent up over the carapace, are tipped with short, needle-like pincers. Inserting these into a piece of sponge or other foreign object (like pins in a cushion), the crab holds it over its back as a disguise. Colonial tunicates, bivalve shells, even sea stars can also be carried. Hawaiʻi has perhaps half a dozen species of sponge crabs. The most common may be *Cryptodromiopsis tridens*, a small, cryptic shallow-water crab sometimes occurring in heads of Cauliflower Coral (*Pocillopora meandrina*). It is not illustrated here.

SLEEPY SPONGE CRAB ▪ **makua-o-ka-līpoa**
Dromia dormia (Linnaeus, 1763)
Family Dromiidae
• This is the world's largest sponge crab. Its limbs and domed carapace are covered by brown, muddy-looking fuzz; the tips of its massive claws, often held under the body out of sight, are bone-white. Small, closely set beady black eyes and a chunk of sponge held over the back complete the picture of this awkward, slow-moving animal. When large sponges are scarce, these crabs will carry rubber slippers, bits of rope, wood or even metal. One found off ʻEwa, Oʻahu, was carrying a Crown-of-Thorns Star, the arms neatly clipped for a perfect fit. Evidence exists that Sleepy Sponge Crabs will feed on these stars, in which case the ʻEwa specimen may have been carrying a meal as well as camouflage. These crabs hide by day in caves and under sponge-coated overhangs; at night they clamber about slowly over the reef. They are considered inedible and may even be poisonous. The species name means "sleepy." To about 8 in. carapace width. Indo-Pacific. Photo: Lānaʻi Lookout, Oʻahu. 20 ft.

Cryptodromiopsis plumosa (Lewinsohn, 1984)
Family Dromiidae

• Working in secret, this small, shaggy-legged crab cuts and shapes a cap of sponge, hollowing out a space that perfectly fits the dome of its carapace. The finished cap completely covers the crab, making it almost impossible to detect. Although superbly camouflaged, this cautious creature takes no chances, emerging only after dark. It does, however, show a bit of flair: its white eyes are flecked with bright crimson polka dots. If bothered, it folds its legs closely against the edge of its carapace, draws its underside flush with the sponge, and waits out the disturbance. To about 1 in. carapace width. Known only from the Seychelles, New Caledonia and Hawai`i. Photo: Palea Point, O`ahu. 50 ft.

FROG CRABS. FAMILY RANINIDAE

In many parts of the world these sand-dwelling crabs with long carapaces are called "spanner crabs" because their odd, right-angled claws resemble monkey wrenches ("spanners" to the English). They often rest froglike on their hind legs; if disturbed they do not spring but dig backwards rapidly into the sand. At least two species occur in Hawai`i.

KONA CRAB • **pāpa`i kua loa**

Ranina ranina (Linnaeus, 1788)
Family Raninidae

• Kona Crabs are the largest and most common frog crabs in Hawai`i. They live on sand bottoms, usually at depths of 30 ft. or more, and are difficult to see and even more difficult to approach. Edible and large, they are trapped commercially in many parts of the world including Hawai`i. The Hawaiian name means "long-backed crab." The common name, "Kona Crab," is strictly local. Both the genus and species names mean "frog." To 9 in. carapace length. Indo-Pacific. Photo: Kahe Point, O`ahu. 50 ft. This species is protected by law. See Appendix B (p. 346).

BOX CRABS. FAMILY CALAPPIDAE

Box crabs of the family Calappidae usually have winglike extensions of the carapace that completely cover the legs. Their large, oddly shaped claws fold snugly along the front forming a continuous shield that covers the mouth region, hence the alternative common name "shame-faced crabs." Box crabs inhabit sandy, muddy or rubbly areas from shore to depths of 150 ft. or more. Typically burying themselves in the substrate, often with just their eyes protruding, they can be exceedingly difficult to see.

Box crabs have a specialized mechanism on one claw (usually the right) for chipping open the shells of marine snails, their principal prey. The movable part of the claw bears a knob at its base, just above the joint, that fits neatly into the space between two projections on the opposing side. These enable the crab to exert strong localized pressure on the lip of a snail shell, thereby chipping it. Maneuvering a shell into position with its legs, a box crab can chip and turn, chip and turn, can-opener fashion to get at the soft animal inside. A box crab can even attack hermit crabs in this manner. These animals can be successfully kept in aquariums and will accept frozen clams, squid or shrimp. At least six species occur in Hawai`i.

SMOOTH BOX CRAB • **pokipoki**

Calappa calappa (Linnaeus, 1758)
Family Calappidae
• This large box crab is not often encountered in Hawai`i. The front of its smooth, domed carapace is sometimes attractively patterned with orange or dark spots of varying sizes. The rear edge forms a beautiful unbroken curve, paralleled on the carapace surface by low, line-like ridges. This crab occurs on sandy bottoms from the shallows to depths of 50 ft. or more. To about 6 in. carapace width. Indo-Pacific. Photo: Ron Holcom. Kahe Point, O`ahu. 50 ft. (with *Terebra strigilata*)

LUMPY BOX CRAB

• **pokipoki**
Calappa gallus Herbst, 1803
Family Calappidae
• This crab resembles a lumpy rock. It prefers seaweedy sand and rubble environments where its rough, knobby carapace (sometimes with bits of algae growing on it) blends perfectly with the surroundings. Sensing danger it freezes, folding its heavy knobbed claws almost seamlessly against its body. Some specimens have a large black spot on each side of the carapace. The orangish legs, although usually out of sight under the winglike extensions of the carapace, are visible when the crab moves. This species sometimes occurs intertidally. It can partially bury itself, but judging by the lack of long eyestalks, it probably spends most of its time on the surface. The species name means "rooster," perhaps because the row of heavy spines on the upper edge of each claw resembles a cockscomb. To about 2 1/2 in. carapace width. Indo-Pacific. Photo: Sandy Beach, O`ahu. 1 ft.

COMMON BOX CRAB • **pokipoki**

Calappa hepatica (Linnaeus, 1767)
Family Calappidae
• Long eyestalks help distinguish this rough-carapaced box crab from *C. gallus* (p. 267). Spending most of its time in sandy or gravelly environments, it remains buried for long periods with only the eyes protruding. The carapace is broad, covered with large rounded tubercles on the front half and smaller ones at the rear. Sharp-eyed snorkelers and divers can find these box crabs from the shallows to depths of at least 100 ft. They are probably the most common box crabs in Hawai`i. To about 4 in. Indo-Pacific. Photo: Kahe Point, O`ahu. 50 ft.

WINGLESS BOX CRAB

Cycloes marisrubri Galil & Clark, 1996
Family Calappidae
• The carapace of this species is longer than it is broad and lacks winglike expansions; the purple-tinged legs are always visible. The carapace and outer surface of the claws are covered with fine granules. The inner surfaces of the claws are fringed with fine golden hair. *Cycloes granulosa* is a synonym. To about 2 in. carapace length. Indo-Pacific. Photo: Kahe Point, O`ahu. 50 ft.

SPIDER CRABS. FAMILY MAJIDAE

Spider crabs have a triangular or teardrop-shape carapace that is drawn out into a pointed rostrum at the front. The legs are long and slender. Many of these slow-moving animals camouflage themselves with bits of algae, small sponges, soft corals or other organisms. Hooked hairlike projections on the carapace and limbs help hold these objects in place. Such crabs are often called collector crabs or decorator crabs. Hawaiian waters are home to more than 20 species of spider crabs.

SIMPLE COLLECTOR CRAB
• **kumulīpoa, pāpa`i limu**

Simocarcinus simplex Dana, 1852
Family Majidae
• These little crabs are common near shore in the brown seaweeds of the genus *Sargassum* (which they match in color). They also occur under stones. They collect bits of algae, sponge and other growths. Males have larger claws than females and a very pointed, teardrop-shape carapace, almost always with a few strands of algae attached near the tip. The female carapace is more oval. This is an extremely slow-moving animal. Both Hawaiian names mean "seaweed crab." To about 1 in. long. Indo-Pacific. Photo: Kewalo, O`ahu. 3 ft. (male)

HILO COLLECTOR CRAB
• **pāpa`i limu**

Schizophroida hilensis (Rathbun, 1906)
Family Majidae
• This is the collector crab most likely to be seen by divers at night. It covers itself with sponges of various colors, algae, and occasionally the blue or purplish octocoral *Anthelia edmondsoni*. Described from a specimen dredged near Hilo, it also occurs in New Caledonia, the Kermedec Islands and Lord Howe Island. This type of distribution, called "antiequatorial" or "antitropical," is common to a number of Hawaiian marine animals that occur only in the cooler waters of the subtropics. To about 1/2 in. carapace length. Photos: a) Palea Point, O`ahu. 30 ft. (with red sponge); b) Honokōhau, Hawai`i. 20 ft. (with blue octocoral) c) Pūpūkea, O`ahu. 30 ft.

Hilo Collector Crab (a)

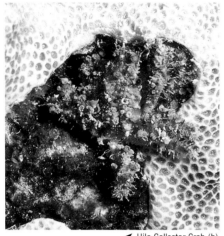

◄ Hilo Collector Crab (b)　　　　　　　　　　◄ Hilo Collector Crab (c)

ELBOW AND URCHIN CRABS. FAMILIES PARTHENOPIDAE AND EUMEDONIDAE

These two families are closely related. Urchin crabs (family Eumedonidae) are small crabs always associated with sea urchins. There is probably only one species in Hawai`i. Elbow crabs (family Parthenopidae) are an odd group of slow-moving crabs whose roughly triangular bodies are often covered with bumps, lumps and tubercles. The claw-bearing limbs (chelipeds) are typically long, spiny or bumpy, and have an elbow-like joint. Many parthenopids are masters of camouflage. In Greek mythology Parthenope was a siren who threw herself into the sea after failing to lure Odysseus onto the rocks with her songs. She was cast ashore at Naples, a city which bore her name in ancient times.

SEA URCHIN CRAB

Echinoecus pentagonus (A. Milne Edwards, 1879)
Family Eumedonidae
• The females of this parasitic crab live in the rectums of sea urchins, in Hawai`i exclusively in the common Banded Urchin, *Echinothrix calamaris* (p. 317). Except in very shallow water, almost every Banded Urchin hosts one or more of these animals. The tiny dark males live among the spines and are usually difficult to see; the larger females are imprisoned inside the rectum (located on the urchin's upper side). Look closely and you might see one busily manipulating the urchin's whitish fecal pellets upon which it feeds; the crab will be recognizable by its oddly curved and pointed carapace. The few Banded Urchins that do not host crabs may contain one or more White-Stripe Urchin Shrimps *(Stegopontonia commensalis)*; the two do not appear to occur together. To about 1/4 in. carapace width. Indo-Pacific. Photo: Palea Point, O`ahu. 30 ft.

FLAT ELBOW CRAB

Aethra edentata Edmondson, 1951
Family Parthenopidae

• This crab resembles a potato chip with legs. The genus name comes from the Greek *aethes*, meaning "unusual" or "strange." The flat wide carapace with its upturned edges is almost concave except for the central mounds that house the crab's internal organs. Covering the entire animal, it is usually overgrown with pinkish coralline algae, blending almost perfectly with coral rubble. Encrusting sponges may colonize the underside. Small or recently molted specimens, such as the one pictured here, are light tan or beige with few growths. If threatened, this crab snugs down into the rubble and virtually disappears. To about 2 in. carapace width. Endemic. (Similar species occur elsewhere in the Indo-Pacific.) Photo: Kahe Point, O`ahu. 50 ft.

Horrid Elbow Crab (a)

HORRID ELBOW CRAB

Daldorfia horrida (Linnaeus, 1758)
Family Parthenopidae

• This perfectly camouflaged, slow-moving crab exactly resembles the rubble in which it lives and can be spotted only when it moves. Its lumpy, knobby carapace and pincer-bearing front limbs may be overgrown with a thin layer of pinkish coralline algae to match the stones around it. Out of its habitat it looks grotesque, hence the species name. It attains a carapace width of about 1 1/2 in. and occurs from shallow water to scuba-diving depths and beyond. Indo-Pacific. Photos: a) Magic Island, O`ahu. 15 ft.; b) Hālona Blowhole, O`ahu. (missing one cheliped)

Horrid Elbow Crab (b)

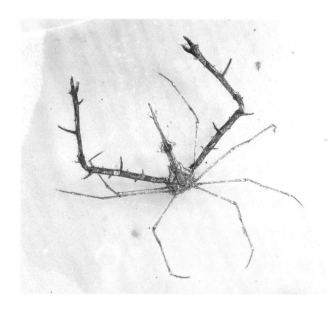

THORNY ELBOW CRAB

Lambracheus ramifer Alcock 1895
Family Parthenopidae
• This bizarre crab has long spindly legs and two large claw-bearing limbs protected by thorn-like spines. The carapace is drawn out between the eyes into an extremely long, sometimes branched point (rostrum). This the crab uses defensively, angling it in the direction of an intruder. The genus and species names both mean "branch" or "branch-ing." To about 1 in. carapace length, including the rostrum. Indo-Pacific. Photo: Waikīkī Aquarium. Crab collected at the "Hale`iwa Trench," O`ahu. 80 ft.

HAIRY ELBOW CRAB

Parthenope sp.
Family Parthenopidae
• The long, flattened, pincer-bearing "arms" of this crab have distinct elbowlike joints and bear comblike sets of lateral spines. Colorless hairs along the margins of the carapace and arms accumulate silt from the surround-ings, making the animal difficult to see unless it moves. It lives in silty sand and rock from the shoreline to depths of at least 150 ft. Although this crab was report-ed from Hawai`i around 1900 as *P. whitei*, it is almost certainly not that species, and is probably undescribed. Photo: Ala Moana Beach Park, O`ahu. 6 in.

SWIMMING CRABS. FAMILY PORTUNIDAE

Many portunids are swift, aggressive predators; their long powerful pincers can give a painful nip. The last two segments on their fifth pair of legs are usually flattened into swimming pad-dles and are effective in propelling the animals through the water. Most common in muddy, sandy or estuarine habitats, crabs of this family are named after Portunus, ancient Greek god of ports and harbors. Some are commercially important, including a few species intentional-ly introduced to Hawai`i. These include Hawai`i's largest swimming crab, the Samoan Crab (*Scylla serrata*), which exceeds 9 in. carapace width. The Blue Crab (*Callinectes sapidus*), a recent and probably illicit introduction from the eastern seaboard of the Americas, has been reported in Hawai`i to date only from Kāne`ohe Bay, O`ahu.

VIOLET-EYED SWIMMING CRAB

Carupa tenuipes Dana 1851
Family Portunidae
• Orange with violet eyes and claw tips, these colorful swimming crabs can occasionally be glimpsed on reefs at night from the low tide mark to depths of at least 50 ft. Fast-moving, they seldom stay still for long. Carapace width is typically about 1 in. *Carupa laeviuscula* is a synonym. Indo-Pacific. Photo: Hekili Point, Maui. 1 ft.

RAINBOW SWIMMING CRAB • **pāpa`i āko`ako`a**

Charybdis erythrodactyla (Lamarck, 1818)
Family Portunidae
• This colorful swimming crab is immediately recognizable by the blue marks on its yellowish orange carapace. Although sometimes called the "Red Legged Swimming Crab," legs of Hawaiian specimens, at least, are usually the same color as the carapace. The large claw-bearing limbs bear numerous spines. These crabs can sometimes be found by day resting in caves, usually on ledges near the ceiling. Some sources give the Hawaiian name `**ala`eke** for this species. It is reported to attain a 7 in. carapace width, but most specimens are smaller. Indo-Pacific. Photo: Pūpūkea, O`ahu. 15 ft.

HAWAIIAN SWIMMING CRAB

Charybdis hawaiensis Edmondson, 1954
Family Portunidae
• This is by far the most common large swimming crab on Hawaiian reefs. Although nocturnal, it can sometimes be observed by day hiding in crevices or between branches of coral (usually *Pocillopora*). The bizarre, vertically striped eyes are distinctive. Its carapace and legs are patchy orange and red with scattered white marks and spots. The swimming paddles at the ends of the last pair of legs are yellowish. The white-tipped claws bear a broad black band, and the pinching fingers are finely grooved. To about 3 in. carapace width. Hawai`i, Tuamotus, and Society Islands. Photo: Pūpūkea, O`ahu, 30 ft.

RED SWIMMING CRAB

Charybdis paucidentata (A. Milne Edwards, 1861)
Family Portunidae
• In size, shape and habits this crab resembles the Hawaiian Swimming Crab (above). It is almost entirely bright red, however, with black claw tips and eyes; the eyes are not striped. The species name means "few teeth" or "small teeth"; Charybdis was the daughter of the Greek sea god Poseidon and also the name of a famous whirlpool between Italy and Sicily. To about 3 in. carapace width. Indo-Pacific. Photo: Hōnaunau, Hawai`i. 40 ft.

274

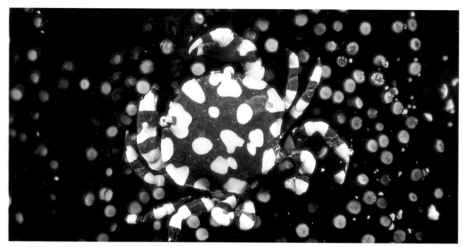

SEA CUCUMBER CRAB

Lissocarcinus orbicularis Dana, 1852
Family Portunidae

• These small crabs live commensally on several species of large sea cucumbers. Although found most often among the tentacles or in the mouth, they occur almost anywhere on the body, including the anus. Extremely variable, they may be almost uniform in color or boldly marked with a harlequin pattern (light with dark spots or the reverse). The dark areas may be brown, reddish, purple or almost black and the light areas yellowish or white. A pair of these crabs (sometimes of different colors) often inhabit a single cucumber (in Hawai'i most commonly *Holothuria atra*, *Bohadschia paradoxa* and *Stichopus* sp. 1). Although a member of the swimming crab family, this species remains on its host and does not swim. To about 1/2 in. carapace width. Indo-Pacific. Photo: Kewalo, O'ahu. 5 ft. (on *Holothuria atra*)

LONG-EYED SWIMMING CRAB ▪ **mo`ala**

Podophthalmus vigil (Weber, 1795)
Family Portunidae

• The eyestalks of this crab are amazingly long and can either be held erect or folded back into grooves along the front of the carapace. Individuals are brown, live on soft, muddy bottoms, and can tolerate brackish conditions. Divers and snorkelers seldom encounter this animal because of its habitat. The species name means "watchful." To about 5 in. carapace width. Indo-Pacific. Photo: John E. Randall, Maumere, Flores, Indonesia.

275

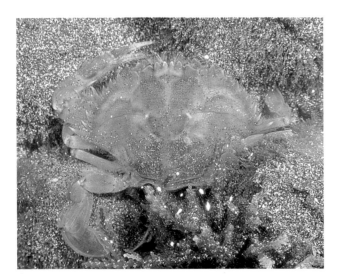

HAIRY SWIMMING CRAB

Portunus pubescens Dana, 1852
Family Portunidae
• The carapace of this crab is covered with short hairs. Its front edge is almost circular in outline and bordered with many teeth, four between the eyes and nine on each side, the last largest. The legs are thickly fringed with hair. It dwells in shallow water and is said to bury itself in sand. To about 2 in. carapace width. Indo-Pacific. Photo: Hekili Point, Maui. Intertidal.

BLOOD-SPOTTED SWIMMING CRAB • **kūhonu**

Portunus sanguinolentus Herbst, 1783
Family Portunidae
• Like many swimming crabs, this species has a strong, sideways-pointing spine on each side of the carapace. The carapace is marked with three red spots. The claws are slender and pointed. Like most portunids, it faces its enemy when threatened, spreading its claws wide in an most effective threat display. With an "arm-spread" of 18 in., this is not an animal to be trifled with. It prefers muddy or sandy habitats from the shoreline to depths of at least 100 ft. and is seldom seen by divers or snorkelers. A few barnacles *(Chelonibia patula)* sometimes occur on its back. The Hawaiian name means "turtle back," referring to the domed carapace. Other local names are "haole crab" or "white crab." It attains up to 7 in. across the carapace (not including the long spines). Indo-Pacific. Photo: Kahe Point, O`ahu. (This beautiful animal was carefully captured and set on sand for the picture.) This species is protected by law. See Appendix B (p. 346).

UNIDENTIFIED SWIMMING
CRAB

Portunus sp.
Family Portunidae
• This unidentified swimming crab is seen at night in caves and crevices. The legs are banded red and white. Photo: Mākaha Caves, Oʻahu. 20 ft.

SAMOAN CRAB

Scylla serrata De Man, 1899
Family Portunidae
• This edible crab was introduced to Hawaiian waters from Samoa in the 1920s and is Hawaiʻi's largest swimming crab. The carapace is smooth and brown. There are four spines between the eyes and nine along each side. It occurs on muddy bottoms in calm bays and brackish river mouths. The common name Samoan Crab is local; this animal occurs from the Red Sea and east coast of Africa to Hawaiʻi and may have been spread by humans through much of its range. Large males have enormous claws and may exceed 9 in. carapace width. Photo: Courtesy Bishop Museum. This species is protected by law. See Appendix B (p. 346).

XANTHID CRABS. FAMILY XANTHIDAE

The xanthids—sometimes called dark-finger crabs, stone crabs or mud crabs—form the largest crab family. Common on Hawaiʻi's reef flats and reefs, they have roughly oval carapaces that are never drawn to point in front like those of spider crabs. The tips of their claws are usually dark and often spoonlike. Although some xanthids are large, brightly colored, hairy or otherwise unusual, most are small, dull-looking crabs that spend their lives in crevices, under stones, or between coral branches. Along some protected shores in Hawaiʻi almost any rock turned will reveal one or more of these tiny creatures scuttling for cover. Relatively slow-moving and unlikely to pinch, the smaller forms are easy to catch and examine. Larger species with strong pincers should be handled with care. More than 100 species of xanthid crabs are known from Hawaiian waters. Seventeen of the largest, most common or unusual are presented here.

CONVEX CRAB • **kukuau**

Carpilius convexus Forsskål, 1775
Family Xanthidae
• A smooth, convex carapace—thick, heavy and bearing no spines—gives this medium-size crab its name. It is solid brownish red, sometimes mottled with gray and white. Two or three small white spots may mark the center of the carapace. It occurs from shallow reef flats to depths of 50 ft. or more. This animal has been seen holding a small urchin *(Echinometra mathaei)* in its claws, methodically snipping off the spines preparatory to eating it. To about 3 in. carapace width. A common Indo-Pacific species. Photo: Kīpapa Island, O`ahu. 3 ft.

SEVEN-ELEVEN CRAB • **`alakuma**

Carpilius maculatus Linneaus, 1758
Family Xanthidae
• Because of its large size and unmistakable color pattern, this is one of the best known Indo-Pacific reef crabs. It is slow-moving and seen most often at night. The massive claw-bearing limbs are of unequal size; the smooth, heavy shell lacks spines. It has seven prominent red spots (two by each eye, three in the center) and four less obvious ones (along the rear margin), making a total of eleven spots. In old Hawai`i the story was told of a hungry god who caught an `**alakuma**. The frantic crab pinched its captor and drew blood, escaping with bloody marks on its back. Again the god grabbed the crab; again the crab pinched and got free. The third time the god was successful, but to this day the crab's descendants bear his bloody finger prints. This crab has been observed carrying urchins and cowries, suggesting a possible food preference. The common name is probably local; the species name means "spotted." To about 6 in. carapace width. Indo-Pacific. Photo: Mākua, O`ahu. 20 ft.

SPLENDID PEBBLE CRAB
• **kūmimi**

Etisus splendidus Rathbun, 1906
Family Xanthidae
• This large, bright red xanthid crab with black, spoon-tipped claws (larger in males) is an uncommon nocturnal animal, sometimes seen by day wedged deep into crevices. The carapace, smooth with broad lobules, is bordered by spines. The legs are fringed with stiff black hairs. The flesh is said to be bitter; in old Hawai`i it was used in medicine, possibly as a heart stimulant. A fossil specimen found near Wai`anae, O`ahu, has been estimated to be 350,000 years old. To about 6 in. or more carapace width. Indo-Pacific. Photo: Lehua Rock, Ni`ihau. 40 ft.

PRETTY LIOMERA

Liomera bella (Dana, 1852)
Family Xanthidae
• Members of the genus *Liomera* are often colorful and may change color somewhat as they grow. This yellow or pale brown species occurs on shallow reef flats. The front of its carapace is divided by furrows; toward the rear the carapace flattens out. The species name means "beautiful." Large specimens are about 1 in. wide. Indo-Pacific. Photo: Hawai`i Kai, O`ahu. Intertidal.

RED LIOMERA

Liomera rubra (A. Milne Edwards, 1865)
Family Xanthidae
• Small specimens of this xanthid are usually a uniform bright red. Larger ones often have white in the center of the carapace and in the furrows. The carapace and limbs are covered with fine granules. To about 1 in. carapace width. Indo-Pacific. Photo: Portlock Point, O`ahu. Intertidal.

CORRUGATED LIOMERA

Liomera rugata (H. Milne Edwards, 1834)
Family Xanthidae
• The intense purple-red of this small crab makes it easy to identify. It occurs under stones from the shoreline down to at least 30 ft. Many xanthid crabs have a carapace divided into lobules, producing a corrugated appearance; this one was named for it— the species name means "wrinkled" or "creased." Maximum size is probably about 1 in. Indo-Pacific. Photo: Magic Island, O`ahu. 10 ft.

KNOTTED LIOMERA

Liomera supernodosa (Rathbun, 1906)
Family Xanthidae
• The carapace of this crab is divided into distinctly pitted lobules separated by deep smooth furrows. It varies from yellowish brown to light lavender. The legs and claw-bearing limbs are covered with knotty nodules. It is known only from the Hawaiian Islands and is most common in the northwestern chain. To about 1 in. carapace width. Photo: O`ahu. (Aquarium specimen; exact location and depth of capture unknown.)

BEARDED CRAB • **kūmimi**

Lophozozymus intonsus (Randall, 1839)
Family Xanthidae
• The legs of this crab are bordered by long yellow hairs. Its claw-bearing limbs are stout and muscular-looking; the pincers are black. The carapace is smooth but clearly divided into regions. It occurs under stones along shallow reef flats where there is good water circulation. The scientific name means "unshorn" or "bearded." To about 2 1/2 in. carapace width. Central Pacific. Photo: Kīpapa Island, O`ahu. 3 ft.

HAWAIIAN POM-POM CRAB • **kūmimi pua**

Lybia edmondsoni Takeda & Miyake, 1970
Family Xanthidae
• Pom-Pom Crabs almost always brandish a stinging anemone in each claw (usually *Triactis producta*) as a defense against predators and possibly as a means of gathering food. They are sometimes called "Boxer Crabs" because of the pugilistic stance they adopt when threatened. Unusual polygonal patterns in pink, brown or yellow mark the carapace; the legs are banded with dark purple. Although striking on a black background, the pattern renders the crab almost invisible in its natural habitat, usually under stones in thin sand or rubble patches over hard substrate from the shallows to depths of at least 100 ft. In captivity, as in the wild, they tend to hide but will emerge from their crevices at feeding time and accept the usual invertebrate foods. Aquarists report them picking off and eating food or detritus that sticks to the tentacles. Recent research by Curt Fiedler at the University of Hawai`i shows that a crab deprived of one anemone will split the remaining anemone, forcing it to regenerate into two complete animals. Crabs of the genus *Lybia* represent an excellent example of symbiosis, tool use, and one animal "farming" another. Although this species is endemic, the Indo-Pacific *Lybia tessellata* is similar. The Hawaiian name **kūmimi** designates an inedible crab; **pua** means "flower." In ancient times this animal was used in sorcery. The scientific name honors Hawai`i's world-renowned zoologist Charles Howard Edmondson (1876-1970). Endemic. To about 1/2 in. carapace width. Photo: "Tracks," Kahe Point, O`ahu.

STRAWBERRY CRAB

Neoliomera pubescens (A. Milne Edwards, 1865)
Family Xanthidae
• Strawberry pink with scattered white spots, this is one of the prettiest Hawaiian xanthids. Small granules and sparse, short yellow hairs cover the carapace. The claw-bearing limbs are large and of equal size, and the inner surfaces of the claws bear thick yellow hair. This crab lives from the shoreline to depths of at least 20 ft. and attains a carapace width of 1 in. or slightly more. It is sometimes collected for the aquarium trade. Indo-Pacific. Photo: Honolulu Harbor. 1 ft.

AREOLATED XANTHID CRAB

Pilodius areolatus (H. Milne Edwards, 1834)
Family Xanthidae
• This is a common crab under stones on shallow Hawaiian reefs. Its carapace is divided into lobules that in turn are covered by light granules. The walking legs are covered with fur and shaggy hair. "Areolated" means having light areas surrounded by dark areas. To about 1 in. carapace width. Indo-Pacific. Photo: Kewalo Park, O`ahu. 3 ft.

RED-EYED XANTHID CRAB

Platypodia eydouxii (A. Milne Edwards, 1865)
Family Xanthidae
• The claws of this crab are densely covered with orange- or yellow-tipped granules. The oval, somewhat convex carapace is more sparsely granulated, sometimes bearing algae or other growths. Reddish eyes make identification easy. It is not uncommon beneath stones on shallow reef flats. The species name (pronounced "ee-dew-eye") honors French naturalist Joseph Fortune Eydoux (1802-1841), who traveled widely collecting and naming animals of many kinds for the National Museum of France. To about 2 in. carapace width. Photo: Kewalo Park, O`ahu. 3 ft.

TEDDY-BEAR CRAB

Polydectus cupulifer
(Latreille, 1812)
Family Xanthidae
• Like the Pom-Pom Crab, this crustacean carries a sea anemone in each claw. Its legs and carapace are covered with a dense coat of hair. The anemone carried is usually *Telmatactis decora* (p. 39), but it will also carry *Anthopleura nigrescens* (p. 37). Edmondson reports: "In picking up a sea anemone, the crab backs up to it and by the activity of its legs rolls the actinian forward under its body, taking hold of it only when it comes in contact with the chelipeds." This crab occurs under stones from the shoreline to at least 50 ft. To about 1 in. carapace width. Indo-Pacific. Photo: Lāna`i Lookout, O`ahu. 30 ft.

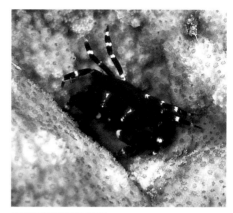

SHOWY XANTHID CRAB

Pseudoliomera speciosa (Dana, 1852)
Family Xanthidae
• This small crab occurs most often between branches of Cauliflower Coral *(Pocillopora meandrina)*. Its carapace is divided into lobules and covered with granules. There is no hair. The species name means "splendid" or "showy." To about 1/2 in. carapace width. Indo-Pacific. Photo: Pūpūkea, O`ahu. 20 ft.

BROAD-FRONTED CRAB

Xanthias latifrons (De Man, 1887)
Family Xanthidae
• These small attractive crabs are most active at night but can sometimes be seen by day along vertical walls near crevices and caves. The carapace is dark maroon. In younger specimens two conspicuous white marks edged in red extend over the front of the carapace between the eyes. Bright white bands adorn the legs and claw tips. The species name means "broad front." This crab occurs to depths of at least 300 ft. and attains about 3/4 in. carapace width. Indo-Pacific. Photo: Waimea, O`ahu. 30 ft.

CORAL GUARD CRABS. FAMILY TRAPEZIIDAE.

Crabs of this family live in association with branching corals. In Hawai`i the preferred hosts are Cauliflower Coral *(Pocillopora meandrina)*, Antler Coral *(P. eydouxi)* and Lace Coral *(P. damicornis)*. Mated pairs live deep between the branches, feeding on coral mucus and tissue. Recent studies indicate that the coral host may concentrate fat globules at the tentacle tips specifically to feed the crabs. Completely dependent on the living coral for food and shelter, they use their massive claws to ward off intruders that threaten their host. If a coral-eating Crown-of-Thorns Star settles on their coral head, the guard crabs nip its tube feet. and drive it away. This works against divers too. Place your hand over a small head of branching coral containing these crabs, and you will soon receive a sharp pinch. All species, except the slightly larger Yellow-Spotted Guard Crab, attain a carapace width of about 1/2 in. All are widely distributed in the Indo-Pacific.

BROWN GUARD CRAB

Trapezia digitalis Latreille, 1823
Family Trapeziidae
• These crabs are deep brown (sometimes half brown, half white on juveniles). There are no spots. They inhabit Cauliflower Coral and Antler Coral. Pairs often share a coral head with the Common Guard Crab, *Trapezia intermedia* (p. 284). To about 1/2 in. carapace width. Indo-Pacific and Eastern Pacific. Photo: Magic Island, O`ahu. 20 ft.

RUSTY GUARD CRAB

Trapezia ferruginea Latreille, 1823
Family Trapeziidae
• This species is a uniform orange or rusty brown without spots. It inhabits Cauliflower and Antler Coral. To about 1/2 in. carapace width. Indo-Pacific and Eastern Pacific. Photo: Kewalo Pipe, O`ahu. 50 ft. (in Cauliflower Coral, *Pocillopora meandrina*)

YELLOW-SPOTTED GUARD CRAB

Trapezia flavopunctata Eydoux & Souleyet, 1842
Family Trapeziidae
• The yellow spotting of this crab blends well with the texture of its coral host, usually Antler Coral. The spots are larger than those of other Hawaiian spotted coral guard crabs. The background color is light brown, and the walking legs are striped white and reddish brown. This is the largest of Hawai`i's coral guard crabs, attaining about 1 in. carapace width. Indo-Pacific. Photo: Mākua, O`ahu. 15 ft. (in Antler Coral, *Pocillopora eydouxi*)

COMMON GUARD CRAB

Trapezia intermedia Miers, 1886
Family Trapeziidae
• The most common of Hawai`i's coral guard crabs, this is usually the first to colonize a growing coral head. It is light brown covered with pale reddish dots. Most heads of Cauliflower Coral will harbor a pair of these crabs, which also colonize Antler Coral and Lace Coral. To about 1/2 in. carapace width. Indo-Pacific. Photo: "Hale`iwa Trench," O`ahu. 15 ft. (in Lace Coral, *Pocillopora damicornis*)

Red-Spotted Guard Crab (a)   Red-Spotted Guard Crab (b)

RED-SPOTTED GUARD CRAB

Trapezia tigrina Eydoux & Souleyet, 1842
Family Trapeziidae
• Many bright red spots on a light brown background identify this crab. It is most common at depths greater than 40 ft. in Cauliflower and Antler Coral. At such depths most red light disappears and they appear brown. *Trapezia wardi* is a synonym. To about 1/2 in. carapace width. Indo-Pacific. Photos: Pūpūkea, O`ahu. 50 ft. a) in Cauliflower Coral, *Pocillopora meandrina*; b) molt.

ROCK CRABS AND BUTTON CRABS. FAMILIES GRAPSIDAE AND PALICIDAE

Rock, or grapsid, crabs live on rocky shores and in nearby shallow waters. A few species inhabit floating logs and other debris; others are semiterrestrial or live in fresh water. Their carapaces vary from square to almost round and are usually flattened, as are the legs. The two claw-bearing limbs are typically short, adapted for picking algae from the rocks. The related Button crabs (family Palicidae) often have round, domed carapaces covered with granules or tubercles. The last pair of legs is slender and short.

THIN-SHELLED ROCK CRAB • `a`ama

Grapsus tenuicrustatus (Herbst, 1783)
Family Grapsidae
• These crabs (often called "Sally Lightfoot Crabs") are ubiquitous on rocky shores, clambering about in the splash zone and retreating into crevices or the water when approached. They are greenish black, marked with faint striations. Their molted shells, found high on the rocks, often turn bright red from the heat of the sun (just as a lobster turns red when boiled). Edible, they were of importance in old Hawai`i and gave rise to a number of stories and sayings. When trouble arose and people gathered out of curiosity, it was said "When the sea is rough the `a`ama crabs climb up on the rocks." The Hawaiian name means "to loosen or relax." To ask a favor of the gods, an `a`ama might be offered, that the gods would "loosen" and grant it. The species name means "thin shell." To about 3 in. carapace width. Indo-Pacific. Photo: Waikīkī, O`ahu.

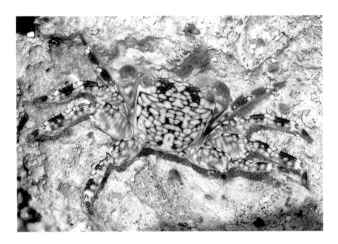

PLEATED ROCK CRAB • `a`ama

Pachygrapsus plicatus (H. Milne Edwards, 1837)
Family Grapsidae
• Common on rocky shores both in and out of the water, this crab has a squarish carapace marked with transverse sets of light squarish or rectangular spots. These spots follow sets of transverse ridges that give the species its name. To almost 1 in. carapace width. Indo-Pacific. Photo: Barber's Point, O`ahu.

BLUE-EYED ROCK CRAB

Percnon affine (H. Milne Edwards, 1853)
Family Grapsidae
• These crabs are common in caves and crevices. They often face an inquisitive diver, their bright blue eyes contrasting sharply with the dun color of the body and legs. The claw-bearing limbs are short and stout. *Percnon pilimanus* is a synonym. To about 3 in. carapace width. Indo-Pacific. Photo: Pūpūkea, O`ahu. 20 ft.

FLAT ROCK CRAB • pāpā

Percnon planissimum (Herbst 1804)
Family Grapsidae
• These flat, active little crabs occur most commonly on smooth rounded boulders, scooting about on the surface and always retreating backwards to the other side when approached. They can be seen in tide pools and to depths of at least 30 ft. The carapace has a blue-green or white line down the center and a thin iridescent blue-green line across the front. The eyes are often yellow. The claw-bearing limbs and pincers are short compared with the walking legs. To about 1 in. carapace width. Indo-Pacific. Photo: Pūpūkea, O`ahu. 20 ft.

SCALY ROCK CRAB

Plagusia depressa tuberculata (Lamarck, 1818)
Family Grapsidae
• The almost circular carapace of this large and unusual grapsid is covered with flat tubercles that resemble forward-point-ing scales. Each "scale" is fringed at the front with short bristles. The overall color is maroon-brown. The animal can be found in tide pools and splash pools in areas of constant wave action. The subspecies *tuberculata* is widespread in the Indo-Pacific. There is also a tropical Atlantic subspecies. To about 2 3/4 in. carapace width. Indo-Pacific. Photo: Lāna`i Lookout, O`ahu. (This fresh molt was found in a splash pool; the crab itself, in the process of hardening its new shell, was in a crevice a few inches away.)

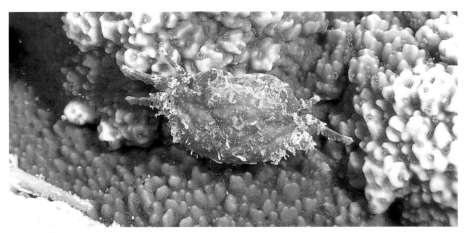

BUTTON CRAB

Pseudopalicus sp.
Family Palicidae
• This little crab is undescribed. It has a rough, domed carapace and the legs are fringed with hair, enabling it to swim short distances when disturbed. Although not uncommon in rocky habitat, it is seen only at night. The similar *P. oahuensis* occurs in the same habitat. The specimen pictured was crawling on Rice Coral (*Montipora capitata*). To about 1/3 in. carapace width. Photo: Magic Island boat channel, O`ahu. 15 ft.

GHOST CRABS AND FIDDLER CRABS. FAMILY OCYPODIDAE

These pale, active crabs live on sandy beaches in burrows above the waterline. Two species occur in Hawai`i: the Horn-Eyed Ghost Crab, *Ocypode ceratophthalma* (Pallas, 1772), is the largest and possibly the most common, but is strictly nocturnal. Its presence on a beach is indicated by the conical piles of sand that mature males build near the entrances of their burrows to attract females. The Pallid Ghost Crab *(O. pallidula)* is described below. In ancient Hawai`i, a poorly placed home, open on all sides to attack, was called **lua `ōhiki**, or "sand crab hole." Fiddler crabs, also of this family, live in great numbers on mud flats in many parts of the world. Two species were recorded from Hawai`i years ago but have not been seen since.

PALLID GHOST CRAB • **`ōhiki**

Ocypode pallidula Jacquinot, 1852
Family Ocypodidae

• "Try and catch this little crab" wrote C. H. Edmondson in his children's book *Hawaii's Seashore Treasures*. "He will lead you a merry chase. His ten legs never seem to get in the way of each other. And forward or sideways is all the same to this speedy sand-racer." This is the smaller of the two ghost crabs found in Hawai`i, and the only one seen by day (although usually only at dusk or dawn). It lives higher on the beach than the Night Ghost Crab and in some locations is more abundant. Unlike the nocturnal Horn-Eyed Ghost Crab, mature males of this species do not build conical piles of sand at the burrow entrances; rather, they flip the excavated sand into a fan-shape pattern of small clumps. The species name means "pale" or "pallid." To about 1 in. carapace width. Indo-Pacific. *Ocypode laevis* is a synonym. Photo: Kapa`a, Kauai.

CORAL GALL CRABS AND CHAMBER CRABS. FAMILY CRYPTOCHIRIDAE

The small crabs of this family live as commensals on stony corals. Some create hollow swellings (galls) at the tips of the branches; others occupy chambers within the coral that are connected to the surface by a small opening. Females are usually imprisoned within their chambers. The usual hosts in Hawai`i are corals of the genera *Porites*, *Cyphastrea* and *Pocillopora*. The carapaces of some chamber-dwelling species are specially shaped at the front to close the small opening.

CORAL GALL CRAB

Hapalocarcinus marsupialis
Stimpson, 1859
Family Cryptochiridae
• This crab causes branching corals to form galls (hollow swellings) at the branch tips in which the females imprison themselves for life. The tiny, free-living males enter and exit through pores at the crest of the gall. Females carry their eggs in a pouch, hence the specific name. They attain almost 1/2 in. Males are far smaller. The host coral in Hawai`i is almost always *Pocillopora damicornis*; other species of *Pocillopora* are used rarely. This photo shows two exceptionally large galls in Antler Coral (*Pocillopora eydouxi*). Indo-Pacific and Eastern Pacific. Photo: Kahe Point, O`ahu. 3 ft.

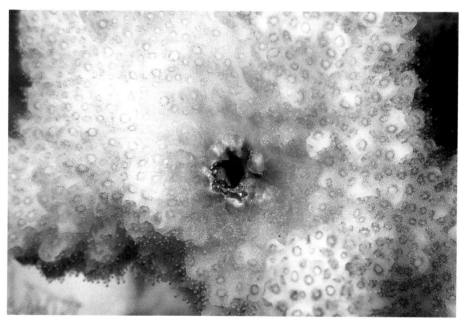

KAHE POINT CRAB

Pseudocryptochirus kahe McCain & Coles, 1979
Family Cryptochiridae
• This small crab creates and inhabits chambers in Cauliflower Coral and Antler Coral (*Pocillopora meandrina* and *P. eydouxi*, respectively). Each chamber has a small pitlike entrance surrounded by pinkish discoloration. Females, larger than males, are imprisoned within their chambers; the tiny males can exit freely, probably mating with multiple females. Up to 32 females have been found in one coral colony. Typically 20 to 30 percent of Cauliflower Coral colonies contain these crabs; those in high energy environments tend to have the most. The species was discovered at Kahe Point, O`ahu, during an environmental survey for the Hawaiian Electric Company, hence the species name. Females attain about 1/5 in. carapace length. Probably Indo-Pacific in distribution. Photo: Kahe Point, O`ahu. 25 ft.

ECHINODERMS
PHYLUM ECHINODERMATA

Cushion Stars (*Culcita novaeguineae*). These two animals were found next to each other exactly as shown. Behind, a pair of Arc-Eyed Hawkfish stand guard in a head of Cauliflower Coral. The Lobe Coral in the foreground contains channels made by pairs snapping shrimps. "Five Graves," Kīhei, Maui. 30 ft.

No land animals remotely resemble echinoderms. The name comes from Greek words for "spiny skin," but perhaps the most unusual characteristic of these animals is their five-part body plan. No other animals are organized in this way, with circulatory, nervous and skeletal systems radiating in five directions from a central axis. Head, brain and heart are absent.

A second unique characteristic of echinoderms is a "water vascular system," a complex of water-filled reservoirs and canals ending in multitudes of hollow tube feet. These tube feet, extended and contracted in part by the changing water pressure within, are used for locomotion, feeding and sometimes respiration. Although some echinoderms have lost their tube feet, all have a water vascular system.

Echinoderms have an internal calcareous skeleton. Actually, it is the skeleton, not the skin, that bears the spines for which echinoderms were named. In some echinoderms, such as sea urchins, the skeleton is fairly obvious; in others it takes the form of disconnected plates or even microscopic ossicles radially arranged within the skin or flesh.

Of the major body systems, only the digestive system is not radial. The mouth and anus generally lie at opposite ends of the body's central axis, on the lower and upper surfaces, respectively (or in sea cucumbers, at the ends); the digestive tract runs between them. Some echinoderms, notably brittle stars, lack an anus.

In most echinoderms the sexes are separate. When spawning, males and females release sperm and eggs into the water. The fertilized eggs hatch into larvae that drift with the plankton before settling in appropriate habitat to mature. In a few species, especially those in the colder oceans, eggs develop directly into miniature adults. The female parent sometimes retains these youngsters on her body until they reach a certain size. Some sea stars can reproduce by detaching an arm, which then regenerates into another complete star.

Many echinoderms are either scavengers or predators on stationary (sessile) organisms such as algae, corals, sponges, clams and oysters. Others filter food particles from sand, mud or water. Echinoderms are exclusively marine, ranging from brackish intertidal zones to abyssal depths. Almost all the large animals known from deep ocean trenches are echinoderms. A few deep-water sea cucumbers are pelagic, but the great majority of echinoderms, except in their larval stages, are strongly associated with the sea bottom or other solid surfaces.

Sea stars, brittle stars, sea urchins, sea cucumbers, crinoids (known from the main Hawaiian Islands only from deep water), and a small, newly discovered group known as "sea daisies" are the only living classes within this phylum. About a dozen more classes are known only from the fossil record. So ancient are the echinoderms that known fossil species outnumber the living ones two to one. Of about 6,500 living species, some 280 occur in Hawaiian waters. Echinoderms vary in size from tiny urchins and stars about 1 in. across to sea cucumbers 6 ft. long; most are in the 4 to 5 in. range.

SEA STARS
PHYLUM ECHINODERMATA. CLASS ASTEROIDEA

Sea stars are familiar shoreline animals throughout much of the world. No cartoon of a beach scene is complete without a five-armed "starfish" lying on the sand. In Hawai`i, however, one must snorkel or dive to find them; they are rare in tide pools (except for one tiny species) and are almost never cast up on the beach.

Most sea stars have an easily recognized five-part body consisting of a central disk and radiating arms (also called rays). The surface can be knobby, spiny, bristly or smooth. Other echinoderm characteristics include hydraulically operated tube feet and a skeleton composed of calcium carbonate ossicles (rods, crosses or plates) buried within the skin. There is neither head nor tail, but a mouth is present on the underside and an anus on the upper. As with sea urchins, tiny pincer-like organs (pedicellariae) may occur on the body surface. A sievelike plate (madreporite), located off-center on the upper side of the disk, allows water to enter the vascular system.

Depending on species, sea stars can have up to 50 arms. The usual number is 5. On the reef these flexible arms drape fluidly over objects of any shape; when handled the star may contract and become rigid. The underside of each arm bears two or four rows of tube feet tipped with suckers. These tube feet arise from furrows (ambulacral grooves) radiating from the central toothless mouth. At each arm's tip is often a light-sensitive tube foot; many sea stars will keep these tips tilted slightly up to better sense their environment. Internally, the arms share parts of the digestive and reproductive systems and other organs. A sea star can regenerate a lost arm; in a few cases, a detached arm can regenerate a complete star.

Sea stars are predators and scavengers. Some attack other echinoderms, including spiny sea urchins, by everting their stomach over their prey and either digesting it outside the body or "swallowing" it whole; some use their pedicellariae to seize bottom-dwelling fishes that inadver-

Green Linckia *(Linckia guildingi)*. "Treasure Islands," Kona, Hawai`i. 50 ft.

tently come to rest on the star. Others evert their stomachs on sessile animals—sponges, anemones, or coral polyps that cannot move away—and digest them on the spot. A few may subsist on detritus or algae. Sea stars that pull apart clam or oyster shells and insert their stomach into the opening are confined to the family Asteriidae, known in Hawai`i only from deep water; shallow-water species in the Islands do not feed this way.

In aquariums sea stars can be voracious. Captive specimens (although probably not Hawaiian species) have even been known to attack and consume sleeping fishes. Many of these animals could wreak havoc in an invertebrate or "living-reef" aquarium. Some stars will survive in a conventional aquarium if fed bits of clam, shrimp or other flesh (even though this might not be their natural diet). To keep them in top condition, water quality must be excellent and nitrate levels low.

Like most echinoderms, sea stars reproduce by the synchronized release of eggs and sperm. Fertilized in the water, the eggs hatch into free-swimming larvae that drift as plankton for days or weeks before settling to the bottom and transforming into juveniles. Many stars, however, forgo sex and reproduce by detaching an arm (autotomy) or splitting apart (fission). Each piece then grows new parts to become a complete animal.

Little is recorded about the uses of sea stars in ancient Hawai`i. Inedible and uncommon, they were probably disregarded. General names used were **pe`a**, **pe`ape`a** and **hōkū-kai** but specific names, if any existed, have been lost.

There are about 1,800 known species of sea stars, ranging in diameter from about 1/2 in. to 3 ft. Although their greatest diversity occurs in the tropical Indo-Pacific, one of the best places to observe these animals is along the Pacific coast of North America; an amazing assortment occurs there, even in tide pools. Hawai`i has only about 20 known shallow-water species with 68 more from deep water. Fifty-three of the deep-dwelling species are known only from Hawai`i, but some are probably more widespread. Fifteen are illustrated here.

CROWN-OF-THORNS STAR

Acanthaster planci (Linnaeus, 1758)
Family Acanthasteridae
• Notorious for its depredation of coral reefs, the Crown-of-Thorns may be the best known of all sea stars. It feeds on living coral polyps (in Hawai`i, usually *Pocillopora meandrina*), leaving behind white patches of dead coral. Periodic population explosions have generated alarm in some areas; in Hawai`i, however, it has not caused extensive damage. Up to 18 in. across, with 12-19 arms and covered with stout sharp spines, this is an unusual star, the more so because the spines are venomous. Puncture wounds cause sharp burning pain that subsides to a dull ache and disappears after a few hours. Swelling, numbness and discoloration of the puncture site are common, usually disappearing within 48 hours. The large Triton's Trumpet snail preys upon this star, and sometimes the Harlequin Shrimp attacks it as well. Lined Fireworms may enter through wounds created by the shrimps to feast on the star from inside. The tiny commensal shrimp *Periclimenes soror* often lives among the spines. Although this star is usually brownish red in Hawai`i, it appears greenish at depths greater than about 30 ft. Indo-Pacific and Eastern Pacific. Photo: Hanauma Bay, O`ahu. 40 ft.

TOENAIL STAR

Asterodiscides tuberculosus
(Fisher, 1906)
Family Asterodiscididae
• Perhaps the most unusual of Hawaiian shallow-water stars, this knobby species bears two or more pinkish marginal plates at the tip of each arm. It is maroon to red; the five arms are stubby and cylindrical in cross-section. This is one of many Hawaiian sea stars discovered by American zoologist Walter K. Fisher during the 1902 cruise of the research vessel *Albatross*. The first specimens were dredged from about 300 ft. Divers occasionally see it at about 100 ft. on coarse sand or hard substrate adjacent to sand. To at least 7 in. diameter. Indo-Pacific. Photo: Ed Robinson. "Monolith," Lāna`i. 100 ft.

Asterina anomala Clark, 1921
Family Asterinidae
• Growing no larger than a quarter, this tiny star has a variable number of short, blunt arms, usually six or seven. It reproduces by splitting in half (fission) and is often incomplete. The entire upper surface is covered with short spines. It occurs under stones on shallow reef flats and in tide pools, sometimes in great numbers. **Limu** pickers sometimes find it in the seaweed they gather. The color is variable, usually yellow-green or greenish brown becoming lavender toward the arm tips. A bright red eyespot is present at the upturned tip of each arm. To about 1/2 in. diameter. Indo-Pacific. Photo: Scott Johnson. O`ahu.

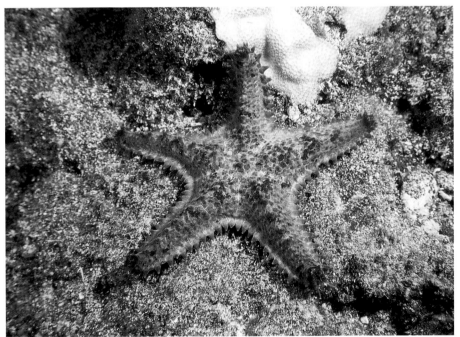

SHERIFF-BADGE STAR

Asteropsis carinifera (Lamarck, 1816)
Family Asteropseidae
• This star can be greenish, yellowish, reddish brown or bright red. Its five arms are triangular (both in shape and in cross-section), and its entire perimeter is rimmed by thick, toothlike spines. Spines also run down the central ridge of each arm. The underside is broad and flat. This animal is seen at night under ledges and on the walls or ceilings of caves. It also occurs under stones on shallow reef flats. Some aquarists report that it preys on sea urchins, but it also seems to do quite well feeding on algae from the sides of the tank. It sometimes hosts the tiny shrimp *Periclimenes soror* (p. 224) and is the only species in its genus. To about 7 in. Indo-Pacific and East Pacific. Photo: Palea Point, O`ahu. 40 ft.

STRIATED STAR
Valvaster striatus (Lamarck, 1816)
Family Asteropseidae
• Similar in size and shape to the Sheriff-Badge Star (previous page), this nocturnal animal is uncommon in Hawaiian waters. Sea stars often bear tiny pincer-like organs called pedecillariae on their surface. Bivalve pedecillariae large enough to be seen with the naked eye surround the perimeter of this species. Typically about 5 in. across. Indo-Pacific. Photo: Lana`i Lookout, O`ahu. 30 ft.

MAGNIFICENT STAR
Luidia magnifica Fisher, 1906
Family Luidiidae
• This 10- or 11-armed giant attains a diameter of at least 2 ft. and is one of the most impressive Hawaiian invertebrates. It lives just beneath the sand, its presence occasionally betrayed by a star-shape mark on the surface. If lifted out it sinks slowly back in, digging busily with its thousands of large tube feet. It can also move with surprising speed over the surface. Species of *Luidia* are voracious predators, feeding principally on other echinoderms such as heart urchins and other stars. Instead of everting their stomachs they distend their mouths and swallow their prey whole. Described in 1906 from one specimen (dredged off Moloka`i from a depth of between 250 and 450 ft.) it remains a poorly known, rarely seen creature, recorded so far only from Hawai`i and the Philippines. The genus is named after an early investigator of sea stars, Edward Lhuyd, who wrote under the Latin name "Luidius." *Luidia aspera*, a similar but smaller species (up to 12 in.) sometimes seen by divers, has only eight arms. Photo: Kahe Point, O`ahu. 50 ft.

FISHER'S STAR

Mithrodia fisheri Holly, 1932
Family Mithrodiidae
• The surface of this star is covered with a coarse mesh of ridges, bumps and projections. Its arms, round in cross- section, are likely to be of unequal length and the edges usually bear a row of long, blunt lateral spines. Color varies from almost white, tan or cinnamon-brown to orange-red, usually with three to four darker bands on the arms. Most active at night, it typically occurs on the ceilings or walls of caves, where it probably feeds on sponges, bryozoans or other sessile animals. The name honors American zoologist Walter K. Fisher (1878-1953), who pioneered the study of Hawaiian sea stars and cucumbers in the early 1900s. This species is usually about 4-6 in. across; Edmondson records specimens to 20 in. Although similar to *M. bradleyi* of the Eastern Pacific and *M. clavigera* of the Indo-Pacific, it is probably endemic to Hawai`i. Photo: Mākua, O`ahu. 20 ft.

CATALA'S STAR

Thromidia catalai Pope & Rowe, 1977
Family Mithrodiidae
• In Hawai`i this star is pinkish to light cinnamon-brown, darkening to reddish brown at the tips. In small specimens the central disk is lighter, the arms reddish brown with one or more light bands. The surface is firm and completely covered with small hard nodules. The species occurs as shallow as 40 ft. on hard substrate adjacent to deep water, such as along the slopes and walls of Molokini Islet, Maui, or Lehua Rock, Ni`ihau. The first known specimens were dredged from a depth of 100-150 ft. off Nihoa Island in 1906. They were misidentified, and the species was not fully described until 1977. The name honors René Catala of the Nouméa Aquarium in New Caledonia. This star grows to at least 2 ft. in diameter and develops arms so round and fat that it has been compared with an overstuffed toy. In Hawai`i, specimens of this size generally occur only at depths of several hundred feet. Tropical Western and Central Pacific. Photo: Lehua Rock, Ni`ihau. 90 ft. (juvenile coloration)

296

FAMILY OPHIDIASTERIDAE

This large family is well represented in Hawai`i. Its members usually have a small central disk with five long arms, cylindrical in cross-section.

RED VELVET STAR

Leiaster glaber Peters, 1852
Family Ophidiasteridae

• This beautiful star, with its small central disk, slender arms and irregular red blotches, is sometimes confused with the more common Spotted Linckia (p. 299). It has more red, however, and its skin is soft and smooth (almost slimy), whereas *Linckia* species have a harder, more granular surface. Also, this animal is nocturnal while the Spotted Linckia is found in the open at all times of day. A juvenile about 2 in. across was light tan with scattered red blotches. The species name means "smooth." To about 8 in. Indo-Pacific and Eastern Pacific. Photo: Pūpūkea, O`ahu. 30 ft.

PURPLE VELVET STAR

Leiaster leachi (Gray, 1840)
Family Ophidiasteridae

• This large reddish purple star appears blue in deep water; like all species of *Leiaster* it has a completely smooth, almost velvety texture. The five arms, nearly round in cross-section, may be of slightly unequal lengths. The central disk is small. Although the first Hawaiian specimen was found in shallow water at Kōloa, Kaua`i, the species usually occurs at depths of at least 70 ft. or more on hard substrate mixed with pockets of sand and rubble. Small specimens do well in captivity and are sometimes exported (although they sometimes shed an arm if stressed). The Hawaiian population, differing in the number and shape of the spines and possibly in color, has been given the subspecies name *hawaiiensis*. To about 15 in. Indo-Pacific. Photo: Mākua, O`ahu. 80 ft.

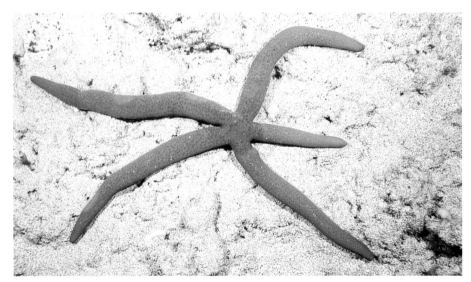

GREEN LINCKIA

Linckia guildingi Gray, 1840
Family Ophidiasteridae

• This star usually has five (sometimes four or six) long arms, cylindrical in cross-section. Although appearing smooth, the skin is hard and granular to the touch and covered with low nodules. The arms narrow where they join the central disk and are easily detached. Most specimens have at least one arm shorter than the rest, evidence that it was recently lost. This is not necessarily due to predation; like the Spotted Linckia (below), it can reproduce by shedding an arm, which then regenerates to form a new animal. In Hawai`i this species is dark grayish green, brown or (rarely) blue; elsewhere it can be tan, beige or even reddish. It occurs on hard substrate, often wedging itself tightly into crevices. What it eats and how it feeds are not completely known, although in captivity it will accept carrion. The genus name honors J.H. Linck, an early investigator who published a monograph on sea stars in 1733. Tropical Atlantic and Pacific. To about 16 in. Photo: Hanauma Bay, O`ahu. 70 ft. (See also p. 292.)

Spotted Linckia (a) ➤

SPOTTED LINCKIA   Spotted Linckia (b)

Linckia multifora (Lamarck, 1816)
Family Ophidiasteridae
• The most commonly seen star in Hawai`i, this species is almost always found on hard substrate where it may feed on coralline algae. Like all *Linckia* species, it has a hard granular skin entirely covered with low inconspicuous nodules. The five (occasionally six) slender arms are cylindrical in cross-section and almost always of unequal length. The color is variable: usually grayish pink to orange-brown with round red or maroon spots, irregularly spaced but often concentrated toward the ends of the arms. The number of spots varies; some individuals lack them almost entirely. The tips of the arms show some blue. To reproduce these stars can shed (autotomize) an arm, which then regenerates a complete animal. Stars in the process of regenerating (with one long arm and four short ones) are called "comets." At least two species of eulimid snails parasitize the undersides of the arms, sometimes digging deep into the flesh; both are uncommon. Because parasitized stars often have five complete arms, the snails are thought to inhibit autotomy. The species name means "many pores." To about 4 1/2 in. Indo-Pacific. Photos: a) Hanauma Bay, O`ahu. 30 ft. (typical coloration, with comet); b) Palea Point, O`ahu. 20 ft.

Cylindrical Star (a)   Cylindrical Star (b)

CYLINDRICAL STAR

Dactylosaster cylindricus (Lamarck, 1816)
Family Ophidiasteridae
• This star and the following species are similar in size, shape and habitat. Variable in color, they are difficult to tell apart in the field. Both have a small disk with five arms, each bearing nine longitudinal rows of skeletal plates. This species has small clumps of granules on the surface of the arms with smooth skin between. (Compare with *Ophidiaster hemprichi*, below.) The arms are round in cross-section, constrict slightly at the base and taper little. The blunt tips terminate in small plates. The overall color varies from red to orange, brown, blue-green, tan, violet and white; it may be uniform or marked with red-dish blotches. These animals occur down to at least 15 ft., usually under rocks. The Hawaiian population lacks pedicellariae and has been given the subspecies name *pacificus*. To about 5 in. Indo-Pacific. Photo: a) Magic Island boat channel, O`ahu. 5 ft. b) Lāna`i Lookout, O`ahu. 15 ft.

Hemprich's Sea Star (a)

Hemprich's Sea Star (b)

HEMPRICH'S STAR

Ophidiaster hemprichi Müller & Troschel, 1842
Family Ophidiasteridae
• This star has five (occasionally four) cylindrical arms, often unequal in length and "striped" with nine longitudinal rows of small plates. The arms taper only slightly near the tips, which terminate in a small smooth plate; their surface is completely covered with granules. (Compare with *Dactylosaster cylindricus*, above.) The disk is small. The overall color is usually reddish brown to gray-brown marked with small irregular blotches, mostly white but also pink, brown and yellow. These stars occur on hard or rubbly substrate under ledges and stones at depths of 3 to at least 900 ft. The name honors German naturalist Wilhelm Friedrich Hemprich (1796-1825), who died of a fever while collecting biological specimens in the Middle East. To about 4 in. across. Tropical Pacific. *Ophidiaster squameus* is a synonym. Photos: a) Waikīkī, O`ahu. 40 ft.; b) Pūpūkea, O`ahu. 20 ft.

FAMILY OREASTERIDAE

Stars of the family Oreasteridae have large domed disks and relatively short arms, or no arms at all.

Cushion Star (a) ➤

Cushion Star (b)

CUSHION STAR

Culcita novaeguineae (Müller & Troschel, 1842)

Family Oreasteridae

• Fully grown Cushion Stars resemble plump, bristly, pentagonal cushions about 10 in. across and 3 to 4 in. thick at the center. They may be red, maroon, tan or yellow. Although adults have no obvious arms, five colorful rows of tube feet radiate from the central mouth on the underside. These stars feed on living coral by everting their stomachs (a colorless gelatinous mass that quickly retracts when the animal is disturbed) and digesting the polyps where they grow. In Hawai`i they prefer small colonies of *Pocillopora meandrina* and *P. damicornis*, which they often kill completely. They attack other corals too. Their bristly surface is frequently home to *Periclimenes soror* (p.224), a tiny commensal shrimp that matches the color of its host. If turned upside down, a Cushion Star is able to right itself by inflating one side of its body with water, arching over to get a grip with its tube feet, and pulling itself over. Very small specimens are flat with five stubby arms and resemble star-shape Christmas cookies. To about 10 in. across. Indo-Pacific. Photos: Hanauma Bay, O`ahu. 80 ft. a) & b) adults; c) juvenile. (See also p. 290.)

Cushion Star (c)

KNOBBY STAR

Pentaceraster cumingi (Gray, 1840)
Family Oreasteridae
• These large brown or maroon stars have five stiff arms, triangular both in outline and cross-section. The massive central disk is high and broad, while the underside is smooth and flat. The entire animal is covered with small orange knobs, many of which are interconnected by ridges. These stars usually occur at depths of 100 ft. or more, either on sand or hard substrate mixed with rubble. *Pentaceraster hawaiiensis* is a synonym. To at least 12 in. diameter. Hawai`i and the tropical Eastern Pacific. Photo: Animal collected off Wai`anae, O`ahu at 140 ft. was placed on sand in shallower water for photo.

BRITTLE STARS
PHYLUM ECHINODERMATA. CLASS OPHIUROIDEA

Brittle stars are probably Hawai`i's most abundant echinoderms. They resemble sea stars in shape but are entirely different animals belonging to a class of their own. Turn over almost any stone below the high water mark and you are likely to see one or more of these spiny creatures thrashing arms in a quick scuttle for cover. Dive at night and you will see numbers of them partly out of their crevices. Keenly sensitive to light, they immediately seek shelter when exposed to it and are probably the fastest moving of all echinoderms. Their flexible arms have given them another common name, serpent stars. The class name, Ophiuroidea, derives from *ophis*, the Greek word for snake.

A brittle star's spiny, segmented arms contain none of its vital organs, all of which are located in the central disk. The organs of sea stars, by contrast, are partly contained within their arms. In addition to providing locomotion, a brittle star's arms often serve as the first line of defense— they simply break off should a predator grasp one. Missing arms are quickly regenerated. Arms are also used in feeding and anchoring the animal inside a crevice. Some species of brittle stars filter plankton from the water using their spiny or finely branched arms; most probably use their

Pied Brittle Star *(Ophiocoma pica)*, Lāna`i Lookout, O`ahu. 30 ft.

arms to scavenge detritus from the bottom (helped by hydraulically operated sucker-less tube feet that secrete a sticky mucus). Brittle stars have a mouth on the underside but no anus.

Like most echinoderms, brittle stars have separate sexes. Eggs and sperm are usually discharged into the water where fertilization takes place. Brittle stars often survive well in captivity but are unrewarding aquarium animals because they are nocturnal. Some emit bioluminescent flashes that have mystified more than one aquarist after lights have been turned off.

There are at least 2,000 species of brittle stars, more than any other class of echinoderms. Their success may be due to their motility and their ability to utilize small spaces for shelter. Hawai`i has 57 known species, 19 from shallow water. (The five largest and one representative small species are shown here.) Six shallow-water and 22 deep-water species are currently considered endemic. Although some tropical brittle stars are colorful, common Hawaiian species tend to be drab. The spectacular, many branched basket stars often found in other parts of the tropics are absent from the Islands.

Little is recorded about brittle stars in ancient Hawai`i. Inedible, they were probably disregarded. Most large species from Hawai`i belong to the family Ophiocomidae. Small brittle stars with five or six banded arms, seen at night and sometimes abundant, probably belong to the family Amphiuridae (genus *Ophiactis*).

RETICULATED BRITTLE STAR

Ophiocoma brevipes Peters, 1851
Family Ophiocomidae
• The genus *Ophiocoma* contains four of the five largest shallow-water brittle stars in Hawai`i. Of these, this species is the lightest in color, the smallest and probably the least common. The disk pattern is variable, usually gray or light tan with fine darker brown reticulations covering all or part of the disk. The five arms are the same light color as the disk, usually with numerous narrow dark bands. The arms are 3-4 in. long and the disk is usually less than 1 in. wide. This animal is found under stones on mixed sand and rubble from the shallows to depths of at least 40 ft., often in the company of the similar *O. dentata* (below). Naturalist Cory Pittman has found an easy way to tell them apart: if lifted and dropped through the water *O. brevipes* will curl its arms over its disk whereas *O. dentata* will trail its arms straight up. Indo-Pacific. Photo: Hekili Point, Maui.

TOOTHED BRITTLE STAR

Ophiocoma dentata Müller & Troschel, 1842
Family Ophiocomidae
• One of the most abundant of Hawaiian brittle stars, adults of this species vary from gray black to gray brown, typically lighter on the underside. There are five arms, usually with diffuse light bands (becoming fainter in large specimens); the dark disk is often marked with even darker spots. These animals are found under stones and in crevices, especially in heads of *Porites lobata*. They feed at night with two or three arms extended, securely anchored by the remaining arms. *Ophiocoma insularia* is a synonym. To about 1 in. disk diameter; arms to about 4 in. Indo-Pacific. Photo: Hekili Point, Maui. 2 ft.

SPINY BRITTLE STAR

Ophiocoma erinaceus Müller & Troschel, 1842

Family Ophiocomidae

• This species is uniformly black by day; gray bands appear at night. There are five arms, and the disk may show a faint five-part radial pattern. The tube feet on the underside are sometimes red. Like *O. dentata* (above) it is common under stones, especially on the leeward sides of the Islands. The species name means "hedgehog," probably because of the spines on the arms. It is no more spiny, however, than many other brittle stars. With a disk diameter of about 1 in. and arms up to 5 1/2 in. long, it is the largest common brittle star in Hawai`i. This is a widely distributed Indo-Pacific species. Photo: Makapu`u, O`ahu. 20 ft.

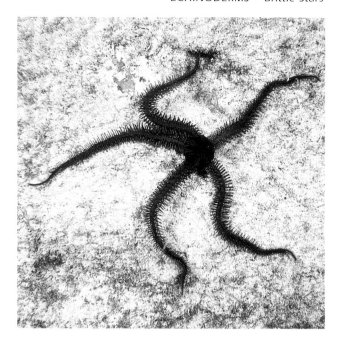

PIED BRITTLE STAR

Ophiocoma pica Müller & Troschel, 1842

Family Ophiocomidae

• This spiny brittle star has fine looping white or gold lines on the black disk. Each of its five relatively short, broad dark arms displays a double row of small, light yellow plates. The spines are long and slender. Often nestling deep among the branches of Cauliflower Coral *(Pocillopora meandrina)*, it is one of the few Hawaiian brittle stars visible during the day. It also occurs under stones but is less common in that habitat than the two species above. The scientific name means "magpie," a bird with striking black and white plumage. The disk is about 1 in. across, the arms about 3 in. long. Indo-Pacific. Photo: Kahe Point, O`ahu. 40 ft. (See also p. 303.)

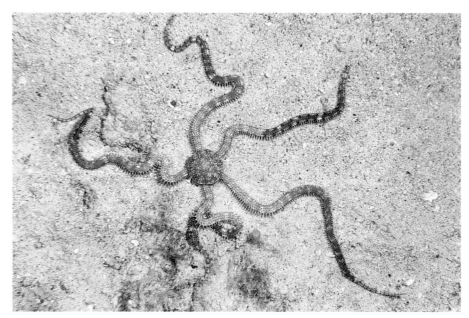

LONG-ARM BRITTLE STAR

Ophionereis porrecta Lyman, 1861
Family Ophionereididae
• This attractive species has slender, serpentine, buff to reddish brown arms, faintly banded with white. The rounded disk is marked with an almost meshlike network of lines. A relatively uncommon brittle star, it occurs from the shoreline to at least 40 ft. The species name means "stretched out" or "extended." The disk is usually less than 1/2 in. across, the arms 4-6 in. long. Indo-Pacific. Photo: Pearl Harbor boat channel, O`ahu. 40 ft.

SPONGE BRITTLE STAR

Ophiactis sp.
Family Amphiuridae
• This small, six-armed brittle star was one of a group found in the red sponge *Mycale (Mycale)* sp., common in the boat channel at Magic Island, O`ahu (p. 14). These animals inhabit cavities in the sponge, extending several arms through openings at its surface to catch drifting food particles. They probably never leave the sponge. Several species of this genus occur in Hawai`i, one often seen at night on vertical rocky walls. They may have five, six, or seven arms. The disk is usually less than 1/4 in. across. Photo: Magic Island boat channel, O`ahu. 15 ft.

URCHINS
PHYLUM ECHINODERMATA. CLASS ECHINOIDEA

Banded Urchin *(Echinothrix calamaris)* surrounded by Rice Coral and Lobe Coral. This specimen is atypical. Despite the name, most adults of this species are a uniform dull black. Molokini Islet, Maui. 40 ft.

Sooner or later, every snorkeler or diver connects with an urchin; the experience is likely to be painful. These echinoderms are characterized by a globular or flattened skeleton (test) bristling with hundreds of movable spines, tube feet, and tiny pincer-like organs called pedicellariae. The mostly hollow inner space contains the digestive and reproductive systems and other organs. Urchins divide easily into two basic groups, "regulars" and "irregulars." Regular urchins, usually called sea urchins, are described immediately below. The irregulars—heart urchins, sand dollars sea biscuits, and the like—are discussed in their own subsection.

Sea urchins usually have either slender sharp spines or massive blunt ones. Most live on the sea bottom or on solid surfaces. Their skeletons, or tests, are roughly spherical and formed of fused plates. If you find a sea urchin test you will see both pores and nodules on the surface, radiating in five bands from the mouth (on the underside) to the anus (at the top.) The animal's tube feet extend through the pores; the spines originate from the nodules.

The mouth of all regular urchins (and some irregulars) is equipped with a complex five-part radial set of jaws and teeth controlled by 60 separate muscles. With these, urchins can scrape algae from the substrate, attack and eat other organisms, or even bore into solid rock. Almost 2,400 years ago the Greek philospher and naturalist Aristotle compared the structure of this unusual organ to that of lanterns in use at the time; an urchin's mouthparts have been known ever since as "Aristotle's lantern."

A sea urchin's tube feet are tipped with suckers that help it to cling, move about and remove unwanted debris. Its pedicellariae (tiny pincers set on stalks) also clear unwanted material and sometimes serve as a second line of defense.

Sea urchin spines can be long, short, slender, stubby or clublike. Most urchins have at least two types. Members of the large tropical family Diadematidae have exceptionally long, sharp, sometimes venomous spines. Puncture wounds from long-spined urchins can be intensely painful. The spines break easily, are almost impossible to extract, and often contain pigments that stain the surrounding tissue an alarming bluish black. The pain usually subsides after about an hour, although nearby joints may remain stiff and sore for weeks. The spines are eventually absorbed by the body and usually need not be removed. Some short-spined Indo-Pacific urchins of the family Toxopneustidae, sometimes called flower urchins, have venomous pedicellariae that can deliver a potentially fatal sting. Fortunately, the most dangerous of these are not found in Hawai`i.

The spines of sea urchins provide protection for a number of other animals. At least four species of shrimps in Hawaiian waters live commensally with urchins, while the rectum of the Banded Urchin *(Echinothrix calamaris)* is home to a tiny crab. Juvenile fishes, especially the Hawaiian Domino Damselfish *(Dascyllus albisella)*, sometimes seek refuge within urchin spines as well.

To reproduce, male and female urchins simultaneously emit a milky fluid containing either eggs or sperm; at such times they may appear to smoke like little chimneys. Some climb to the tops of coral heads or rocky outcroppings to better disperse their gametes. The fertilized eggs hatch into larvae that drift with the plankton for days or weeks before transforming into juvenile urchins and settling on the bottom. Some deep-sea or cold-water species skip the larval stage and hatch directly from eggs deposited in special pouches on the urchin's surface.

How did sea urchins get their name? Although an urchin today is "a small, mischievous boy," in old England the word meant "hedgehog." The ancient Hawaiians had distinct names for at least four broad categories of urchins: **wana** (those with long slender spines), **`ina** (medium-length slender spines), **hāwa`e** (short slender spines) and **hā`uke`uke** (thick, flattened or stubby spines). They considered the gonads of many urchins a delicacy, eating them raw, cooked, or dried, and also prepared sauces and condiments using the urchin's inner liquids. Certain Hawaiian families and individuals, however, revered **wana** as **`aumakua** (the embodiment of ancestors). Such people neither harmed nor ate urchins; an urchin appearing in a dream or vision was held by them to have special meaning.

In the aquarium, sea urchins with short spines are easiest to keep and often live a long time. It is fascinating to watch the coordinated action of their myriad tube feet as they crawl up the glass. Such urchins generally graze on algae, but their unusual five-part mouth (Aristotle's lantern) is capable of devouring almost anything that doesn't run away, including dead fish, tube worms, molluscs and even other urchins. Obviously these voracious omnivores are a poor choice for "mini-reef" type invertebrate aquariums. In fish tanks, however, they can help control unwanted algae. Echinoderms in general require excellent water quality. If your urchins' spines droop, conditions may not be to their liking. Change the water or return them to the sea.

About 900 urchin species exist worldwide with some 75 known in Hawai`i. Of these, at least 22 occur in shallow water. Thirty-five of the deep-dwelling species are known at present only from Hawai`i but are probably more widespread.

REGULAR URCHINS

Nine orders of urchins (of a total of 14) have radial symmetry and are known as regular urchins. Four orders of regular urchins occur in Hawai`i's shallow seas: the Cidaroida (heavy, widely spaced spines; ancestral to all other living urchins), Diadematoida (rigid or flexible test with long hollow spines), and Temnopleuroida and Echinoida (both with rigid test and solid spines).

FAMILY CIDARIDAE

Members of the primitive family Cidaridae (order Cidaroida) have massive, blunt, well-spaced primary spines, ringed at the base by smaller secondary spines. Lacking the skin of living tissue present on the spines of other urchins, their primary spines are usually covered with a layer of algae and detritus.

ROUGH-SPINED URCHIN

Chondrocidaris gigantea A. Agassiz, 1863
Family Cidaridae
• These large urchins are occasionally found in the open, usually at depths greater than 30 ft. Their thorny spines provide an excellent substrate for growths of algae, bryozoans and sponges, which in turn camouflage the urchin. (Contrast this with the smooth, brightly colored spines of the Red Pencil Urchin [p. 317] on which no fouling organisms settle.) Small specimens are not uncommon in some areas but conceal themselves under coral slabs or in crevices by day and are seldom seen. The test attains a diameter of 4 in. and the spines a length of 6 in. Known only from Hawai`i and New Caledonia. Photo: Mākua, O`ahu. 100 ft.

TEN-LINED URCHIN • **ha`ue`ue**

Eucidaris metularia (Lamarck, 1816)
Family Cidaridae
• Small, attractive and common, this is sometimes called the "sputnik urchin." Its smooth upper surface displays a beautiful five-part geometric pattern that differs somewhat from individual to individual. The thick spines, arranged in ten vertical rows, are broader in the middle than at base or tip and are marked with alternating bands of pale red and yellow. Unless newly grown, however, the spines are usually covered with algae or other marine organisms that obscure the colors. This urchin hides by day under rocks or in crevices, from the shoreline to depths of at least 50 ft. The species name derives from the Latin *metula*, meaning a small cone or pyramid. The test grows to about 1 in. diameter. Indo-Pacific. Photo: Kahe Point, O`ahu. 20 ft.

Thomas's Urchin ➤

◄ THOMAS'S URCHIN

Actinocidaris thomasi (A. Agassiz & Clark, 1907)
Family Cidaridae
• Long known only from deep-water specimens dredged by the steamer *Albatross* around the turn of the century, this rarely seen urchin sometimes occurs as shallow as 30 ft. It hides by day under coral slabs or in crevices, probably emerging at night. The thick primary spines are marked lengthwise with closely spaced lines of raised dots. Newly grown spines are pale maroon; older ones are covered with algae and other growths. The secondary spines and test are a rich maroon. At the center of the test a beautiful five-part reticulate pattern of lines marks the position of the apical plates. A few small limpet-like molluscs may attach to the spines in shallow-water specimens; those from deep water may host barnacles. Maximum size is probably about 5 or 6 in. across, including the spines. Known only from Hawai`i. Photo: Palea Point, O`ahu. 80 ft.

FAMILY DIADEMATIDAE

Urchins of this large family (order Diadematoida) have hollow, slender, needle-like spines. They often wave them menacingly at approaching divers. Despite the intense pain caused by punctures, the long spines bear no venom. The shorter secondary spines, however, may be coated with venomous glandular material (genera *Echinothrix* and *Leptodiadema* only). Diadematids are characterized by the presence of an anal sac, a balloon-like organ at the top center of the test that acts somewhat like a chimney, channeling wastes away from the urchin's body.

BLUE-SPOTTED URCHIN

Astropyga radiata (Leske, 1778)
Family Diadematidae
• Uncommon in Hawai`i, this reddish brown, long-spined urchin inhabits sand or rubble bottoms at about 100 ft. or more. The spineless areas of its test contain rows of iridescent blue spots that can expand and contract, sometimes almost disappearing. The spines are usually banded brownish red and white, but coloration is variable. The test is somewhat flexible. In other areas of the Indo-Pacific this species inhabits shallow sea-grass beds. To about 8 in. across, including the spines. Indo-Pacific. Photo: Mike Severns. Mākena, Maui. 85 ft.

LONG-SPINED URCHIN • **wana hālula**

Diadema paucispinum (A. Agassiz, 1863)
Family Diadematidae
• These urchins are remarkable for the length of their purplish black spines—up to several times the diameter of the test. Juveniles have purplish red spines, sometimes with faint banding. The spines of adults are never banded. These urchins are uncommon in the reef environment, occurring most predictably in harbors and other sheltered locations, typically on vertical surfaces. In exposed habitat they prefer depths below 60 ft. (the precipitous back wall of Molokini Islet, Maui, is a good place to see them) but sometimes occur in surprisingly shallow water. The spines, although long, are relatively sparse, accounting for the species name meaning "few spines." To about 12 in., including the spines. Hawai`i and the islands of the South Pacific. (*Diadema savignyi*, an Indo-Pacific urchin with similarly long spines and electric blue lines on the test, has also been recorded from Hawai`i but is extremely rare.) Photo: Magic Island boat channel, O`ahu. 15 ft.

Banded Urchin (a) ➤

Banded Urchin (b)

Banded Urchin (c)

BANDED URCHIN • **wana**

Echinothrix calamaris (Pallas, 1774)
Family Diadematidae

• This is the most common long-spined urchin in Hawai`i. The spines of adults are usually dull black but may be banded or even grayish white. Hollow and fragile, they are easily broken. Clumps of shorter venomous secondary spines nestle between the primaries—golden brown in small specimens, black in adults. Younger urchins have banded white and dark green primary spines, and the primaries of very small specimens are blunt, flaring toward the tips. The test may have bluish green lines. The balloon-like anal sac, when distended, is marked with black and white spots. The rectum of older specimens, however, almost invariably contains a small parasitic crab, *Echinoecus pentagonus* (p. 270), that obliterates the anal sac. The crab's pointed carapace is usually visible through the anal opening. A prettier but rarer companion is the commensal shrimp *Stegopontonia commensalis* (p. 226), which clings head-inward to the long spines. Examine enough urchins and you will eventually find a pair. To about 8 in., including the spines. Indo-Pacific. Photos: a) Palea Point, O`ahu. 30 ft. (black spines); b) Maile Ledge, O`ahu. 70 ft. (white spines); c) Mākua, O`ahu. 30 ft. (subadult, showing secondary spines and anal sac) (See also p. 307)

BLUE-BLACK URCHIN • **wana**

Echinothrix diadema (Linnaeus, 1758)
Family Diadematidae
• The black primary spines of this urchin have a beautiful dark blue satiny sheen. They are often slightly thicker than those of *E. calamaris* (p. 313). Juveniles of the two species look similar, with white and dark green banded primary spines (flaring at the tips in very small specimens). The secondary spines of juveniles are also banded, whereas those of *E. calamaris* are uniform in color. The distended anal sac is small, black and inconspicuous, and the rectum is generally too small to accommodate the tiny parasitic crab that lives in *E. calamaris*. Holthuis' Urchin Shrimp (p. 227) may live among the spines. These urchins occur down to at least 70 ft. but are most abundant above 15 ft. along turbulent rocky shores where their thicker, stronger spines may give them an advantage. Scavengers (as are many urchins), they sometimes feed on the cast-off shells of spiny lobsters. To about 6 in., including the spines. Indo-Pacific. Photo: Lāna`i Lookout, O`ahu. 15 ft.

Fine-Spined Urchin ➤

FINE-SPINED URCHIN

Leptodiadema purpureum A. Agassiz & Clark, 1907
Family Diadematidae

• These unusual and little-known urchins have extremely fine, flexible golden or brown spines reminiscent of spun glass. Active at night, they occur on rocky substrate at depths of 20 to 150 ft. or more. Small individuals hide under slabs and stones by day; large ones withdraw into the deepest recesses of caves and crevices. Moving surprisingly fast for urchins, they retreat quickly when exposed to light. The long fine primary spines are flexible but can still prick a finger; the shorter secondary spines, tipped with dark glandular material, cause intense pain and may bear venom. These urchins often host a pair of the commensal shrimps *Stegopontonia commensalis* (p. 226), which cling to the light-colored spines. Both shrimps and urchins do well in aquariums, the latter grazing "hair algae" off the walls of the tank. The species was described from one tiny specimen dredged off Moloka`i in 22-24 fathoms. The test was dull purple, hence the species name. Although recorded only from Hawai`i, this urchin or something very much like it occurs also in the Marshall Islands. Specimens are usually about 1 1/2 in. across, including the spines, but may grow to more than twice that size. Photo: Pūpūkea, O`ahu. 20 ft.

FAMILY ECHINOMETRIDAE

This diverse family includes four of Hawai`i's most abundant urchins. It belongs to the order Echinoida, characterized by solid spines and a rigid test.

HELMET URCHIN ▪ **hā`uke`uke kaupali**

Colobocentrotus atratus (Linnaeus, 1758)
Family Echinometridae

• These unusual urchins cling to exposed rocky shores where few other animals survive. Their shallow domed tests are covered with stubby, shingle-like spines that offer little resistance to the surf. Longer flat spines ring the test like a skirt of armor. (Other common names are Shingle Urchin and Armored Urchin.) Covered with water they are deep purple; exposed to air they are almost black. Despite the constant pounding of enormous waves, they creep about almost imperceptibly on their strong tube feet, scraping thin films of algae from the rocks. At low tide look for them from shore especially on smooth lava rock that is tan or pinkish from the growth of coralline algae. Admire from a distance, however; sudden large waves can be dangerous. The Hawaiian name **"kaupali"** means "cliff-clinging." The species name means "dressed in black" (as if in mourning). A tiny snail with beautiful spiral banding *(Vexilla vexillum)* sometimes feeds on these animals. So do humans; in old Hawai`i it was said "When the **hala** fruit ripens, the **hā`uke`uke** is fat," meaning that the tests are full of eggs or sperm. To about 3 in. Indo-Pacific. Photo: Makapu`u, O`ahu. 1 ft.

ROCK-BORING URCHIN • `ina kea

Echinometra mathaei (de Blainville, 1826)
Family Echinometridae
• This is the most common urchin in Hawai`i and possibly the Indo-Pacific. By continually scraping with its short spines and five-toothed "lantern" it is able to excavate into solid rock, usually limestone or tuff. Most shallow reef flats in Hawai`i are riddled with channels and holes bored by these animals over the years. (Black urchins occurring in the same shallow habitat are a different species, described below.) They also inhabit deeper water, typically wedging themselves under branches of Finger Coral at depths of 50 or 60 ft. The species ranges from greenish white to reddish. In Hawaiian, `ina denotes an urchin with medium-length spines, and **kea** means "white." Reddish specimens were called `ina `ula. To about 2 1/2 in., including the spines. Indo-Pacific. Photo: Hanauma Bay, O`ahu. 50 ft.

OBLONG URCHIN • `ina

Echinometra oblonga (de Blainville, 1826)
Family Echinometridae
• Like the lighter-colored species above, this dull black urchin uses its spines and teeth to bore into softer rocks such as limestone or volcanic tuff. It is often the dominant urchin (and a principal agent of erosion) on shallow rocky shores exposed to constant wave action; it also occurs on reef flats intermingled with *E. mathaei*. Unlike the latter, it is restricted to depths of less than about 10 ft. Apart from the black color and a slightly more oblong test, the physical differences between the two are slight. *Vexilla vexillum*, a tiny snail with beautiful spiral banding, feeds on both of these animals. To about 2 1/2 in., including the spines. Photo: Makapu`u, O`ahu. 1 ft.

NEEDLE-SPINED URCHIN

Echinostrephus aciculatus A. Agassiz, 1863
Family Echinometridae
- These rock-boring urchins have fine lavender, bronze or black spines, long only on the exposed upper surface; spines around the perimeter are short. The test is light in color. Occupying deep round cavities in coral limestone or basalt, these animals never leave their holes, whose entrance diameters are often smaller than their own. It has been suggested that they subsist on bits of floating algae caught on their long upper spines. In Hawai`i, however, this seems unlikely; their diet remains a mystery. The species is fairly common in the main islands at depths greater than 10 ft. but is seldom as abundant as the shorter-spined, rock-boring species of *Echinometra*. At Midway Atoll and perhaps elsewhere in the northwestern chain it is superabundant and the dominant urchin on hard substrate down to at least 90 ft. The species name means "having small needles." The test attains a diameter of about 1 1/2 in. Indo-Pacific. Photo: Pūpūkea, O`ahu. 30 ft.

RED PENCIL URCHIN ▪ **hā`uke`uke `ula`ula**

Heterocentrotus mammillatus (Linnaeus, 1758)
Family Echinometridae
- These striking urchins have bright red, clublike spines that add a characteristic color note to many Hawaiian reefs. The shorter stubby spines covering the test vary from almost white to very dark red. At night, surprisingly, the long red spines often turn chalky pink. A thin layer of live tissue covering the spines inhibits the growth of algae and other marine organisms. Years ago, when blackboards were made of slate, the long spines could be used in place of chalk, thus the alternate common name Slate Pencil Urchin. In old Hawai`i the spines were decoratively carved. The small red shrimp *Levicaris mammillata* (p. 229) sometimes lives among the spines. These urchins seldom survive long in captivity unless plentiful algae is provided. Specimens from other parts of the Pacific may be yellowish or chocolate brown. The Hawaiian name `ula`ula means red. Another Hawaiian name is **pūnohu**. The species name means "having breasts" or "nipples," referring to the nodules prominent on the tests of dead specimens. In the main Hawaiian Islands it seldom exceeds 8 in., including the spines. At Midway Atoll, Northwestern Hawaiian Islands, it attains 12 in. Indo-Pacific, but abundant only in Hawai`i. Photo: Kahe Point, O`ahu. 15 ft.

Top picture on next page ➤

◄ Red Pencil Urchin

FAMILY TOXOPNEUSTIDAE

Members of the family Toxopneustidae (order Temnopleuroida) have short spines and often cover themselves by day with algae, bits of shell, rubble or other debris. Some species (not found in Hawai`i) have venomous pedicellariae that can give a painful or even fatal sting to humans.

WARTY SAND URCHIN

Cyrtechinus verruculatus
(Luetken, 1864)
Family Toxopneustidae
• This tiny urchin lives under rocks, gravel or sand from the shoreline to depths of at least 30 ft. Its short whitish, yellowish or tan spines are often covered with sand grains. Although not uncommon, it is seldom seen because of its burrowing habits. The species name means "small warts." To about 1 in. Indo-Pacific. Photo: Pūpūkea, O`ahu. Tide pool.

PEBBLE COLLECTOR URCHIN
• **hāwa`e po`o hina**

Pseudoboletia indiana (Michelin, 1862)
Family Toxopneustidae

• Grayish white to purple, this urchin occurs from tide pools to depths of at least 50 ft. By day it covers or may even bury itself. At night it throws off its camouflage and roams the reef. Tiny shrimps of two species *(Periclimenes insolitus* and *Gnathophylloides mineri)* sometimes live among the spines, usually on the under-surface of the urchin. The Hawaiian name **po`ohina** means "gray head." In old Hawai`i short-spined, shallow-water urchins (**hāwa`e**) were considered nearly useless as food; by extension, the word **hāwa`e** denoted a useless person. To about 5 in., including the spines. Indo-Pacific. Photo: Kahe Point, O`ahu. 45 ft.

COLLECTOR URCHIN • **hāwa`e maoli**

Tripneustes gratilla (Linnaeus, 1758)
Family Toxopneustidae

• This is a short-spined black urchin that occurs in tide pools and to depths of at least 40-50 ft. Its spines (not particular-ly sharp) radiate in five double rows from the center of the test. Some whitish spines may intermingle with the black ones (rarely, almost all spines are white); the smooth areas between the rows of spines are often a deep blue. This animal col-lects bits of algae, shell or other material on its spines, possibly for disguise, protection from bright light, or even as a means of storing food. The small shrimp *Gnathophylloides mineri* (p. 227) sometimes lives among the spines. The species name refers to a kind of cake eaten in ancient Rome; the Hawaiian name means "natural" or "native." To about 5 in., including the spines. Indo-Pacific. Photos: a) Lāna`i Lookout, O`ahu. 30 ft,; b) Makapu`u, O`ahu. 20 ft.

Brissus latecarinatus (test). Makapu`u, O`ahu. 30 ft. The empty tests of irregular urchins are more often seen than the living animals, which usually live under the sand.

IRREGULAR URCHINS

Five orders of urchins have bilateral rather than radial symmetry and are known collectively as irregular urchins. These include the heart urchins (order Spatangoida); arrowhead urchins, sand dollars, sea biscuits and sea pancakes (order Clypeasteroida); and bean urchins (order Holectypoida). Most irregular urchins live their entire lives under sand. Unlike the radially symmetrical regular urchins, these animals have definite front and rear ends with the anus shifted off-center toward the back. Often the mouth is shifted forward. An "Aristotle's lantern" may or may not be present. The spines, usually short and hairlike, typically point backward to facilitate the animal's passage through the sand.

Heart urchins have somewhat elongated globular tests and lack a lantern. Their slender spines may be short or long. Sand dollars usually have thin flat tests. Sea biscuits and sea pancakes belong to the same order as sand dollars but have somewhat thicker tests. All three are usually covered with short, velvety spines. The tests of most irregular urchins display a beautiful five-part petaloid pattern on the upper surface formed by pores through which specialized tube feet (used for respiration) once protruded on the living animal. Irregular urchins are often abundant under the sand and constitute an important food source for many marine animals in Hawai`i. The tiny *Mortonia australis*, only 1/4 in. long, occurs in great numbers just below the waterline on some beaches. A similar species, *Echinocyamus incertus*, can also be superabundant. Some species grind coarse sand into fine, thereby contributing to the formation of our beautiful Hawaiian beaches.

ROUND-MOUTH BEAN
URCHIN

Echinoneus cyclostomus
Leske, 1778
Family Echinoneidae
• Almost a living fossil, this is one of only two surviving urchins of the order Holectypoida, whose members lack the petaloid pattern common to other irregular urchins. Five bands extending around the sides of this urchin's oval test make it easy to recognize; the bands are also visible on the living animal. The short spines are whitish to light brown and the tube feet, ocurring at the edges of the bands, are red. The oval mouth is in the center of the test's underside, the anus posterior to it. This urchin occurs from the shoreline (usually under pieces of coral rubble in sandy areas) to depths of almost 2,000 ft. The species name means "round mouth." To about 1 in. All warm seas. Photo: Mike Severns. Molokini Islet, Maui. 90 ft.

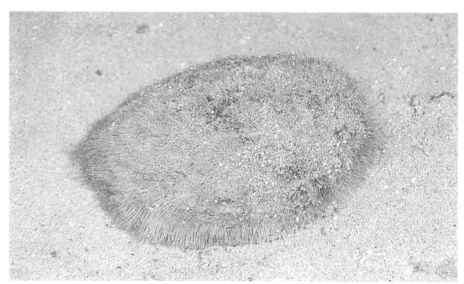

KEELED HEART URCHIN

Brissus latecarinatus (Leske, 1778)
Family Brissidae
• This is the largest and best-known Hawaiian heart urchin. As it moves under the sand it often leaves a depression on the surface by which it can be located and lifted out for inspection. Live specimens are covered with short brownish, pinkish or greenish spines. An empty test turned upside down has a crescent shape mouth at the front and a large anal opening at the rear. The test's upper surface displays a distinct five-part petaloid pattern; the four rearmost rays are prominent and a fainter fifth one points forward. Large specimens attain a length of about 7 in. The species is common throughout the Indo-Pacific. Photo: "Yellow Brick Road," south shore, Kaua`i. 40 ft. (See also p. 320.)

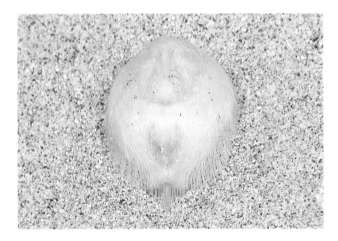

HALF-STAR HEART URCHIN

Rhinobrissus hemiasteroides A. Agassiz, 1879
Family Brissidae
• Although poorly represented in museum collections, this species appears to be fairly common in shallow Hawaiian waters. It is also recorded from northern Australia and islands of the South Pacific. This live specimen was found by combing the sand with the fingers. Placed on the surface, it soon buried itself. Maximum size not known; the specimen was about 1 1/2 in. long. Photo: Kahe Point, O`ahu. 20 ft.

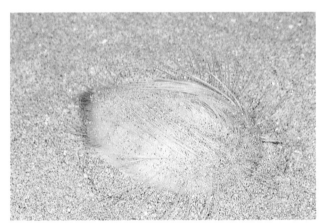

HAWAIIAN LOVENIA

Lovenia hawaiiensis Mortensen, 1950
Family Loveniidae
• Members of the family Loveniidae (and also the family Spatangidae) have long, thin flexible spines. These and other irregular urchins can be quite abundant under the sand and are an important food source for a variety of animals, including sea stars of the genus *Luidia* and Spotted Eagle Rays. This specimen was about 2 in. long. Known only from Hawai`i. Photo: Kahe Point, O`ahu. 50 ft.

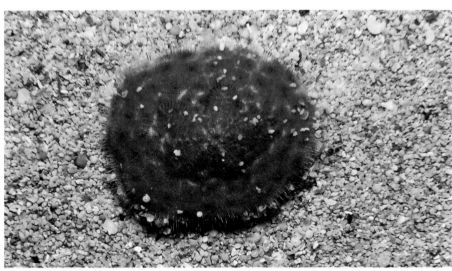

Reticulated Sea Biscuit ➤

RETICULATED SEA BISCUIT

Clypeaster reticulatus (Linnaeus, 1758)
Family Clypeasteridae
• Sea biscuits, sometimes called cake urchins, have oblong, somewhat flattened tests and short velvety spines. They are thicker than sand dollars but not as globular as heart urchins. This small species occurs in sandy tide pools and down to at least 80 ft. Living specimens have short spines and are light brown, greenish or red. They live just under the sand, emerging at night and even clambering slowly on and over stony rubble. Empty tests are seen more often than living animals. Longer than they are broad, with roughly parallel sides, the tests bear the usual five-part petaloid pattern on the upper surface. The underside is markedly concave with the mouth at the center, the anus at the edge. The family and genus names derive from *clypeus*, the Latin word for "shield." To about 1 in., but usually smaller. Indo-Pacific. Photo: Bruce Mundy. Pūpūkea, Oʻahu. 80 ft.

SEA CUCUMBERS
PHYLUM ECHINODERMATA. CLASS HOLOTHUROIDEA

Unidentified sea cucumber, probably of the genus *Stichopus*. Molokini Islet, Maui. 190 ft. (Mike Severns)

Sea cucumbers are easy to find in Hawaiʻi. Clinging to rocky reefs or lying like great sausages on the sand, the larger species are a familiar sight to all snorkelers and divers. Many smaller sea cucumbers live hidden lives beneath stones, in crevices or partially buried in sand or mud.

Unlike other echinoderms, sea cucumbers lie on their sides and have distinct front and rear ends. Most have definite upper and lower sides as well. Although their five-part radial symmetry is not obvious, an endwise view of a sea cucumber usually reveals a ring of tentacles surrounding the mouth; the number (typically 20) is almost always divisible by five. In some species, five inward-pointing teeth guard the anus. Tube feet (usually present) may be in five double rows. Other manifestations of the echinoderm's radial structure are internal.

Almost all sea cucumbers are soft-bodied. Arms, hard spines and pedicellariae are lacking. The skeleton is reduced to microscopic ossicles (spicules) in the skin and a few internal calcareous plates. Examination of the ossicles (shaped like rods, anchors, wheels, buttons, tables, etc.) is often necessary for positive identification of specimens.

Most large sea cucumbers feed by slowly sweeping their down-turned mouth back and forth across the bottom, much like a vacuum cleaner. Some crawl about sweeping as they go; others remain anchored in a crevice, stretching out to sweep the area in front. Swallowing sand, sediment or mud, these animals filter out the organic matter and excrete the remainder in strands or piles of pellets at the other end. Many other sea cucumbers gather food with their tentacles, either sweeping them across the substrate or using them as filters to catch matter suspended in the water. Small burrowing species exist as well.

The sea cucumber anus serves not only to excrete waste, but also as the opening through which many of these animals "breathe." Expanding and contracting their muscular body wall in slow rhythm, they alternately draw in and expel water; organs called "respiratory trees" extract the oxygen. Although lacking eyes, many sea cucumbers are sensitive to light and quickly retract their tentacles when approached. All sea cucumbers can creep about slowly, and a few can swim.

As might be imagined, these sluggish creatures have interesting defenses. Many possess an internal mass of strong, sometimes toxic white threads (Cuvierian tubules) that become amazingly sticky on contact with water. Extruded through the anus, they adhere strongly to anything except the cucumber itself and can easily entangle and immobilize a crab or other predator. (Fishermen in Micronesia sometimes use these matted threads as impromptu foot protection.) Some species produce a potent poison called holothurin, which can be lethal to fishes and other predators, including humans. Some Pacific peoples, however, use sea cucumber extracts to stun or kill fish, which they eat without apparent ill effect. Auto-evisceration is yet another possible mechanism of defense: the cucumber ejects all its internal organs through the anus, presumably to sidetrack a predator while the cucumber itself escapes. The ejected organs take up to several months to grow back, during which time the cucumber cannot feed, although it may absorb some nutrients through its skin.

Reproduction in sea cucumbers follows the usual echinoderm pattern: males and females release sperm and eggs into the water. Fertilized eggs hatch into free-swimming larvae that go through several stages before settling to mature. Some cold-water species skip the larval stage and develop directly, incubated by one of their parents. Spawning sea cucumbers can sometimes be observed on Hawaiian reefs, expecially during high tide on a full moon. For about 20 minutes animals of the same species, wherever on the reef they happen to be, raise their front ends and sway gently, sometimes almost in unison, emitting milky gametes through a pore near the mouth.

Sea cucumbers are a valuable resource in some parts of the world. Boiled and dried, the body walls of certain species are exported to the Far East and the Mediterranean area under the names "trepang" or "bêche-de-mer." Others are eaten raw and undried. You can sample sea cucumber in many Chinese restaurants. Except for a few colorful filter-feeding species from the Western Pacific, sea cucumbers are not important in the aquarium trade. The smaller species might make good night scavengers but will probably hide during the day. Only species without Cuvierian tubules should be kept.

In the Hawaiian language, sea cucumbers are known as **loli**. In olden days many kinds were recognized and named, but we no longer know to which species all the names apply. Some **loli** were eaten and others perhaps used as medicine. Sea cucumbers were also used in love magic. "When **loli** is the offering," it was said, "passionate is the love."

Two of the four holothurian orders common on Indo-Pacific reefs are represented in shallow Hawaiian waters. The order Aspidochirotida—to which most Hawaiian species belong—is characterized by sausage-shape animals, often with thick tough skins. Their bodies typically bear numerous soft fleshy projections called papillae that are actually modified tube feet. Normal tube feet are also present. These cucumbers often eject sticky Cuvierian tubules. Cucumbers of the order Apodida are long and wormlike, with thin skins and collapsible bodies. They lack tube feet and do not eject sticky threads.

There are approximately 1,200 living sea cucumber species ranging in length from about 1 in. to 5 or 6 ft. fully extended. About 50 are known from Hawai`i, roughly half inhabiting shallow water. Four deep-water species are considered endemic. Many Hawaiian sea cucumbers are nocturnal; some remain unidentified.

FAMILY HOLOTHURIIDAE

Members of the large family Holothuriidae (order Aspidochirotida) are common in shallow tropical seas. They typically have thick, muscular bodies equipped with tube feet used for clinging to the substrate and moving about. Pointed fleshy projections (papillae) often cover the body. Most feed on organic material in the sand or mud. With the aid of their oral tentacles they collectively pass many tons of bottom material through their tubular bodies, undoubtedly playing an important ecological role. Many of these animals can eject quantities of sticky white threads to entangle and confuse predators.

WHITE-SPOTTED SEA CUCUMBER • **loli**

Actinopyga mauritiana (Quoy & Gaimard, 1833)
Family Holothuriidae
• Members of the genus *Actinopyga*, sometimes called toothed sea cucumbers, have an anus ringed with five inward-pointing teeth. These may give some protection against pearlfishes and other animals that live in the intestines of sea cucumbers. The White-Spotted Sea Cucumber is firm, its thick-walled brown body dimpled with small white spots and marks. It clings tightly to the reef with its tube feet in areas of strong surge where other species might be swept away. Snorkelers at Hanauma Bay, O`ahu, can almost always find these animals attached to boulders on the seaward side of the reef. They do not expel internal organs or sticky threads. To about 8 in. Indo-Pacific. Photo: Molokini Islet, Maui. 20 ft.

PLUMP SEA CUCUMBER • loli

Actinopyga obesa (Selenka, 1867)
Family Holothuriidae
• The body of this toothed cucumber is a rich, solid brown; the five anal teeth are light yellow. It is cylindrical in cross-section and its skin, although covered with small projections and creases, appears relatively smooth. It clings to hard, often vertical surfaces but is easily dislodged, thus requiring a more sheltered habitat than *A. mauritiana* (above). It occurs to depths of over 100 ft. Like all members of its genus, it does not expel sticky threads. To about 1 ft. Western and Central Pacific. Photo: Honolua Bay, Maui. 20 ft.

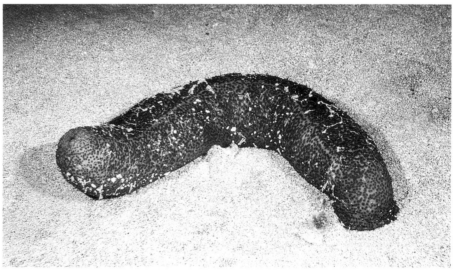

PARADOXICAL SEA CUCUMBER • loli

Bohadschia paradoxa (Selenka, 1867)
Family Holothuriidae
• This animal lies fully exposed in sand patches near the reef and on reef slopes at depths of about 50 ft. or more, sometimes covering itself with bits of algae and detritus. Its color varies from dark brown to golden yellow-brown. The body is cylindrical in cross-section and covered with fine dark papillae. Like most cucumbers it feeds by pressing its mouth to the bottom and ingesting quantities of sand, which it passes periodically in little puffs from the anus. If disturbed it readily releases quantities of sticky white Cuvierian tubules, reported to cause a rash or mild sting on humans. Strongly adhesive, these threads stick to anything but the cucumber itself and are extremely difficult to remove (especially if caught in body hair). The genus name honors Johann Bohadsch, an 18th-century professor of natural history in Prague. To about 20 in. Indo-Pacific Photo: "Treasure Islands," Kona coast, Hawai`i. 50 ft.

BLACK SEA CUCUMBER • **loli okuhi kuhi**

Holothuria atra Jaeger, 1833
Family Holothuriidae
• This is Hawai`i's most common large sea cucumber. It lies fully exposed on sand or rubble bottoms from the shallows to depths of at least 100 ft., its smooth black surface usually coated with a fine layer of sand. (Sometimes circular sandless "holes" occur in pairs along the sides.) Round in cross-section, it is softer and more slender than the Teated Sea Cucumber (below). When handled, its skin may "bleed" a clear red fluid that quickly washes away. It emits no sticky threads, but if sufficiently disturbed it ejects its internal organs. A small commensal crab, *Lissocarcinus orbicularis* (p. 275), often inhabits the tentacles and mouth. A pearlfish *(Onuxodon fowleri)* may live in the intestines, entering and exiting through the anus. Tiny parasitic snails (family Eulimidae) attach to its outside surface. Although this sea cucumber is eaten and has commercial value in some parts of the Pacific, it contains holothurin and is toxic unless properly prepared. The species name means "black." To about 20 in. Indo-Pacific. Photo: Molokini Islet, Maui. 80 ft.

TEATED SEA CUCUMBER • **loli**

Holothuria whitmaei Bell, 1887
Family Holothuriidae
• This hard, black sea cucumber lives from the shallows to depths of 70 ft. or more on rocky substrate mixed with sand. The wide body, flat on the bottom, resembles a low loaf of bread and is usually covered with a thin layer of sand. Teatlike projections along the base have earned it the common names "mammy-fish" or "teat-fish." The thick body wall is prized in some parts of the world for producing "bêche-de-mer" or "trepang." Until recently this species was known as *Holothuria nobilis*; the name was changed in 1997 because of a nomenclatural technicality. To at least 1 ft. Indo-Pacific. Photo: Hanauma Bay, O`ahu. 50 ft.

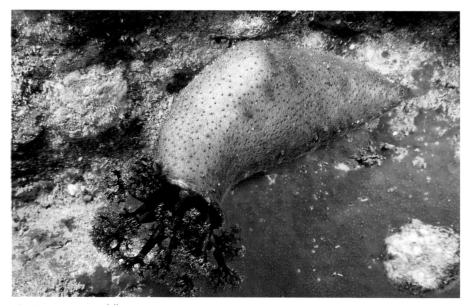

ASHY SEA CUCUMBER • **loli pua**

Holothuria cinerascens Brandt, 1835
Family Holothuriidae
• Ashlike flecks of shell or sand usually cover this purplish-black to gray animal, giving both its common and scientific names. Common in tide pools, on reef flats and on wave-washed shorelines, it wedges itself into cracks or under boulders, extending its 20 branched tentacles into the current to catch drifting food particles. The body is often hidden, but on some reef flats its tufts of black tentacles are visible every few feet. This animal also clings completely exposed to the roofs and sides of boulders, caves and overhangs. The Hawaiian name means "flower **loli**." This may also have been the species called "**loli-lu`au**," or "taro-tops **loli**" and in olden days was the sea cucumber most favored for eating. To about 6 in. Indo-Pacific. Photo: Lāna`i Lookout, O`ahu. 10 ft.

EDIBLE SEA CUCUMBER • **loli**

Holothuria edulis Lesson, 1830
Family Holothuriidae
• This unusual holothurian looks like a burnt hot dog turned over on the grill: black on the upper side, pink on the lower. The skin is thick, smooth and soft and often covered with fine silt. Common on the reef flats and slopes of Kāne`ohe Bay, O`ahu, and perhaps Pearl Harbor, the species is rare or absent elsewhere in the Islands. In other parts of the world its coloration may differ. Although edible, as the common and species names suggest, it is not highly esteemed. Indo-Pacific. Typically about 6-8 in. Photo: Kāne`ohe Bay, O`ahu. 15 ft.

DIFFICULT SEA CUCUMBER • **loli**

Holothuria dificilis (Semper, 1868)
Family Holothuriidae
• This small, black, shallow-water sea cucumber is covered with fine, fleshy projections (papillae). Its body is soft and yield-ing to the touch. Remaining in crevices or under stones by day, it emerges at night, often clinging to vertical surfaces along protected rocky shorelines. It is common, for example, along sea walls in Waikīkī at depths of only a few inches. If handled it spurts numerous exceedingly fine sticky threads from its anus that are difficult to remove. The threads are so fine they resemble a milky fluid. To about 5 in. Indo-Pacific. Photo: "Toilet Bowl," Hanauma Bay, O`ahu. 10 ft.

LIGHT-SPOTTED SEA CUCUMBER • **loli**

Holothuria hilla Lesson, 1830
Family Holothuriidae
• This soft flaccid sea cucumber is covered with limp fleshy spikes much lighter in color than the body. The animal can inflate, becoming slightly more firm. This causes the spikes to disappear, leaving only yellow or whitish spots that contrast nicely with the brown body. The animal is distinctive but remains concealed most of the time. At night it may partly extend its front end from under its rock to sweep for food. It is common in some areas. The species name means "sausage" or "intestine." To about 1 ft. Indo-Pacific. Photo: Black Point, O`ahu. 3 ft.

IMPATIENT SEA CUCUMBER • **loli koko**

Holothuria impatiens (Forsskål, 1775)
Family Holothuriidae
• This sausage-shape sea cucumber is plump and firm to the touch. Gray to pale brown and sometimes lightly spotted or mottled, it is ringed by darker brown bands, especially toward the anterior end (most evident when the animal is expanded). The wrinkly, gritty skin is covered on all sides with tiny projections. The animal occurs under stones on shallow protected reef flats. As might be guessed by the name, it ejects a mass of long sticky Cuvierian tubules at the slightest provocation. The threads are fine, but less so than those of the Difficult Sea Cucumber (p. 329). To about 8 in. Tropical seas around the world. Photo: Kāhala Beach, O`ahu. 3 ft.

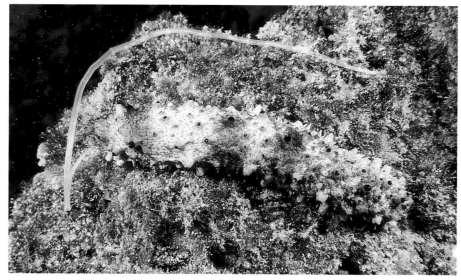

STUBBORN SEA CUCUMBER • **loli ka`e**

Holothuria pervicax (Selenka, 1867)
Family Holothuriidae
• This common sea cucumber hides under stones by day. Up to five or six diffuse brown bands may encircle its tan body. Blunt projections on the body wall are tipped by soft black thornlike points which collapse when the animal is taken from the water. On the slightest provocation it ejects thick blue-white Cuvierian threads. Difficult to remove, these strong sticky filaments are probably responsible for this species' name, meaning "obstinate" or "stubborn." To about 1 ft. Indo-Pacific. Photo: Kahe Point, O`ahu. 25 ft.

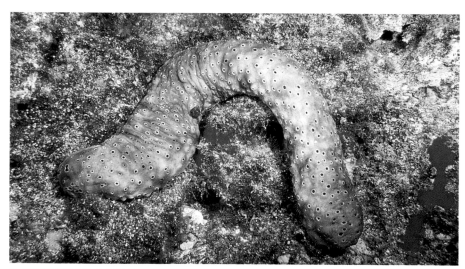

BLACK-SPOTTED SEA CUCUMBER

Holothuria sp.
Family Holothuriidae
• Light brown with pale-edged black spots, this animal remains unidentified. It resembles the Indo-Pacific cucumber *H. turriscelsa* but might be an undescribed Hawaiian endemic. One authority suggests that it could be a deep-water form of *H. pervicax* (p. 330). Like the latter, it ejects thick, blue-white Cuvierian threads when disturbed. The animal is active by day and most sightings have been below 75 ft. Specimens are about 12-15 in. long. Photo: Lehua Rock, Ni`ihau. 90 ft.

LEOPARD SEA CUCUMBER • **loli**

Holothuria pardalis (Selenka, 1867)
Family Holothuriidae
• This is the commoner of the two Hawaiian cucumber species marked with a double row of dark spots; it has the larger spots. (Compare with the Sand Sea Cucumber, below.) The dark spots become indistinct when the animal fully expands or contracts. The body is light tan and firm to the touch. This animal occurs under stones from near shore down to at least 50 ft. It does not eject sticky threads. To about 4 in. Indo-Pacific and Eastern Pacific. Photo: Kewalo Beach Park, O`ahu. 3 ft.

Holothuria arenicola
Semper, 1868
Family Holothuriidae
• This animal has a whitish or light brown body usually marked with a double row of small reddish spots and bearing small hard papillae. It resembles the more common Leopard Sea Cucumber (above) but has distinctly smaller spots and a lighter-color body. It is usually found under stones or buried in sand or sediment, sometimes with a mound at the entrance to its burrow. It does not eject sticky threads. To about 10 in. Indo-Pacific and tropical Western Atlantic. Photo: Black Point, O`ahu. 3 ft.

WHITE SEA CUCUMBER • **loli**

Family Holothuriidae
Labidodemus semperianum Selenka, 1867
• These pale white sea cucumbers are found under rocks in shallow water, their firm plump bodies partly or almost totally buried. Sometimes only the tip of the anus (tinged with purplish blue) protrudes, enabling the creature to breathe. (Respiration in sea cucumbers takes place largely through the anus.) The species name means "always violet," possibly in reference to the color of its posterior. To about 5 in. Indo-Pacific. Photo: Kewalo Park, O`ahu. 2 ft.

FAMILY STICHOPODIDAE

Cucumbers of the family Stichopodidae (order Aspidochirotida) are square or trapezoidal in cross-section and the majority (those of the genus *Stichopus*) have tube feet in rows on their flat underside. Most are warty or lumpy. Many are nocturnal. These animals do not eject sticky threads but if handled roughly they may eject their internal organs (auto-eviscerate). Members of the genus *Stichopus* have an even more unusual defense mechanism: they can "melt," becoming completely limp and eventually disintegrating if taken from the water. If not too far gone they can reverse the process. Such changes in connective tissue consistency occur on the molecular level and are unique to echinoderms. Using a similar mechanism, some sea stars stiffen when disturbed.

HAWAIIAN SPIKY SEA CUCUMBER

Stichopus sp. 1
Family Stichopodidae
• This large firm sea cucumber is studded with fleshy spikes up to 1 1/2 in. long, some double (joined at the base). The body is roughly trapezoidal in cross-section. Dark reddish brown, it lies in the open on mixed sand and rubble adjacent to the reef and on steep rubbly slopes from about 50 to at least 150 ft. Although long known to divers, it has not yet been scientifically described. To about 20 in. Occurs only in Hawai`i. (*Stichopus chloronotus*, a somewhat similar dark green or black species reported only rarely from the Islands, is square in cross-section, with blunt spikes only along the corner edges.) Photo: Kahe Point, O`ahu. 50 ft.

HAWAIIAN YELLOW-TIP SEA CUCUMBER

Stichopus sp. 2
Family Stichopodidae
• This undescribed animal bears numerous wartlike bumps resembling those of the common Warty Sea Cucumber (below). Thin black lines may criss-cross the whitish body and encircle the bases of the warts, which are often orange-red, becoming yellowish at the tips. Small dark spots may cover the body. This cucumber is strictly nocturnal, found under stones by day. It and *Stichopus* sp. 1 (above) are under study by Dr. Gustav Paulay of the University of Guam. To about 4 in. Known only from Hawai`i. Photo: Kahe Point, O`ahu. 40 ft.

WARTY SEA CUCUMBER • `unae

Stichopus horrens Selenka, 1867
Family Stichopodidae
• Covered with large, irregular, wartlike bumps in four loose rows, this cucumber varies from uniform dark brown to mottled grayish brown or orange red. The body is usually firm; when handled roughly it becomes limp and flabby. The animal is often seen clinging to rocky walls at night, to depths of at least 50 ft. By day it is found under stones or wedged into crevices. The principal prey of the Partridge Tun Snail (p. 126), it can detach a flap of skin when attacked, often escaping. The species' name means "dreadful," probably in reference its appearance. To about 10 in. Indo-Pacific. Photo: Palea Point, O`ahu. 10 ft.

FAMILY SYNAPTIDAE

The serpentine sea cucumbers of the family Synaptidae (order Apodida) have extremely thin body walls and lack tube feet. Hooklike spicules embedded in the skin help them adhere to the reef and make them curiously sticky to the touch. Oral tentacles gather food particles from the substrate, furling and unfurling in slow rhythm to bring food to the mouth. Light-sensitive organs at the base of the tentacles sense the approach of potential predators, and the animals contract quickly when disturbed. When removed from the water, their long, wormlike bodies (containing rows of round or squarish swellings) collapse like a plastic bag. They do not eviscerate or eject sticky threads.

LION'S PAW SEA CUCUMBER • weli

Euapta godeffroyi (Semper, 1868)
Family Synaptidae
• This sea cucumber is nicely patterned with creams, browns, yellows and black. It occurs from the shallows to depths of at least 40 ft. A nocturnal animal, it can sometimes be found by day under stones along the shore. It is most common, however, in slightly deeper water. Like others of its kind, it moves about slowly by contracting and expanding its accordian-like body. Stretched out, a large specimen may measure up to 4 ft. Indo-Pacific. Photo: Palea Point, O`ahu. 20 ft.

CONSPICUOUS SEA CUCUMBER • **weli**

Opheodesoma spectabilis Fisher, 1907
Family Synaptidae
• Sprawled along calm, protected shallow reef flats and often entwined in seaweed, these highly visible animals vary from bright pinkish orange to almost white. Their long, wormlike bodies, segmented by a series of swellings, resemble a human colon. Occurring in Kāne`ohe Bay and Pearl Harbor, O`ahu, they are rare or absent elsewhere in the Islands. To about 3 ft. Indo-Pacific. Photo: Kāne`ohe Bay, O`ahu, 1 ft.

KEFERSTEIN'S SEA CUCUMBER

Polyplectana kefersteini (Selenka, 1867)
Family Synaptidae
• This is the smallest of the three thin-walled sea cucumbers in Hawai`i. It is uniform brown with fine light speckles; the body wall may have a segmented appearance due to rows of round swellings, or it may appear smooth. By day it hides under stones or in crevices; at night it emerges to feed. It is most common at depths of 20 ft. or more but can also be found on shallow reef flats. To about 2 ft. if stretched out, but usually smaller. Indo-Pacific. Photo: Palea Point, O`ahu. 25 ft.

TUNICATES
PHYLUM CHORDATA. SUBPHYLUM UROCHORDATA (OR TUNICATA)

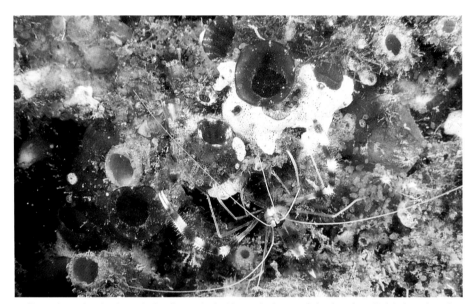

Five or six species of tunicates, both solitary and colonial, grow in close proximity along the wall of the boat channel at Magic Island, O`ahu. The bright red siphons of *Herdmania momus* dominate, some overgrown by a white colonial species, probably of the family Didemnidae. The shrimp is *Stenopus hispidus*.

Tunicates are filter-feeding animals that as adults either attach permanently to hard substrates or drift in the open ocean. They may be solitary or colonial. Although simple-looking, tunicates are surprisingly complicated and evolutionarily advanced. Most have larvae that resemble tiny tadpoles, complete with gill slits, a muscular tail, a "backbone" consisting of a stiff rod, and a nerve cord—structures that place tunicates in the same phylum as human beings! As adults, most tunicates lose these vertebrate-like characteristics. Nonetheless, they are of special interest to zoologists studying the ancestral roots of fishes, mammals and other familiar vertebrates.

There are four classes of tunicates. Nearly all of these animals belong to the class Ascidiacea and are known as ascidians. Adult ascidians are sessile (fixed permanently to the sea bottom or other hard surface). Solitary species are commonly called "sea squirts." Colonial ascidians have no common name. In the abyssal oceanic depths, ascidian-like Sorberaceans, all solitary, also attach to the bottom. Members of the remaining two classes, Thaliacea and Larvacea, are pelagic, swimming weakly or strongly in the open ocean. Thaliaceans, better known as "salps," sometimes enter coastal areas where they can be seen by snorkelers and divers. Larvaceans, tiny transparent animals that never lose their tadpole shape, drift with the plankton both in deep water and at the surface of the ocean.

Solitary ascidians (*askidio* means "bag" in Greek) have pouchlike or tubelike bodies with two openings raised on siphons. If handled they close their siphons and contract, often squirting little jets of water. They typically attach to the undersurfaces of rocks, seaweeds, harbor pilings, floats and the like. A few (mostly deep-sea forms) live in sand or mud. Solitary species range from about 1/8 in. to roughly 5 in. across. Some cold-water forms grow larger.

Colonial ascidians are composed of numerous rather elongated individual "modular" bodies called zooids. These units might be less than 1/20 in. to 1 in. long. Zooids may be joined at the base to form groups resembling bouquets of flowers, or they may be deeply embedded in common tissue (the so-called compound ascidians). Groups of zooids in a colony often share common excurrent (cloacal) openings. Colonies range from a fraction of an inch to over 3 ft. across and often resemble encrusting sponges.

All ascidians have soft bodies. They secrete a protective outer test, or tunic, formed of scattered living cells and a great deal of noncellular material containing substantial amounts of cellulose, a compound more often found in plants than animals. Tunics can be hard and tough, gelatinous, or soft and slimy. In solitary species the tunic may function somewhat like an external skeleton; in colonial ones it typically forms a common mat of supporting tissue. Under the tunic a muscular body wall encloses a gill sac, or basket, called the pharynx. The pharynx is surrounded by a cloacal chamber, gut tract, heart, blood vessels, nerves, reproductive organs, endocrine organs, etc. The heart is unusual in that it reverses every 50 beats or so, changing the direction of blood flow throughout the body.

A tunicate's basket-like pharynx contains numerous perforations (in fact, gill slits) lined with beating hairs (cilia) that create a water current. Entering through the oral siphon, water pours into the pharynx where oxygen and food are extracted. It then passes through the gill slits into the cloacal chamber, and finally out of the animal via the excurrent siphon. Sheetlike webs of mucus lining the pharynx catch fine food particles and are periodically rolled into cords and "swallowed." Digestion takes place in the stomach and intestine; wastes are discharged with the outflowing water.

Ascidians are hermaphrodites. Solitary species usually release both eggs and sperm into the sea, where fertilization takes place. Eggs hatch into larvae (called tadpoles) that quickly settle in a suitable location to mature. Colonial species often brood their eggs internally and release the tadpoles after they hatch.

Salps (pelagic tunicates of the class Thaliacea) have cylindrical bodies with oral and excurrent siphons on opposite sides. Pumping water by rhythmic contractions of the body wall, they move by gentle jet propulsion as they feed. These animals are usually transparent and many are luminescent. Night divers in the Caribbean sometimes encounter chains of salps whose members light up in sequence likes strings of Christmas lights.

Cylindrical salp colonies called pyrosomas (the name means "fire body") are shaped like giant flexible test tubes and may grow to astounding lengths—up to 30 ft.! These colonies comprise hundreds or thousands of zooids, whose currents combine to drive the colonies briskly through the sea. They light up brilliantly at night. It has been suggested that flashing pyrosomas mistaken for military torpedos were responsible for the 1964 "attacks" on two American ships in the Gulf of Tonkin, precipitating U.S. involvement in the Vietnam conflict. On a smaller scale, a pyrosoma observed drifting near the surface in Hanauma Bay, O`ahu, was about 3 ft. long, 5 in. in diameter, and bright red.

Of some 1,300 species of tunicates, about 70 are known from Hawai`i. Forty-five of these are ascidians. Most recorded Hawaiian species live in calm harbors and lagoons, a large number probably transported here as fouling organisms on ships. Investigators using scuba will probably find dozens more native species outside the harbor environment. No Hawaiian name is recorded for these animals.

CRATERED APLIDIUM
Aplidium crateriferum (Sluiter, 1909)
Family Polyclinidae
• This compound ascidian is white with irregularly shaped, transparent, common cloacal chambers that receive the excurrent flow of many individual zooids. It grows in crevices and under overhangs along exposed vertical walls and is especially common along O`ahu's north shore. The species was first described from Gardner Pinnacles, Northwestern Hawaiian Islands. Large colonies attain about 2 in. Known only from Hawai`i. Photo: Pūpūkea, O`ahu. 30 ft.

GOLD RING APLIDIUM

Aplidium sp.
Family Polyclinidae
• Close inspection of these white colonies reveals raised excurrent openings that are almost transparent. The irregular cloacal chambers they empty receive the outflow from many individual zooids. The openings of the smaller oral siphons are delicately rimmed in gold. Compound ascidians of the genus *Aplidium* usually occur only at scuba depths in Hawai`i and appear to be somewhat site-specific. This species is found at Mākena and Molokini Islet, Maui. The Ivory Tunicate (below) occurs around Lāna`i, and the Cratered Aplidium (above) off O`ahu. Colonies are 1 or 2 in. across. Photo: Mike Severns. Mākena, Maui.

IVORY TUNICATE

Aplidium(?) sp.
Family Polyclinidae
• This ivory-color compound ascidian is common in caves and under ledges along the south shore of Lāna`i. Lines of oral siphons, often double, join the large common cloacal orifices. Photo: "First Cathedral," Lāna`i. 40 ft.

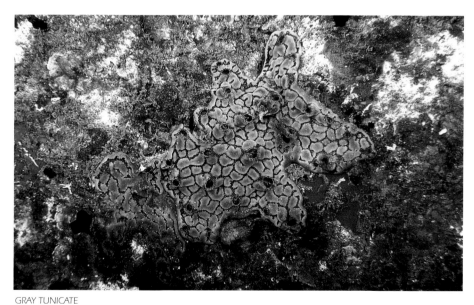

GRAY TUNICATE

Aplidium(?) sp.
Family Polyclinidae
• Similar in appearance and habitat to the species above, this ascidian colony was photographed at the entrance to an enormous underwater cave off the rugged north shore of Moloka`i. To confirm its identify even to genus would require dissection by a specialist. Photo: Mōkapu Rock, Moloka`i. 70 ft.

ORANGE-RED DIDEMNID

Didemnum sp.
Family Didemnidae
• The large family Didemnidae is characterized by tiny zooids (rarely exceeding 1/25 in. long) embedded in a thin, firm, encrusting tunic. Their small size coupled with the presence in the tunic of masses of calcareous spicules makes dissection tedious and very difficult. Yet without dissection, identification to species or even genus is almost impossible. The potent antiviral and antitumor compound didemnin B was isolated from a Caribbean tunicate of this family. This bright orange-red colony was photographed in Maunalua Bay, O`ahu, at a depth of 40 ft. It grows in the open and can be abundant, but on O`ahu occurs in only a few locations.

WHITE DIDEMNID

Didemnum sp.
Family Didemnidae
• The gold tints of this predominantly white animal are due to symbiotic algae of the genus *Prochloron* (a type of alga once thought to live only in tunicates but now collected as tiny nanoplankton from the open sea.) Oral siphons evenly puncture the surface. Tiny tadpole larvae are barely visible in the scattered common excurrent orifices. Colonies several inches across occur in crevices along exposed vertical walls and also under coral slabs. Photo: Pūpūkea, O`ahu. 30 ft.

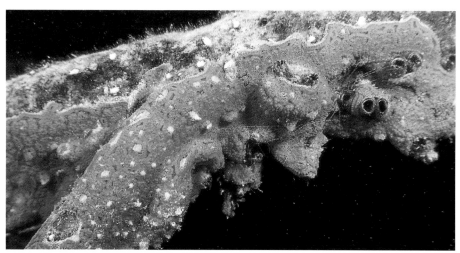

LADDER TUNICATES

Botryllus sp.
Family Styelidae
• These colonial tunicates are common in Kāne`ohe Bay, O`ahu. Groups of zooids form meandering double lines sometimes described as "ladder-like." (In some other species they organize themselves into star-shape circles or ovals.) Each zooid has its own oral siphon on the colony surface; excurrent siphons open into common cavities. Colonies are soft and brightly colored: usually red, orange, purple, green or brown, but sometimes with bands of white or gold between the siphons. How many species occur in Hawai`i is not known. The genus name *Botrylloides* is still widely used for colonies such as the one pictured here, while *Botryllus* traditionally refers to colonies with star-shape groups of zooids. Recent investigators regard the two genera as synonymous, under the older name *Botryllus*. Both names derive from the Greek word *botrys*, meaning "bunch of grapes." Photo: Moku o Lo`e (Coconut Island), Kāne`ohe Bay, O`ahu. 3 ft. (on mangrove root)

HERDMAN'S SEA SQUIRT

Herdmania momus (Savigny, 1816)
Family Pyuridae
• This common tunicate has a globular body with long, flaring red or orange siphons, usually streaked and spotted with white. Its tunic tends to be overgrown by other organisms and the body wall beneath it is full of microscopic needle-like spicules. In some parts of the world commensal shrimps live within the gill sac. Tunicates of the family Pyuridae are known to concentrate iron in their tissues. The species name, from the Greek word *momos*, means "blame." The genus name honors 19th century tunicate specialist W.A. Herdman. To about 1 1/2 in. Worldwide in warm waters, but more often in harbors and lagoons than on the reef. Probably an introduced species in Hawai`i. Photo: Pūpūkea, O`ahu. 30 ft. (with several species of compound ascidians) (See also p. 336.)

BLACK SEA SQUIRT

Phallusia nigra Savigny, 1816
Family Ascidiidae

• This common ascidian varies from velvety black to translucent gray (the latter usually in heavily shaded areas, such as under docks) and may also be blue-black, red-purple or brown. The body is cylindrical with the oral siphon near one end and the excurrent siphon opening parallel to it on the side and bending a little towards the oral siphon. It lives in harbors and also on mud flats in Kāne`ohe Bay, O`ahu, attached to rocks and dead coral. Experiments show that it can filter about 45 gallons of water in 24 hours. The specimen pictured was covered with minute arthropod symbionts, possibly larval isopods. These sea squirts grow about 4 in. tall. Tropical Atlantic, Red Sea, Western Indian Ocean, Hawai`i, and Micronesia. Probably introduced. Photo: Magic Island, O`ahu. 15 ft.

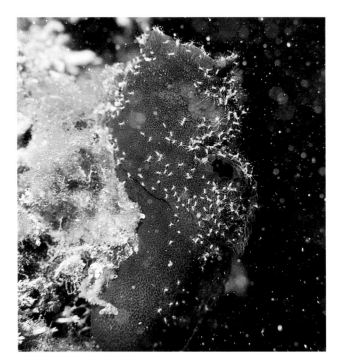

YELLOW-GREEN SEA SQUIRT

Ascidea sydneiensis Stimpson, 1855
Family Ascidiidae

• This is one of the few solitary sea squirts in Hawai`i occurring in the exposed rocky habitats favored by scuba divers (usually in crevices along vertical walls with only the siphons exposed). It also occurs in protected waters on pilings and floats. Specimens were dredged from 150-250 ft. by the steamer *Albatross* in the early 1900s. In silty areas this animal is usually black, dark brown, orange-brown or salmon red due to foreign particles incorporated in the tissues. In clear water it is light yellow-green, as shown here. Species of several ascidian families, among them the Ascidiidae, have the curious ability to concentrate the metal vanadium in greenish yellow cells called vanadocytes. Vanadium may facilitate the synthesis of cellulose and deter predators, but in truth its role remains a puzzle. Adults attain about 4 in. This species occurs worldwide in warm waters, with considerable variability. It may have been introduced to Hawai`i. Photo: Pūpūkea, O`ahu. 30 ft. (with an encrusting bryozoan on the rock above and a pink didemnid and transparent ascidians growing on its tunic)

CHAINED SALPS

Salpa(?) sp.
Order Thaliacea
• Salps (pelagic tunicates) often occur in chainlike colonies, as in this photo. Each individual modular body has a brown nerve bundle visible through its transparent body wall. Drifting inshore from the open ocean, these animals are only occasionally encountered in Hawai`i. Each body in the chain is about 1/2 in. long. Photo: Honokōhau, Hawai`i.

Hydrolithon gardineri forms both encrustations and masses of knobs and slender branches that may be round or flattened in cross-section. Common on shallow reef flats where there is strong water movement, it is usually pale pinkish or greenish. It also occurs down to at least 70 ft. Heads can be 1 ft. or more across. It is pictured here with Finger Coral *(Porites compressa)* and small encrusting patches of Rice Coral *(Montipora capitata)* on the reef flat at Ali`i Beach Park, O`ahu. 5 ft.

Some marine plants, notably species of red algae *(Rhodophyta),* deposit calcium carbonate in their cell walls as they grow. This is limestone, the same substance secreted by stony corals. Like cnidarian corals, a few of these plants become hard and stony. Known as crustose coralline algae, they play an important role in the building of coral reefs, cementing together coral rubble and also laying down significant quantities of limestone in their own right. For this reason some scientists prefer the term "biotic reefs" in place of "coral reefs." Coral limestone in Hawai`i is largely a product of these plants, some of which thrive in shallow water along rocky turbulent shores where cnidarian corals cannot easily grow. The growing edge of the reef where the waves break is formed by coralline algae rather than cnidarian corals and is called the "algal ridge."

Some species of coralline algae form pink, tan or purplish crusts on the rocks; others form leafy plates or branching structures somewhat similar to those created by cnidarian corals. To help the reader distinguish between coralline algae and true corals, three species of coralline algae likely to be seen by snorkelers and divers are shown below. Not pictured is *Hydrolithon onkodes,* an encrusting, reef-building alga that forms extensive chalky pink or tan encrustations on either side of the low tide line along surf-swept rocky shores. It is very common in Hawai`i; look for it under Helmet Urchins and `**opihi** (see p. 98). Such algae typically secrete chemicals that attract the larvae of algae-eating marine invertebrates. These keep the coralline algae grazed clean of other algae that might grow over it. Note that coralline algae never have calyces or polyps.

Peyssonelia sp. forms thin, leaflike, purplish-red plates under overhangs, in the entrances of caves and in other shaded areas, generally in quiet water. Although this red alga deposits some limestone, it is not a true coralline. Photo: "Tunnels," north shore, Kaua`i. 20 ft.

Right: Hydrolithon reinboldi forms lumpy bluish or lavender encrustations on rubble or dead coral. It often forms loose nodules that roll with the surf and is found mainly on reef flats in shallow water but also down to at least 60 ft. *Sporolithon erythraea*, a similar species, is reddish. Photo: Black Point, O`ahu. 2 ft.

Bottom: Unidentified pinkish purple and green coralline algae encrust a smooth rocky wall of a cave entrance at the Lāna`i Lookout, O`ahu. 30 ft..

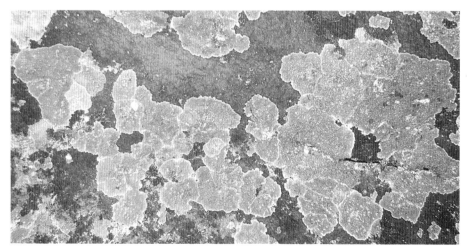

Legal restrictions on the taking of marine invertebrates in the State of Hawai`i.

STONY CORALS
It is unlawful to intentionally take, break or damage any live stony coral.

"LIVE ROCK"
It is unlawful to take, break or damage any rock or coral to which marine life is visibly attached.

SPINY LOBSTERS and SLIPPER LOBSTERS
Closed season May through August.
No spearing. No taking with eggs.

Carapace length (from the base of the eyestalks to the rear margin of the carapace) must be at least 3 1/4 in. for the following two species:
>   Banded Spiny Lobster • **ula** *(Panulirus marginatus)*
>   Tufted Spiny Lobster • **ula** *(Panulirus penicillatus)*

Tail width (between 1st and 2nd tail segments) must be at least 2 3/4 in. for the following two species:
>   Scaly Slipper Lobster • **ula pāpapa** *(Scyllarides squammosus)*
>   Ridgeback Slipper Lobster • **ula pāpapa** *(Scyllarides haanii)*

CRABS
No spearing. No taking with eggs.

Closed season May through August. Carapace must be at least 4 in. long or wide.
>   Kona Crab • **pāpa`i kua loa** *(Ranina ranina)*

Carapace must be at least 4 in. long or wide.
>   Blood-Spotted Swimming Crab • **kuhonu** *(Portunus sanguinolentus)*
>   Samoan Crab *(Scylla serrata)*

OCTOPUSES
The following two species must weigh at least 1 lb.
>   Day Octopus • **he'e mauli** *(Octopus cyanea)*
>   Ornate Octopus • **he`e pūloa, he`e mākoko** *(Octopus ornatus)*

LIMPETS (`opihi)
Shell diameter must be at least 1 1/4 in. for the following three species.
>   Black-Foot `Opihi • **`opihi makaiaūli** *(Cellana exarata)*
>   Yellow-Foot `Opihi • **`opihi `ālinalina** *(Cellana sandwicensis)*
>   Giant `Opihi • **`opihi kō`ele** *(Cellana talcosa)*

CLAMS, OYSTERS and OTHER SHELLFISH
Taking prohibited at any time.

For more detailed information contact:

DIVISION OF AQUATIC RESOURCES
Dept. of Land and Natural Resources
1151 Punchbowl Street
Honolulu, Hawaii 96813
(808) 587-0100

SOURCE MATERIALS

The following materials were used in the preparation of this book. This is by no means a complete bibliography of Hawaiian marine invertebrates. Entries are arranged alphabetically by author within categories that follow the contents of the book. General references are listed first.

GENERAL REFERENCES

Allen, Gerald R. & Roger Steene, 1994. **Indo-Pacific coral reef field guide.** Singapore: Tropical Reef Research. 378 p. (Well illustrated but lacks species notes. Includes vertebrates and invertebrates.)

American Fisheries Society, 1989. **Common and scientific names of aquatic invertebrates from the United States and Canada: decapod crustaceans**. American Fisheries Society Special publication 17. Bethesda, Maryland.

Brown, Roland Wilbur, 1956. **Composition of scientific words: a manual of methods and a lexicon of materials for the practice of logotechnics**. Washington D.C.: Smithsonian Institution Press. 882 p.

Carlton, James T. & Lucius G. Eldredge, in preparation. **Introduced and cryptogenic marine and brackish water invertebrates of the Hawaiian Islands.**

Chave, E.H. & Alexander Malahoff, 1998. **In deeper waters: photographic studies of Hawaiian deep-sea habitats and lifeforms**. Honolulu: University of Hawai`i Press. 125 p. (Fascinating photographs taken from submersibles. Ample text.)

Coles, S.L, R.C. DeFelice, L.G. Eldredge and J.T. Carlton, 1997. **Biodiversity of marine communities in Pearl Harbor, Oahu, Hawaii, with observations on introduced species.** Bishop Museum Technical Report no. 10.

Colin, Patrick L. & Charles Arneson, 1995. **Tropical Pacific invertebrates: a field guide to the marine invertebrates occurring on tropical Pacific coral reefs, seagrass beds and mangroves**. Beverly Hills, California: Coral Reef Research Foundation. 296 p. (A guide with over 1,300 color plates and notes for each species. Well worth buying but contains many misspelled names and other errors.)

Debelius, Helmut & Hans A. Baensch, 1994. **Marine atlas: the joint aquarium care of invertebrates and tropical marine fishes**. Melle, Germany: Mergus. 1,215 p. (Well-illustrated and full of information for the hobbyist.)

Devaney, Dennis M. & Lucius G. Eldredge, eds., 1977. **Reef and shore fauna of Hawaii. Section 1: Protozoa through Ctenophora**. Honolulu: Bishop Museum Press. 278 p.

————, 1987. **Reef and shore fauna of Hawaii. Section 2: Platyhelminthes through Phoronida and Section 3: Sipuncula through Annelida**. Honolulu: Bishop Museum Press. 461 p.

Doty, Maxwell, S. Arthur Reed and Julie Bailey-Brock, 1983. **Field keys to common Hawaiian marine animals and plants.** Honolulu: State of Hawaii, Dept. of Education. Office of Instructional Services.

Edmondson, Charles Howard, 1946. **Reef and shore fauna of Hawaii.** Bernice P. Bishop Museum. Special Publication 22. 381. p. (For many years the "bible" of marine invertebrates in Hawai`i.)

————, 1979. **Hawaii's seashore treasures**. Hilo: Petroglyph Press. (For children. Originally published in 1949 as **Seashore treasures**.) 144 p.

Edmondson, Charles Howard, et al., 1925. **Marine zoology of the tropical Central Pacific.** Bernice P. Bishop Museum. Bulletin 27. 148 p. (Crustacea, echinoderms, polychaete worms, foraminifera collected by the Tanager Expedition. Contains description of the hermit crab *Dardanus sanguinocarpus* by Degener.)

Eldredge, Lucius G. & Scott E. Miller, 1995. **How many species are there in Hawai`i?** Bishop Museum Occasional Papers number 41. p. 3-19. (A list of taxa with total numbers of known species.)

————, 1997. **Numbers of Hawaiian species: supplement 2, including a review of freshwater invertebrates**. Bishop Museum Occasional Papers number 48. p. 3-22.

Fielding, Ann. 1979. **Hawaiian reefs and tidepools**. Honolulu: Oriental Publishing Company. 103 p.

Flora, Charles J., with Corydon C. Pittman and friends. **The S.H.R.I.M.P.ER's Maui, a guide to coral reef biology**. Everson, Washington: Jero. 347 p. (A unique, amazingly informative and well-illustrated [b.w.] guide to Hawaiian reef life at Hekili Point, Maui, informally published for use by marine biology classes at Camp Pecusah, Maui. For information, contact the author at 6618 Lunde Rd, Everson, WA 98247. Highly recommended.)

George, J. David & Jennifer J. George, 1979. **Marine life: an illustrated encyclopedia of invertebrates in the sea.** New York: Wiley. 288 p. (1,300 photographs and detailed descriptions of 27 phyla broken down to the family level. Species information for all illustrations. Out of print.)

Gosliner, Terrence M., David W. Behrens & Gary C. Williams, 1996. **Coral reef animals of the Indo-Pacific: animal life from Africa to Hawai`i exclusive of the vertebrates**. Monterey, California: Sea Challengers. 314 p. (An excellent Indo-Pacific invertebrate field guide. Over 1,100 species illustrated with underwater photographs. Includes brief descriptive information for each.)

Gulko, Dave, 1998. **Hawaiian coral reef ecology**. Honolulu: Mutual Publishing, 1998. 256 p. ("A comprehensive, fun, educational book with tons of information on coral reef ecology...." Gulko does a fine job of making a complex subject understandable. A unique book, profusely illustrated.)

Halstead, Bruce W. 1988. **Poisonous and venomous marine animals of the world (2nd rev. ed.)** Princeton, N.J.: Darwin Press. 1,168 p.

Hiatt, Robert Worth, 1954. **Hawaiian marine invertebrates, a guide to their identification**. Honolulu: University of Hawaii (mimeographed). 140 p. (The most complete set of keys in one volume, but outdated.)

Humann, Paul, 1992. **Reef creature identification: Florida, Caribbean, Bahamas**. Jacksonville, Florida. New World Publications. 320 p.

Jokiel, Paul L. et al., 1986. **Coral reef population biology : results of the 1983 Summer Research Training Program held at the Hawaii Institute of Marine Biology**. Technical report / Hawaii Institute of Marine Biology. 501 p.

Kaplan, Eugene H., 1982. **A field guide to coral reefs of the Caribbean and Florida**. Boston: Houghton Mifflin. 289 p.

Kay, E. Alison & Stephen R. Palumbi, 1987. **Endemism and evolution in Hawaiian marine invertebrates.** Trends in Ecology & Evolution (2)7: 183-186

Kent, Harold Winfield, 1986. **Treasury of Hawaiian words in one hundred and one categories.** Honolulu: Masonic Public Library of Hawaii. 475 p.

Kerstitch, Alex, 1995. **Chemical warriors: marine contributions to human medicine.** Freshwater and Marine Aquarium (18)10: 206-211.

Love, Robert Milton, 1991. **Love's way**. Ocean Realm. Winter 1990-91: 16-21. (On scientific names.)

Mapes, John, 1981. **Ever wonder why reefs are barren?** Hawaiian Shell News 29(9): 9. (Ultraviolet radiation).

Margulis, Lynn and Karlene V. Schwartz, 1982. **Five kingdoms: an illustrated guide to the phyla of life on earth, 2nd ed**. New York: W.H. Freeman. 376 p.

Miller, Scott E. & L.G. Eldredge, 1996. **Numbers of Hawaiian species: supplement 1**. Bishop Museum Occasional Papers number 45. p. 8-17.

Moe, Martin A., 1992. **The aquarium reference: systems and invertebrates.** Plantation, Florida: Green Turtle Publications. 509 p. (Contains a useful overview of the known phyla.)

Pearse, Vicki, et al., 1987. **Living invertebrates**. Palo Alto, Calif.: Boxwood Press. 848 p. (A lively, easy-to-read text with many illustrations.)

Ruppert, Edward E. & Robert D. Barnes, 1993. **Invertebrate zoology, 6th ed**. Saunders College Publishing. 1056 p. (This standard text was used as the main general reference in the preparation of this book.)

Russo, Ron, 1994. **Hawaiian reefs, a natural history guide.** San Leandro, Calif.: Wavecrest Publications. 174 p.

Schoenberg, Olive, 1986. **The night that Haleiwa reef burned**. Hawaiian Shell News 34(6): 12.

Thomas, Craig and Susan Scott, 1997. **All stings considered: first aid and medical treatment of Hawai`i's marine injuries**. Honolulu: University of Hawai`i Press. 231 p.

Titcomb, Margaret, et al., 1978. **Native use of marine invertebrates in old Hawaii**. Pacific Science 32(4): 325-391.

Walls, Jerry G., 1974. **Starting with marine invertebrates.** Hong Kong: T.F.H. Publications. 160 p.

Walls, Jerry G., (ed.), 1982. **Encyclopedia of marine invertebrates**. Neptune, N.J.: T.F.H. Publications. 736 p. (Detailed discussion of all major phyla with numerous color plates.)

SPONGES

Bergquist, Patricia R. **Porifera**. (in D.M. Devaney & L.G. Eldredge, eds., **Reef and Shore Fauna of Hawaii.** Section 1). Honolulu: B.P. Bishop Museum Press. p. 53-69. (A key to sponges of the Hawaiian Islands, but now outdated.)

De Laubenfels, M.W,, 1950. **The sponges of Kaneohe Bay, Oahu**. Pacific Science 4(1): 3-36.

————, 1951. **The sponges of the island of Hawaii**. Pacific Science 5(3): 256-271.

————, 1957. **New species and records of Hawaiian sponges**. Pacific Science 11(2): 236-251.

Hooper, John N.A., 1996. **Guide to sponge collection and identification**. Published on the internet. ("This is an early version of a sponge identification guide that is still being developed by the Queensland Museum....")

Love, Robert Milton, 1992. **Sponges get no respect**. Ocean Realm. February 1993: 58-67.

Mancini, Allessandro. 1990. **Sponges in the marine aquarium.** Freshwater and Marine Aquarium 13(7): 8; July 1990.

CNIDARIANS

HYDROIDS, SIPHONOPHORES, SCYPHOZOANS, CUBOZOANS

Cook, William J., 1977. **Order Hydroida** (in D.M. Devaney & L.G. Eldredge, eds., **Reef and Shore Fauna of Hawaii. Section 1**). Honolulu: B.P. Bishop Museum Press. p. 71-147. (Describes hydroids of the Hawaiian Islands.)

————, 1982. **New scyphozoan records for Hawaii: *Anomalorhiza shawi* Light, 1921, and *Thysanostoma loriferum* (Ehrenberg, 1835): with notes on several other rhizostomes.** Proceedings of the Biological Society of Washington 97(3): 583 588.

Eldredge, L.G. and Dennis M. Devaney, 1977. **Other Hydrozoans** (in their **Reef and Shore Fauna of Hawaii. Section 1**). Honolulu: B.P. Bishop Museum Press. p. 71-147. (Describes siphonophores and chondrophores of the Hawaiian Islands.)

Eldredge, L.G. and Dennis M. Devaney, 1977. **Class Scyphozoa** (in their **Reef and Shore Fauna of Hawaii. Section 1**). Honolulu: B.P. Bishop Museum Press. p. 71-147. (Describes scyphozoan jellyfish of the Hawaiian Islands.)

Love, Robert Milton, 1994. **Jellyfish**. Ocean Realm. June 1994: 68-75.

Young, Forrest A., 1995. **Aquarium cultivation of Cassiopeia**. Freshwater and Marine Aquarium 18(6): 62.

Anemones, tube anemones and zoanthids

Ates, Ron, 1992. **Tube Anemones.** Freshwater and Marine Aquarium 15(5): 8 (A thorough and informative article on these fascinating animals)

————, 1995. **Aggression in flower animals**. Freshwater and Marine Aquarium 18(2): 8

————, 1998. **Glass anemones, feared and hated**. Freshwater and Marine Aquarium 21(2): 190

Cutress, Charles E., 1977. **Subclass Zoantharia** (in D.M Devaney & L.G. Eldredge, eds., **Reef and Shore Fauna of Hawaii. Section 1**). Honolulu: B.P. Bishop Museum Press. p. 130-147. (Describes anemones, corallimorphians and cerianthids from the Hawaiian Islands.)

Dunn, Daphne Fautin. 1974. *Radianthis papillosa* **(Coelenterata, Actinaria) redescribed from Hawaii**. Pacific Science 28(2): 171-179.

————, 1974. *Actiniogeton sesere* **(Coelenterata, Actinaria) in Hawaii**. Pacific Science 28(2): 181-188.

————, 1974. **Redescription of *Anthopleura nigrescens* (Coelenterata, Actinaria) from Hawaii.** Pacific Science 28(4): 377-382.

Fautin, Daphne G. & Gerald R. Allen, 1992. **Field guide to anemonefishes and their host anemones.** Perth: Western Australian Museum. 160 p.

Moore, R.E. & P.J. Scheuer, 1971. **Palytoxin: a new marine toxin from a coelenterate.** Science 172: 495-498.

Reimer, A.A., 1971. **Feeding behavior in the Hawaiian zoanthids *Palythoa* and *Zoanthus*.** Pacific Science 25(4): 512-520.

Rosin, R., 1969. **Escape response of the sea anemone *Anthopleura nigrescens* (Verrill) to its predatory eolid nudibranch *Herviella baba spec. nov.*** Veliger 12: 74-77.

Ross D.M., 1970. **The commensal association of *Calliactis polypus* and the hermit crab *Dardanus gemmatus* in Hawaii.** Canadian Journal of Zoology 48: 351-357.

Schlais, Jim. 1979. **Aiptasia, nature's filter.** Freshwater and Marine Aquarium, May 1979.

Steenson, R.A., 1963. **Behavior of the pomacentrid reef fish *Dascyllus albisella* Gill in relation to the anemone *Marcanthea cookei*.** Copeia no. 4: 612-614.

Walsh, G.E. & R.L. Bowers, 1971. **A review of Hawaiian zoanthids with descriptions of three new species**. Zoological Journal of the Linnaean Society 50(2): 161-180.

————. **Order Zoanthinaria** (in D.M. Devaney & L.G. Eldredge, eds., **Reef and Shore Fauna of Hawaii. Section 1**). Honolulu: B.P. Bishop Museum Press. p. 71-147. (Describes all known zoanthids from the Hawaiian Islands. However, the species presented have changed significantly in light of recent research.)

STONY CORALS

Coles, Steve, L., [1995]. **Corals of Oman.** Muscat Printing Press. 106 p.

Grigg, Richard W., 1965. **Ecological studies of black coral in Hawaii**. Pacific Science 19(2): 244-260.

————, 1981. *Acropora* **in Hawaii. Part 2. Zoogeography**. Pacific Science 35(1): 15-24.

————, 1983. **Community structure, succession and development of coral reefs in Hawaii.** Marine Ecology Progress Series, 11:1-14.

Grigg, Richard W. and J. Maragos, 1974. **Recolonization of hermatypic corals on submerged lava flows in Hawaii.** Ecology 55: 387-395.

Grigg, Richard W., et al., 1981. *Acropora* **in Hawaii. Part 1. History of the scientific record, systematics, and ecology.** Pacific Science 35(1): 1-13.

Jokiel, Paul L., 1987. **Ecology, biogeography and evolution of corals in Hawaii.** Trends in Ecology & Evolution 2(7): 179-182.

Maragos, J.E., 1977. **Order Scleractinia: stony corals** (in D.M. Devaney & L.G. Eldredge, eds., **Reef and Shore Fauna of Hawaii. Section 1**). Honolulu: B.P. Bishop Museum Press. p. 158-241.

————, 1995. **Revised checklist of extant shallow-water stony coral species from Hawaii (Cnidaria: Anthozoa: Scleractinia)**. Bishop Museum Occasional Papers no. 42 p. 54-55.

SOFT CORALS AND BLACK CORALS

Devaney, Dennis M., 1977. **Subclass Octocoralia** (in D.M. Devaney & L.G. Eldredge, eds., **Reef and Shore Fauna of Hawaii. Section 1**). Honolulu: B.P. Bishop Museum Press. p. 119-125. (Describes all known soft corals and gorgonians from the Hawaiian Islands.)

Grigg, Richard W., 1977. **Hawaii's precious corals**. Norfolk Island, Australia: Island Heritage. 64 p. (A well-illustrated popular account of deep-water precious corals.)

Grigg, Richard W. and Dennis Opresko, 1977. **Order Antipatharia** (in D.M. Devaney & L.G. Eldredge, eds., **Reef and Shore Fauna of Hawaii. Section 1**). Honolulu: B.P. Bishop Museum Press. p. 242-261. (Describes all known black corals from the Hawaiian Islands.)

WORMS

Bailey-Brock, Julie H., 1987. **Phylum Annelida**. (in D.M. Devaney & L.G. Eldredge, eds., **Reef and Shore Fauna of Hawaii. Section 3**). Honolulu: B.P. Bishop Museum Press. p. 213-454.

Bailey-Brock, Julie H. 1976. **Habitats of tubicolous polychaetes from the Hawaiian Islands and Johnston Atoll.** Pacific Science 30(1): 69-81.

Johnson, Scott. 1983. **The colorful flatworms**. Sea Frontiers 29(1): 2; Jan-Feb 1983.

Lyman, Libbie H., 1950. **A new Hawaiian polyclad, _Stylochoplana inquilina_, with commensal habits**. Occasional Papers of the Bernice P. Bishop Museum 20(4): 53-58; June 21, 1950.

Mancini, Allessandro. 1990. **Tropical tubeworms in the marine aquarium.** Freshwater and Marine Aquarium 13(8): 32; August 1990.

Poulter, Jean L. 1975. **Hawaiian polyclad flatworms: Prosthiostomids**. Pacific Science 29(4): 317-339

————, 1987. **Phylum Platyhelminthes** (in D.M.Devaney & L.G. Eldredge, eds., **Reef and Shore Fauna of Hawaii. Section 2**). Honolulu: B.P. Bishop Museum Press. p.13-58.

Prudhoe, Stephen, 1985. **A monograph on polyclad turbellaria.** London: British Museum (Natural History) and Oxford University Press.

Schroeder, Helen, 1991. **Polychaetes: worms in the marine aquarium**. Freshwater and Marine Aquarium 6:160

BRYOZOANS

Dade, W.B. and T. Honkalehto, 1986. **Common ectoproct bryozoans of Kaneohe Bay, Oahu.** (in P.L. Jokiel, et al., eds., 1986. **Coral Reef Population Biology** p. 52-65.)

Soule, John D., Dorothy F. Soule & Henry W. Chaney, 1987. **Phyla Entorprocta and Bryozoa (Ectoprocta)**. (in D.M. Devaney & L.G. Eldredge, eds., **Reef and Shore Fauna of Hawaii. Section 2**). Honolulu: B.P. Bishop Museum Press. p. 83-170.

MOLLUSCS

GENERAL

Kay, E. Alison, 1979. **Hawaiian marine shells: reef and shore fauna of Hawaii, Section 4: mollusca.** Honolulu: Bishop Museum Press. 653 p. (The authoritative reference on all known Hawaiian molluscs.)

Kay, E. Alison & Olive Schoenberg-Dole, 1991. **Shells of Hawai`i.** Honolulu: University of Hawai`i Press. 89 p. (Small color photos; no species descriptions. Includes some opisthobranchs.)

Moretzsohn, Fabio & E. Alison Kay, 1995. **Hawaiian marine mollusks, an update to Kay, 1979**. 23 p. (Unpublished.)

Morris, Percy A., 1966. **A field guide to Pacific coast shells, including shells of Hawaii and the Gulf of California. 2nd ed.** Boston: Houghton Mifflin. 297 p.

Paulay, Gustav, 1996. **New records and synonymies of Hawaiian bivalves (Mollusca)**. Records of the Hawaii Biological Survey for 1995. Bishop Museum Occasional Papers 45: 18-29.

Tinker, Spencer Wilkie, 1958. **Pacific sea shells: a handbook of common marine molluscs of Hawaii and the South Seas. Rev. ed**. Rutland, Vermont: C.F. Tuttle, 1958. 240 p. (Although dated and out of print, still a useful book. Detailed descriptions of each species. Black and white illustrations.)

SNAILS

Anon., 1971. **Hawaii's giant shell species**. Hawaiian Shell News 19(1): 3

————, 1973. **Hawaii's giant tiger cowry**. Hawaiian Shell News 21(2): 1

————, 1990. **Recent finds: the biggest tiger of them all?** Hawaiian Shell News, 38(12): 8.

Badcok, Eric C., 1981. **Hawaii's *Cypraea rashleighana* Melvill and the collector for whom it was named**. Hawaiian Shell News 29(6): 3.

Beals, Marty, 1976. ***Cypraea mauiensis*—an endangered species?** Hawaiian Shell News 24(1): 10.

Beckwith, Vela & Ron, 1993. **The little jewels of Kamehameha Reef** (*Cypraea semiplota*). Hawaiian Shell News 41(2).

Burch, Beatrice, 1982. **Who's this man, John S. Gaskoin?** Hawaiian Shell News 30(9): 1

Burgess, C.M., 1975. **The "carneola complex," some further thoughts on three (or maybe four) related cowries.** Hawaiian Shell News 28(7): 1

————, 1985. **Cowries of the world.** Cape Town: Gordon Verhoef, Seacomber Publications 268 p. This beautiful book (based on the author's 1970 work **The Living Cowries**) is the "bible" of cowry collectors worldwide. The author lives in Hawai`i and offers a great deal of local information.

Corpuz, Gladys C., 1980. **The stormy life of the limpet**. Hawaiian Shell News 29(11): 5

Cross, E.R., 1973. ***Charonia tritonis* egg laying**. Hawaiian Shell News 21(5): 1

Dayle, Bob, 1991. **Makua Reef's snakehead cowries—then and now**. Hawaiian Shell News 39(3): 9.

Earle, John L., 1980. **A look at Hawaii's rarest murex**. Hawaiian Shell News 28(10): 1.

————, 1981 **The 37th murex, or how to find a rare specimen**. Hawaiian Shell News 29(3): 1 (*Chicoreus insularum*)

————, 1983. **The Great Hawaiian shell shortage**. Hawaiian Shell News 31(8):1

Hayes, Therese, 1983. **The influence of diet on local distribution of *Cypraea***. Pacific Science 37(1): 27-36.

Holland, Maurice, ed. 1959. **Helpful hints for shell hunters**. Honolulu: Hawaiian Malacological Association. 77 p.

Johnson, Scott, [n.d.] **Living seashells**. Honolulu: Oriental Publishing Co., 117 p. (Beautiful photos of the living animals. Indo-Pacific in scope.)

Love, Robert Milton, 1992. **Living shells**. Ocean Realm. January 1992: 66-73

Kay, E. Alison & William Magruder, 1977. **The biology of opihi**. Honolulu: State of Hawaii, Dept. of Planning and Economic Development. 46 p.

Kerstitch, Alex, 1991. **Living molluscs**. Freshwater and Marine Aquarium 12: 32.

————, 1994. **Farmers of the sea, the blue revolution. Part III, molluscs**. Freshwater and Marine Aquarium August 1994. p. 144

Lee, Barbara Churchill, 1986. **Some high living by Hawaii's tiger cowries**. Hawaiian Shell News 33(11): 4

Leehman, Elmer G., 1974. **The puka shell craze in Hawaii**. Hawaiian Shell News 22(11): 1.

————, 1978. **Tale of the tiger: demand is long, but supply is short**. Hawaiian Shell News 26(10): 1.

Lillico, Stuart, 1973. ***Cypraea ostergaardi***. Hawaiian Shell News 21(6): 5.

————, 1984. ***Cypraea alisonae*, the shell that used to be called *C. teres***. Hawaiian Shell News 32(4): 3.

Kohn, Alan J., 1959. **The ecology of *Conus* in Hawaii**. (in E. Alison Kay, ed., **A Natural History of the Hawaiian Islands, selected readings II**). Honolulu: University of Hawai`i Press. p. 210-253. (Detailed information on shallow-water habitats of Hawai`i is included as part of this in-depth discussion.)

Mayor, Rick, 1975. **Food preference of a captive *Murex insularum***. Hawaiian Shell News 22(8): 3.

Mienis, Henk K., 1983. **How many species of Nerita live in Hawaii?** Hawaiian Shell News 31(2): 13.

Nelson, T.J., 1973. **Hawaiian *Conus* scarcity ratings**. Hawaiian Shell News 21(2): 6.

Pip, Eva and Donald Dan, 1990. **Re: Why we do it, and should we?** Hawaiian Shell News 38(10): 10 (Arguments for and against the taking of live shells.)

Purtymun, Bob, 1973. **Oahu observations**. Hawaiian Shell News 21(5): 3. (*Conus striatus* in captivity)

[Quirk, Stephen], 1972. **Hawaiian seashells**. Honolulu: Robert Boom Co. 30 p. (Although brief and slightly dated, still the most useful and inexpensive popular guide to common Hawaiian shells.)

Richert, Tom, 1980. **Ouch!** Hawaiian Shell News 29(11): 8. (On cone shell stings.)

Rohrbach, Jim, 1989. **Cruising with the Moana Wave**. Hawaiian Shell News 37(7): 1. (Trawling for deep water squid.)

Severns, Mike, 1979. **The collector's collector**. Hawaiian Shell News, May, 1979: 10. (Mollusc shells ingested by *Calliactis polypus*.)

————, 1986. **Old lava flows, helmet shells and green lips.** Hawaiian Shell News 34(11): 8.

Schoenberg, Olive, 1971. **The return of *Haminoea aperta*.** Hawaiian Shell News 19(9): 3.

Schoenberg, Olive, 1971. **The care and feeding of textile cones in captivity** Hawaiian Shell News 19(9): 7 (Continued in the subsequent issue.)

————, 1976. **News from the tank: *Cymatium* bites cone.** Hawaiian Shell News 24(4): 7.

————, 1976. **The oldest textile in captivity?** Hawaiian Shell News 24(11): 4

————, 1978. **What shall I feed to my *Terebra*?** Hawaiian Shell News 26(2): 11.

————, 1982. **Hawaii's puzzling *Fusinus*.** Hawaiian Shell News 30(4): 1.

————, 1986. ***Neritina granosa* Sowerby on the menu.** Hawaiian Shell News 34(6): 3.

Smith, Vernon E. 1973. **The best from Hawaiian Shell News.** Hawaiian Shell News 21(12). (Reprint of an article from March 1962 on collecting *Hydatina amplustre*.)

Takahashi, Chris, 1981. **Watch the fingers! *Strombus maculatus* can be hazardous**. Hawaiian Shell News 29(9): 10.

————, 1985. **A 1985 guide to Hawaiian cowries.** Hawaiian Shell News 33(5): 5

————, 1986. **Tales of Honolulu's Fort Kamehameha reef.** Hawaiian Shell News 34(1): 5.

————, 1986. **It's Hawaii's year of the murex.** Hawaiian Shell News 34(11): 14.

————, 1987. **Hawaiian specimens: *Lima fragilis*.** Hawaiian Shell News 35(10): 7.

Tau`a, Keli`i, 1990. **The story of the pupu shell.** Hawaiian Shell News 38(2): 10.

Thorsson, Wesley M., 1979. **Separating *Cypraea teres* from *C. rashleighana*.** Hawaiian Shell News 27(9): 4.

————, 1982. **All about Hawaii's *Cypraea gaskoini* Reeve.** Hawaiian Shell News 30(9): 1.

————, 1990. **Hawaiian shell observations.** Parts 1 & 2. Hawaiian Shell News 38(9), 38(10).

Wells, William Bruce, 1978. **A "new cone" and an old mystery.** Hawaiian Shell News 26(7): 1.

————, 1980. **Hawaii's ancient cone finally has a name.** Hawaiian Shell News 28(11): 12.

Wendt, Dorothy, 1984. **Puttering on the beach: think purple.** Hawaiian Shell News 32(6): 7 (*Janthina* & *Physalia*)

————, 1984. **Puttering on the beach: watch for the sea's gifts.** Hawaiian Shell News 32(7): 9 (*Glaucus*, *Porpita* & *Velella*)

————, 1984. **Some marine "weed-eaters."** Hawaiian Shell News 32(9): 11 (*Aplysiidae*)

Wolf, Charles S., 1973. **More on Hawaiian tigers.** Hawaiian Shell News 21(2): 7. (Arguments in favor of subspecies status)

SLUGS

Bertsch, Hans & Terrence M. Gosliner, 1989. **Chromodorid nudibranchs from the Hawaiian Islands.** Veliger 32(3): 247-265.

Bertsch, Hans & Scott Johnson, 1979. **Three new opistobranch records for the Hawaiian Islands.** Veliger 22(1): 41-44

————. **Hawaiian nudibranchs: a guide for scuba divers, snorkelers, tidepoolers and aquarists.** Honolulu: Oriental Publishing Co., 112 p.

————, 1982. **Three new species of dorid nudibranchs from the Hawaiian Islands.** Veliger 24(3): 208-218.

Brunckhorst, David J., 1989. **Fabulous finds of phyllidiids.** Hawaiian Shell News 37(6): 7.

————, 1993. **The systematics and phylogeny of Phyllidiid nudibranchs (Doridoidea)**. Records of the Australian Museum. Supplement 16. 28 January 1993. 107 p. (Descriptions and color plates of all known species.)

Gosliner, Terrence M., 1979. **The Systematics of the Aeolidacea (Nudibranchia: Mollusca) of the Hawaiian Islands, with descriptions of two new species.** Pacific Science 33(1): 37-77.

————, 1987. **Nudibranchs of southern Africa: a guide to opistobranch mollusks of southern Africa.** Monterey, Calif.: Sea Challengers. 136 p.

Johnson, Scott, 1981. **Those stinky *Phyllidia* nudibranchs.** Hawaiian Shell News 29(5): 5.

Kay, E. Alison & David K. Young, 1969. **The Doridacea (Opistobranchia; Mollusca) of the Hawaiian Islands**. Pacific Science 23(2): 172-231.

Kerstitch, Alex, 1988. **Living rainbows of the sea.** Freshwater and Marine Aquarium 3: 15.

Love, Robert Milton, 1992. **Nudibranchs**. Ocean Realm. October 1992: 60-71.

Pawlik, J.R. et al., 1988. **Defensive chemicals of the Spanish dancer nudibranch *Hexabranchus sanguineus* and its egg ribbons**. Journal of Experimental Marine Biology and Ecology 119(2): 99-109.

Rudman, W.B. 1986. **The Chromodorididae (Opisthobranchia: Mollusca) of the Indo-West Pacific: the genus *Glossodoris* Ehrenbergh.** Zoological Journal of the Linnaean Society 86: 101-184.

Rudman, W.B. 1977. **Chromodorid opisthobranch Mollusca from East Africa and the tropical West Pacific**. Zoological Journal of the Linnaean Society 61: 351-397.

Takahashi, Chris, 1987. **Hawaiian specimens: _Umbraculum sinicum_**. Hawaiian Shell News. 35(2): 8.

Tarr, Rob, 1989. **The umbrella shell**. Hawaiian Shell News 37(4): 12.

Thompson, T.E & I. Bennett, 1972. **Observations of hexabranchus**. Veliger 15(1): 1-5 July 1972.

Thorsson, Wesley M., 1991. **Observations on Aplysiidae**. Hawaiian Shell News. 39(2): 10. (Notes on _Stylocheilus longicauda_.)

Willan, Richard C. & Neville Coleman, 1984. **Nudibranchs of Australasia**. Sydney, Australia: Australasian Marine Photographic Index. 56 p.

BIVALVES

Earle, John, 1985. **Nine colorful pectens from Hawai`i**. Hawaiian Shell News 33(7): 1.

Moffitt, Robert B., 1994. **Pearl oysters in Hawai`i**. Hawaiian Shell News 42(4): 3-4.

Severns, Mike, 1980. **Lahaina's elusive _Spondylus linguaefelis_**. Hawaiian Shell News 28(7): 7.

Sims, Neil Anthony & Dale J. Sarver, 1994. **Bringing back Hawaii's black pearls**. Hawaiian Shell News 42(1): 5-6.

Thorsson, Wesley M., 1980. **Observations on Hawaii's _Pinna_ population**. Hawaiian Shell News 28(7): 1.

———, 1987. **Things that snap on the cliffs I _Arca ventricosa_**. Hawaiian Shell News 35(9): 8.

———, 1987. **Things that snap on the cliffs II _Spondylus tenebrosus_** Reeve 1856. Hawaiian Shell News 35(9): 14.

———, 1989. **_Pinna_ observations**. Hawaiian Shell News. 37(8): 5.

OCTOPUS AND SQUID

Bailey, Steve, et al., 1996. **Husbandry of one species of squid in captivity**. Freshwater and Marine Aquarium 19(1): 196-202. (Oval Squid _Sepioteuthis lessoniana_.)

Berry, S. Stillman, 1914. **The Cephalopoda of the Hawaiian Islands**. Washington D.C. Bureau of Fisheries Bulletin, 32: 255-362.

Cousteau, Jacques-Yves and Philippe Diole, 1973. **Octopus and squid: the soft intelligence**. Garden City, New York: Doubleday. 304 p.

Houck, Becky A., 1977. **A morphological and behavioral study of an extra-ocular photoreceptor in octopods**. Ph.D. dissertation, University of Hawai`i. (Contains short informal description of the undescribed "crescent octopus.")

Kerstitch, Alex, 1992. **Socorro**. Freshwater and Marine Aquarium 15(3): 32. (Diver describes being attacked by a 7-foot squid.)

Lane, Frank W., 1960. **Kingdom of the octopus: the life history of the Cephalopoda**. New York: Sheridan House. 300 p. (Although published 30 years ago, still a wonderful popular account, crammed with fascinating information.)

McFall-Ngai, Margaret, 1996. **Essential partners, studies of the symbiotic relationship between animals and microorganisms**. Science Spectra 1(4): 14-19. _(Euprymna scolopes)_

Norman, Mark D., 1991. **_Octopus cyanea_ Gray, 1849 (Mollusca: Cephalopoda) in Australian waters: description, distribution and taxonomy.** Bulletin of Marine Science 49(1-2): 20-38.

Roper, Clyde F.E. & F.G. Hochberg, 1988. **Behavior and systematics of cephalopods from Lizard Island, Australia, based on color and body patterns**. Malacologia 29(1): 153-193.

Takahashi, Chris, 1985. **Hawaii's eight-arm shell collector**. Hawaiian Shell News 33(11): 3.

Travis, John, 1996. **An illuminating partnership for squid**. Science News Online. Sept. 14, 1996. _(Euprymna scolopes)_ http://38.214.184.12/sn_arch/9_14_96/fob2.htm

Van Heukelem, W.F., 1966. **Some aspects of the ecology and ethnology of _Octopus cyanea_ Gray**. M.S. thesis, University of Hawai`i.

Van Heukelem, W.F., 1983. **_Octopus cyanea_**, (in P.R. Boyle, ed., **Cephalopod life cycles, volume 1**. p. 267-276. (Description of the life cycle of _O. cyanea_.)

Voss, Gilbert L., 1981. **A redescription of _Octopus ornatus_ Gould, 1852 (Octopoda: Cephalopoda) and the status of _Callistoctopus_ Taki, 1964.** Proceedings of the Biological Society of Washington 94(2): 525-534.

Wendt, Dorothy, 1985. **BigEye, the unzippered octopus**. Hawaiian Shell News 33(6): 12.

Wood, James B., 1994. **Don't fear the raptor: an octopus in the home aquarium**. Freshwater and Marine Aquarium 17(4); April 1994. (Keeping a small undescribed Hawaiian octopus in captivity.)

CRUSTACEANS

GENERAL

Barry, C.K., 1965. **Ecological study of the decapod crustaceans commensal with branching coral *Pocillopora meandrina* var. *nobilis* Verrill**. M.S. Thesis, University of Hawai`i.

Clarke, Thomas A. 1972. **Collections and submarine observations of deep benthic fishes and decapod crustacea in Hawaii**. Pacific Science 26(3): 310-317.

Coles, S.L., 1980. **Species diversity of decapods associated with living and dead reef coral *Pocillipora meandrina*.** Marine Ecology Progress Series 2: 281-291.

————, 1986. **A guide to animals symbiotic with reef corals in Hawaiian waters.** in P.L. Jokiel, et al. eds, 1986. **Coral Reef Population Biology.**)

Debelius, Helmut, 1984. **Armoured knights of the sea**. [Germany] Kernen Verlag, distributed by Quality Marine, Los Angeles, Calif. 120 p. (An engaging and well-illustrated account of crabs, shrimps and lobsters of interest to aquarists.)

Headstrom, Richard, 1979. **Lobsters, crabs, shrimps and their relatives**. South Brunswick, N.J.: A.S. Barnes. 143 p. (The best popular introduction to the crustaceans.)

Healy, Anthony & John Yaldwyn, 1970. **Australian crustaceans in colour**. Rutland, Vermont: C.E. Tuttle Co. 112 p.

Kerstitch, Alex, 1988. **Samurai in painted armors**. Freshwater and Marine Aquarium 7: 80.

————, 1988. **Jaws and claws**. Freshwater and Marine Aquarium 10: 8.

————, 1994. **Armored predators of the reef**. Ocean Realm. April 1994: 19-22.

Kosaki, Randall K., 1987. **Hawaiian cave crustaceans**. Freshwater and Marine Aquarium 10(4): 12; April 1987.

Tinker, Spencer Wilkie, 1965. **Pacific crustacea: an illustrated handbook on the reef-dwelling crustacea of Hawaii and the south seas.** Rutland, Vermont: C.E. Tuttle. 134 p.

Titgen, Richard H., 1987. **New decapod records from the Hawaiian Islands (Crustacea, Decapoda)**. Pacific Science 41(1-4): 141-147.

SHRIMPS

Bailey-Brock, Julie H. & Richard E. Brock, 1993. **Feeding, reproduction, and sense organs of the Hawaiian anchialine shrimp *Halocaridina rubra (Atyidae)*.** Pacific Science 47(4): 338-355.

Banner, Albert H. 1953. **The Crangonidae, or snapping shrimp, of Hawaii**. Pacific Science 7(1): 3-144.

Banner, Albert H. & Dora M. Banner. 1974. **Contributions to the knowledge of the alpheid shrimp of the Pacific Ocean Part XVII. Additional notes on the Hawaiian alpheids: new species, subspecies, and some nomenclatorial changes.** Pacific Science 28(4): 423-437.

Bruce, A.J., 1974. ***Periclimenes insolitus* sp. nov. (Decapoda Natantia, Pontoniidae), a new commensal shrimp from Waikiki Beach, Oahu, Hawaii.** Crustaceana 26(3): 293-307.

Castro, Peter, 1971. **The natantian shrimps (Crustacea, Decapoda) associated with invertebrates in Hawaii.** Pacific Science 25(3): 395-403.

Goy, Joseph W. & Dennis M. Devaney, 1980. ***Stenopus pyrsonotus*, a new species of stenopodidean shrimp from the Indo-West Pacific region (Crustacea: Decapoda)**. Proceedings of the Biological Society of Washington 93(3): 781-796; 1980.

Goy, Joseph W. & John E. Randall, 1986. **Rediscription of *Stenopus devaneyi* and *Stenopus earlei* from the Indo-West Pacific region (Decapoda: Stenopodidae).** Bishop Museum Occasional Papers 26: 81-101.

Kraul, Sid and Alan Nelson, 1986. **The life cycle of the Harlequin Shrimp**. Freshwater and Marine Aquarium 9(9): 28.

Lau, Colin J., 1988. **The ecology of an almost anchialine shrimp, *Parhyppolyte uveae* in Hawaii**. Pacific Science 42(1-2): 125.

Moehring, Janath Lynn, 1972. **Communication systems of a goby-shrimp symbiosis**. Ph.D. Dissertation, University of Hawai`i.

Nomura, Keiichi & Ken-Ichi Hayashi, 1992. ***Rhynchocinetes striatus*, a new species (Decapoda, Caridea, Rhynchocinetidae) from southern Japan.** Zoological Science 9: 199-206.

Okuno, Junji, 1994. ***Rhynchocinetes concolor*, a new shrimp (Caridea: Rhynchocinetidae) from the Indo-West Pacific.** Proceedings of the Japanese Society of Systematic Zoology 52: 65-72

————, 1997. **Review on the genus *Cinetorhynchus* Holthuis, 1995 from the Indo-West Pacific**. In B. Richer der Forges, ed. Les Fondes Meubles des Lagons de Nouvelle-Calédonie (Sédimentologie, Benthos). Etudes et Theses, 3 Paris, ORSTOM, 31-58.

Okuno, Junji & John P. Hoover, 1998. ***Cinetorhynchus hawaiiensis*, a new shrimp forming a cryptic species pair with C. reticulatus Okuno, 1997, and new records of three congeneric species (Decapoda: Caridea: Rhynchocinetidae).** Natural History Research 5(1): 31-42.

Okuno, Junji & Hiroyuki Tachikawa, 1997. *Cinetorhynchus fasciatus*, a new shrimp from the western and central Pacific (Decapoda: Caridea: Rhynchocinetidae). Crustacean Research 26: 16-25

Strynchuk, Justin, 1990. **An insight into the mating habits of the banded coral shrimp.** Freshwater and Marine Aquarium 13(10): 42.

Titgen, Richard H., 1989. **Gnathophyllid shrimp of the Hawaiian Islands, with the description of a new species of *Gnathophyllum* (Decapoda, Gnathophyllidae).** Crustaceana 56(2): 200-210.

————, 1991. **A summary of Albert H. & Dora M. Banner's contributions to the knowledge of the family Alpheidae.** Pacific Science 45(3)

Vaughan, R.A., 1973. **Aspects of the ecology of *Alpheus deuteropus* (Crustacea, Decapoda), a boring shrimp.** Honolulu: University of Hawaii. Dept. of Zoology. Master of Science Research Report.

Wilkerson, Joyce D., 1994. **Scarlet cleaner shrimp: care and reproductive habits of *Lysmata amboinensis*.** Freshwater and Marine Aquarium, 17(8); August 1994.

LOBSTERS

Chong, Karynne, 1980. **Spiny lobsters in Hawai`i.** University of Hawaii Sea Grant College Marine Advisory Program. UNIHI-SEAGRANT-AB-80-04. 4 p.

Edmondson, Charles Howard, 1944. **Callianassidae of the Central Pacific**. Occasional papers of Bernice P. Bishop Museum 18(2): 35-61.

Holthuis, L.B., 1983. **Notes on the genus *Enoplometopus*, with descriptions of a new subgenus and two new species (Crustacea Decapoda, Axiidae)**. Zoologische mededelingen 56(22) 13 April 1983.

————, 1991. **Lobsters of the world: an annotated and illustrated catalogue of species of interest to fisheries known to date.** Rome: Food and Agriculture Organization of the United Nations. 292 p.

Kensley, Brian, 1981. **Notes on *Axiopsis (Axiopsis) serratifrons* (A. Milne Edwards) (Crustacea: Decapoda: Thalassinidea)**. Proceedings of the Biological Society of Washington 93(4) 1253-1263.

McGinnis, Fred, 1972. **Management investigation of two species of spiny lobsters, *Panulirus japonicus* and *P. penicillatus*.** Honolulu: State of Hawaii, Dept. of Land and Natural Resources, Division of Fish and Game. 47 p.

Titgen, Richard H. and Ann Fielding, 1986. **Occurrence of *Palinurellus wieneckii* (De Man, 1881) in the Hawaiian Islands (Decapoda: Palinura: Synaxidae)** Journal of Crustacean Biology 6(2): 294-296.

HERMIT CRABS AND RELATIVES

Haig, Janet & Eldon E. Ball, 1988. **Hermit crabs from north Australian and eastern Indonesian waters (Crustacea Decapoda: Anomura: Paguroidea) collected during the 1975 Alpha Helix Expedition.** Records of the Australian Museum 40: 151-196.

Haig, Janet & Patsy A. McLaughlin, 1983. **New *Calcinus* species (Decapoda: Anomura: Diogenidae) from Hawaii, with a key to the local species.** Micronesica vol. 19, December, 1983. p. 107-121.

Hazlett, Bruce A., 1970. **Interspecific shell fighting in three sympatric species of hermit crabs in Hawaii.** Pacific Science 24(4): 472-482.

McLaughlin, Patsy A., 1986. **Three new genera and species of hermit crabs (Crustacea: Anomura: Paguridae) from Hawaii.** Journal of Crustacean Biology 6(4): 789-803.

McLaughlin, Patsy A. & Janet Haig, 1989. **On the status of *Pylopaguropsis zebra* (Henderson), *P. magnimanus* (Henderson), and *Galapagurus teevanus* Boone, with descriptioins of seven new species of *Pylopaguropsis* (Crustacea: Anomura: Paguridae)**. Micronesica 22(2): 121-171.

McLaughlin, Patsy A. & John Hoover, 1996. **A new species of *Aniculus* Dana (Decapoda: Anomura: Diogenidae) from Hawaii.** Proceedings of the Biological Society of Washington 109(2): 299-305.

Nielson, P.M., 1969. **Hermit crabs and shells.** Hawaiian Shell News 7(8): 1

Ross, D.M., 1970. **The commensal relationship of *Calliactis polypus* and the hermit crab Dardanus gemmatus in Hawaii.** Canadian Journal of Zoology 48:351-357.

Tudge, Christopher C., 1995. **Hermit crabs of the Great Barrier Reef and coastal Queensland.** Brisbane: School of Marine Science, University of Queensland. 40 p.

Wenner, Adrian M. 1972. **Incremental color change in an anomuran decapod *Hippa pacifica* Dana.** Pacific Science 26(3): 346-353.

————, 1977. **Food supply, feeding habits, and egg production in Pacific mole crabs.** Pacific Science 31(1): 39-47.

Wooster, Daniel S., 1982. **The genus *Calcinus* (Paguridea, Diogenidae) from the Mariana Islands including three new species.** Micronesica vol. 18, December, 1982. p. 121-162.

TRUE CRABS

Castro, Peter, 1969. **Symbiosis between *Echinoecus pentagonus* (Crustacea, Brachyura) and its host in Hawaii, *Echinothrix calamaris* (Echinoidea)**. Ph.D. dissertation, University of Hawai`i. 173. p.

————, 1978. **Settlement and habitat selection in the larvae of *Echinoecus pentagonus* (A. Milne Edwards), a brachyuran crab symbiotic with sea urchins.** Journal of Experimental Marine Biology and Ecology 34: 259-270.

Coles, S.L., 1982. **New habitat report for *Maldivia triunguiculata* (Barradaile) (Brachyura, Xanthidae), a facultative symbiont of *Porites lobata* Dana in Hawaii.** Pacific Science 36(2): 203-209.

Edmondson, Charles Howard, 1954. **Hawaiian Portunidae**. Occasional papers of Bernice P. Bishop Museum 21(12): 217-274.

————, 1959. **Hawaiian Grapsidae**. Occasional papers of Bernice P. Bishop Museum 22(10): 153-202.

————, 1962. **Xanthidae of Hawaii**. Occasional papers of Bernice P. Bishop Museum 22(13): 215-309.

Fielding, Anne & Samuel R. Haley. 1976. **Sex ratio, size at reproductive maturity, and reproduction of the Hawaiian Kona crab, *Ranina ranina* (Linnaeus) (Brachyura, Gymnopleura, Raninidae)**. Pacific Sciene 30(2): 131-145.

Huber, Michael E. & Stephen L. Coles, 1986. **Resource utilization and competition among the five Hawaiian species of *Trapezia* (Crustacea, Brachyura)**. Marine Ecology Progress Series 30: 21-31.

Karplus, I., G.C. Fiedler, and P. Ramcharan, 1998. **The intraspecific fighting behavior of the Hawaiian boxer crab, *Lybia edmondsoni*—fighting with dangerous weapons?** Symbiosis 24(1998): 287-302.

Love, Robert Milton, 1995. **Decorator crabs**. Ocean Realm. February 1995: 50-56.

Onizuka, Eric W., 1972. **Management and development investigations of the Kona crab, *Ranina ranina* (Linnaeus). Final report.** Honolulu: State of Hawaii. Dept. of Land and Natural Resources. Division of Fish and Game. 28 p.

MANTIS SHRIMPS

Caldwell, Roy L. & Hugh Dingle, 1976. **Stomatopods**. Scientific American 234(1): 80-89.

Kinzie, Robert A. III, 1968. **The ecology of the replacement of *Pseudosquilla ciliata* (Fabricius) by *Gonodactylus falcatus* (Forskal) (Crustacea: Stomatopoda) recently introduced into the Hawaiian Islands.** Pacific Science 22(4): 465-475.

————, 1984. **Aloha also means goodbye: a cryptogenic stomatopod in Hawaii.** Pacific Science 38(4): 298-311

Townsley, Sidney Joseph, 1953. **Adult and larval stomatopod crustaceans occurring in Hawaiian waters**. Pacific Science 7(4): 399-437.

BARNACLES

Gordon, Joleen Aldous, 1970. **An annotated checklist of Hawaiian barnacles (class Crustacea; subclass Cirripedia) with notes on their nomenclature, habitats and Hawaiian localities**. Honolulu: University of Hawaii. Hawaii Institute of Marine Biology technical report no. 19. 130 p.

Matsuda, C., 1973. **A shoreline survey of free-living intertidal barnacles on the island of O`ahu, Hawai`i.** MSc thesis, University of Hawai`i.

Southward, A.J., et al., 1998. **Invasion of Hawaiian shores by an Atlantic barnacle.** Marine Ecology Progress Series. 165; 119-126.

ECHINODERMS

GENERAL

Agassiz, Alexander & Hubert Lyman Clark, 1907. **Preliminary report on the echini collected in 1902, among the Hawaiian Islands, by the U.S. Fish Commission Steamer "Albatross"...** Bulletin of the Museum of Comparative Zoology at Harvard College. 50(8): 231-259.

Clark, Ailsa M. & Francis W.E. Rowe, 1971. **Monograph of shallow-water Indo-West Pacific echinoderms**. London: British Museum (Natural History). 238 p.

Ebert, Thomas A., 1971. **A preliminary quantitative survey of the echinoid fauna of Kealakekua and Honaunau Bays, Hawaii.** Pacific Science 25(1): 112-131.

Hendler, Gordon, John E. Miller, David L. Pawson & Porter M. Kier, 1995. **Sea stars, sea urchins, and allies: echinoderms of Florida and the Caribbean.** Washington: Smithsonian Institution Press. 390 p. (Beautifully illustrated and an excellent introduction to echinoderms in general, although restricted to Florida and the Caribbean.)

Herwig, Nelson, 1980. **Starfish, sea urchins and their kin**. Sierra Madre, California: Freshwater and Marine Aquarium Magazine. 64 p.

Schoenberg, Olive, 1980. **Living together—an old molluscan custom**. Hawaiian Shell News 29(11): 3. (discusses molluscan parasites of echinoderms)

SEA STARS AND BRITTLE STARS

Edmondson, Charles Howard, 1935. **Autotomy and regeneration in Hawaiian starfishes.** Bernice P. Bishop Museum. Occasional Papers (9)8. 20 p.

Ely, Charles A., 1942. **Shallow-water asteroidea and ophiuroidea of Hawaii**. Bernice P. Bishop Museum Bulletin 176. 63 p.

Fisher, Walter K., 1906. **Starfishes of the Hawaiian Islands**. U.S. Fisheries Commission. Bulletin 23(3): 987-1130.

————, 1925. **Sea stars of the tropical central Pacific**. B.P. Bishop Museum Bulletin 27: 63-88

Furlong, Marjorie and Virginia Pill, 1972. **Starfish, guides to identification and methods of preserving. 2nd ed.** [s.l.] Ellis Robinson Publishing. ("Pacific coast, Alaska, Mexico, Hawaii." 8 Hawaiian species discussed, with illustrations of preserved specimens. May contain errors.)

Glynn, Peter W., 1981. ***Acanthaster* regulation by a shrimp and a worm.** The Reef and Man. Proceedings of the Fourth International Coral Reef Symposium. 2: 607-612.

————, 1984. **An amphinomid worm predator of the crown-of-thorns sea star and general predation on asteroids in eastern and western Pacific coral reefs.** Bulletin of Marine Science 35(1): 54-71.

Glynn, Peter W. & D.A. Krupp, 1986. **Feeding biology of a Hawaiian sea star corallivore, *Culcita novaeguineae*.** Journal of Experimental Marine Biology and Ecology 96(1): 75-96.

Guille, Alain, et al., 1986. **Guide des etoiles de mer, oursins et autres echinodermes du lagon de Nouvell-Caledonie.** Paris: Editions de l'ORSTOM. (Beautifully illustrated guide to the echinoderms of New Caledonia, including many that occur in Hawai`i.)

Love, Robert Milton, 1991. **Asteroids, stars of the sea.** Ocean Realm. Summer 1991: 26-31.

Odom, Charles B. & Edward A. Fischermann, 1972. **Crown-of-thorns starfish wounds—some observations on injury sites.** Hawaii Medical Journal 31(2): 99-100.

SEA URCHINS

Love, Robin Milton. 1993. **Urchins**. Ocean Realm July 1993. p 81-87.

Ogden, Nancy B., J.C. Ogden and I.A. Abbott, 1989. **Distribution, abundance and food of sea urchins on a leeward Hawaiian reef.** Bulletin of Marine Science 45(2): 539-549.

SEA CUCUMBERS

Cannon, L.R.G. & H. Silver, 1986. **Sea cucumbers of northern Australia**. Brisbane: Queensland Museum. 60 p. (An excellent account with color illustrations of many species occurring in Hawai`i.)

Fisher, Walter K., 1907. **Holothurians of the Hawaiian Islands**. U.S. National Museum. Proceedings 32: 637-744.

Leonardo, Lydia R. & Marti Ellen Cowan, 1984. **Shallow-water holothurians of Calatagan, Batangas, Philippines**. [Manila]: University of the Philippines Marine Sciences Center. 56 p.

Rowe, F.W.E & J.E. Doty, 1977. **The shallow water holothurians of Guam**. Micronesica 13: 217-250.

TUNICATES

Antrim, Kay, 1970. **Hey look at that!** Hawaiian Shell News, August, 1970. (Observation of a pyrosoma in Hanauma Bay, O`ahu.)

Monniot, Claude, Francois Monniot & Pierre Laboute, 1991. **Coral reef ascidians of New Caledonia**. Paris: Editions de l'ORSTOM. Institut francais de recherche scientifique pour le developpement en cooperation. 247 p.

Abbot, Donald P., A. Todd Newberry, Kendal M. Morris, 1997. **Reef and shore fauna of Hawaii. Section 6B: Ascidians (Urochordata)** Honolulu: Bishop Museum Press. 64 p.

CORALLINE ALGAE

Keats, Derek W., **An Introduction to nongeniculate coralline algae.** http//www/botany.uwc.ac.za/clines.htm

Magruder, William H. and Jeffrey W. Hunt, 1979. **Seaweeds of Hawaii, a photographic identification guide**. Honolulu: Oriental Publishing Co. 116 p.